Ludus Coventriæ

OR

The Plaie called Corpus Christi

COTTON MS. VESPASIAN D. VIII.

EARLY ENGLISH TEXT SOCIETY

Extra Series No. 120

1922 (*for* 1917), reprinted 1960, 1974

Ludus Coventriæ

or

The Plaie called Corpus Christi

COTTON MS. VESPASIAN D. VIII

EDITED BY

K. S. BLOCK

Published for
THE EARLY ENGLISH TEXT SOCIETY
by the
OXFORD UNIVERSITY PRESS
LONDON   NEW YORK   TORONTO

FIRST PUBLISHED 1922
REPRINTED 1960, 1974

REPRINTED LITHOGRAPHICALLY
BY FLETCHER AND SON LTD, NORWICH

PREFACE

THE *Ludus Coventriæ*, unlike the other extant mystery cycles, is not connected with the trade guilds of a town, and the nature and purpose of this collection have been recognized as a problem of special interest. Investigation has been hampered by the fact that the only edition of the plays (J. O. Halliwell, printed for the Shakespeare Society, 1841), though admirably accurate as to the text, obscures, by changes in arrangement and numbering, certain significant features of the MS. It was therefore desirable that the MS. should be re-edited in order that the bibliographical factors of the problem should be clear.

The plays have, moreover, in themselves some claim to be made more accessible. They illustrate the advance in dramatic representation that was taking place in the fifteenth century. Plays like Nos. 18 and 20 (*Adoration of the Magi* and *Massacre of the Innocents*, with the *Death of Herod*), or No. 24 (*The Woman taken in Adultery*), indicate an improvement in the art of the religious drama; passages like the satiric address of Demon to the audience, p. 225, indicate a widening of its range. The *Passion* plays, which differ from those of the other cycles in being acted, not on a sequence of pageants in procession, but on a standing group, show for this reason more attempt at construction; and there are suggestions here and in some of the other plays of specially effective acting (cf. *Introduction*, p. lvi). In the stage directions of these *Passion* plays (cf. p. 267 and p. 273) we have the most graphic picture of a fifteenth-century performance that has come down to us.

The title *Ludus Coventriæ* has been retained partly because it has been associated with these plays since the first cataloguing of the MS. by Richard James; but partly also because there is, in the opinion of the present editor, some ground for thinking that the title was used generically by James, and has therefore some importance in the history of the drama (cf. *Introduction*, p. xl).

This edition was begun in 1913, and marks of protracted and intermittent work are, I fear, obvious. My thanks are due to Sir Israel Gollancz for constant encouragement, and to the Oxford University Press for much consideration and assistance. I am indebted to my colleague Miss G. D. Willcock for help in compiling the glossary.

K. S. BLOCK.

ROYAL HOLLOWAY COLLEGE.
March, 1922.

CONTENTS

	PAGE
INTRODUCTION	xi
LUDUS COVENTRIÆ	1–377
LIST OF PROPER NAMES	379
GLOSSARY	383

PLAYS AS NUMBERED IN THE MS., IN THE PROCLAMATION, AND IN HALLIWELL'S EDITION

	MS. No.	Pageant No.	Halliwell.	Page.
The Creation of Heaven and the Angels	1	1	1	16
Fall of Lucifer				17
The Creation of the World and Man	2	2		19
Fall of Man			2	21
Cain and Abel	3	3	3	29
Noah and	4	4	4	35
The Death of Lamech		No Pag^t.		
Abraham and Isaac	5	5	5	43
Moses	6	6	6	51
The Prophets	7	7	7	57
The Conception of Mary (Barrenness of Anna)	8	No Pag^t.	8	62
Mary in the Temple	9	No Pag^t.	9	71
The Betrothal of Mary	10	8, 9 (corr. to 10)	10	83
The Parliament of Heaven / The Salutation and Conception	11	No Pag^t. 10 (corr. to 11)	11	97
Joseph's Return	12	11 (corr. to 12)	12	109

viii   Contents

	MS. No.	Pageant No.	Halliwell.	Page.
The Visit to Elizabeth	13	No Pag^t.	13	115
Prologue of Summoner	None	No Pag^t.		123
The Trial of Joseph and Mary	14	12 (corr. to 14)	14	124
The Birth of Christ	15	13 (corr. to 15)	15	135
The Adoration of the Shepherds	16	14 (corr. to 16)	16	146
The Adoration of the Magi	18	15	17	151
The Purification	19	No Pag^t.	18	162
The Massacre of the Innocents	20	16	19	169
The Death of Herod		17		174
Christ and the Doctors	21	18	20	178
The Baptism	None	19	21	188
The Temptation	23	20	22	193
The Woman taken in Adultery	24	21	23	200
The Raising of Lazarus	25	22	24	210
Passion Play. I				
Prologues of Demon and of John the Baptist		No Pag^t.	25	225
The Council of the Jews	26	No Pag^t.		230
The Entry into Jerusalem		23	26	237
The Last Supper	27	24	27	242
and				
The Conspiracy of the Jews and Judas				
The Betrayal	28	25	28	262
[Prologue of the Doctors	None	No Pag^t.	29]	269
Passion Play. II				
King Herod		No Pag^t.		271
The Trial of Christ before Annas and Caiaphas	29	26		274
Peter's Denial				277
The Death of Judas		28	30	278
The Trial of Christ before Pilate	30	27		280
The Trial of Christ before Herod		No Pag^t.		283
Pilate's Wife's Dream	31	27, 29	31	287
The Trial of Christ and the Thieves before Pilate		27, 29		289
The Condemnation and Scourging		30		294
The Procession to Calvary	32	No Pag^t.	32	294
The Crucifixion		30		296
The Descent into Hell of Anima Christi	33	31	33	305

Contents

	MS. No.	Pageant No.	Halliwell.	Page.
The Embassy to Pilate of Joseph of Arimathea		32		307
The Episode of Longeus	34	31	34	310
The Descent from the Cross and Burial		32		311
The Guarding of the Sepulchre				312
The Harrowing of Hell		23 (scribal slip)		318
The Resurrection and Appearance to the Virgin	35		35	320
The Compact of the Soldiers and Pilate		No Pag^t.		322
The Announcement to the Three Maries	36	34	36	327
The Appearance to Mary Magdalen	37	35	37	333
The Appearance on the way to Emmaus	38	36	38	337
The Appearance to Thomas		37		346
The Ascension and	39	38	39	349
The Choice of Matthias		No Pag^t.		351
The Day of Pentecost	40	39	40	352
The Assumption of the Virgin	41	No Pag^t.	41	354
Doomsday	42	40	42	373

ERRATA

Some lines in the second Passion play between lines 985 and 1035, and again between lines 1295 and 1355, are wrongly numbered. The total number of lines in the play is correctly given.

INTRODUCTION

THE MS. Vespasian D. viii of the Cotton Collection in the British Museum is a small thick volume containing 225 leaves 8″ × 5½″. In the process of binding the leaves have all been cut down and remounted, and their breadth varies slightly: folios 91 and 92, containing the opening of the *Magi* play, are noticeably narrower than the rest. *Description of MS.*

A modern hand has numbered the folios. The reverse sides are numbered by tens only, 9ᵛ, 19ᵛ ... 49ᵛ being numbered 10, 20 ... 50; 60ᵛ, 70ᵛ ... 110ᵛ, 60, 70 ... 110; 121ᵛ, 131ᵛ ... 161ᵛ, 120, 130 ... 160; 172ᵛ, 182ᵛ ... 212ᵛ, 170, 180 ... 210; and 223ᵛ, 220. *Pagination.*

The lettering of the quires, A to W, is of about the same date. In four cases the original catchwords remain: on fo. 40ᵛ, the last page of quire B; on fo. 148ᵛ, the last page of quire N; on fo. 179ᵛ, the last page of quire S; on fo. 189ᵛ, the last page of quire T. In other cases the quire letters correspond with a change of paper. *Quires.*

Seven different kinds of paper are found in this MS. A and B (ff. 1–40ᵛ) consist of paper of the *Bunch of Grapes* watermark (*Les Filigranes*, C. M. Briquet, Paris 1907, No. 3055). Quires C to M (ff. 41–135ᵛ), with the exception of quire E (ff. 51 and 52) and of the interpolated folios 95, 96, and 112, consist of paper of the *YHS in a Sun* watermark (*Les Filigranes*, No. 9477). The interpolated quire E has the *Pitcher*[1] watermark of the *pot d'étain* type (*Les Filigranes*, No. 12498 or 12501), not found, according to M. Briquet, before the last decade of the fifteenth century. The interpolated folios 95, 96 have the *Hand* watermark of the gloved type marked with a *3* on the palm and surmounted by a pentagon.[2] Unfortunately not enough of the *Paper.*

---

[1] *Mind, Will and Understanding* in the Macro MS. is written on paper bearing a *Pot* watermark. E.E.T.S., E.S. xci, p. xxx.

[2] The three plays in the Digby MS., *The Conversion of St. Paul, The*

lower part of the mark on fo. 95 remains to make it clear whether the wrist is laced. No example exactly corresponding with this mark is given by M. Briquet. It belongs to the general class described by him as *Main aux quatre doigts serrés, le pouce seul écarté.* Of the subdivision *lacée au poignet,* M. Briquet gives 1526 as the date of the earliest example known to him. If the lines which can just be discerned across the gauntlet could be taken as lacing, these folios, according to the evidence of the handwriting, would be earlier. Fo. 112, also an interpolation, has no watermark, but the wire lines correspond with those of folios 95, 96. Quire N (ff. 136–48ᵛ), with the exception of fo. 143, and quires P, Q, R (ff. 152–63ᵛ), have the *Bull's Head* watermark (*Les Filigranes,* No. 14184). Folio 143 has no watermark, but the wire lines correspond with those of the *Bunch of Grapes* or of the *YHS* watermark. Quire O (ff. 149–51ᵛ) has the *Bunch of Grapes* watermark. Quires S and T (ff. 164–89ᵛ), with the exception of folios 184 and 185, have the *Two Crossed Keys* watermark (*Les Filigranes,* No. 3887). Folios 184 and 185 have again unfortunately no watermark, but the wire lines are not those of the *Two Crossed Keys* paper, but correspond with those of the paper having the *Bunch of Keys* or the *YHS* watermark. Quires V and W (ff. 190–225), with the exception of folios 213–22, have the *YHS* watermark. The interpolated folios containing the play of the *Assumption* have the *Two-Wheeled Cart* watermark (*Les Filigranes,* No. 3528).

Besides these seven watermarks in the MS. itself, another—a shield bearing a crook—is found on the fly-leaf on which Dr. James, Sir Robert Cotton's first librarian, has written a note of the contents. Only a small portion is shown, and it does not correspond exactly with any of M. Briquet's examples. It comes nearest to what he describes as *La crosse de Bâle dans un écu,* a sixteenth-century mark. Paper of apparently the same watermark is used for part (ff. 31–107) of the 1621 catalogue of the Cottonian library.

Evidence of composi- This variety in the actual material of the MS. corresponds

Massacre, and *Mind, Will and Understanding,* are written on paper bearing different forms of the *Gloved Hand* watermark, but none exactly corresponding with that in the Vesp. D. viii. MS. The paper of *Mankind* in the Macro MS. has a *Gloved Hand* watermark. Cf. E.E.T.S., E.S. xci, p. xxviii.

in an interesting way with differences and discrepancies in the plays themselves, and throws light on the process of compilation. Some of the interpolations, folios 51 and 52, 95 and 96 and 112, are in a different handwriting from the body of the MS. and are later additions to the collection, in the case of 95 and 96 and 112 certainly, in the case of 51 and 52 possibly, replacing original material; and the play of the *Assumption*, though part of the set made up by the original compiler of this MS., is also in a different handwriting. tion of MS as to method of compilation.

In the case, however, of fo. 143 and quire O, and of folios 184 and 185, the handwriting is the same as that of the quires N, P, Q, R and S, T in which they are interpolated, and the added passages indicate the methods of the compiler.

The quires N, P, Q, R contain a separate group of plays, those that are described later in Contemplacio's Speech (fo. 165) as having been *shewyd the last ȝere*; the subject-matter is complete without fo. 143 or quire O, and the interpolation of these gives rise to duplication. From a cancelled speech in the MS. at the bottom of fo. 142 it is clear that fo. 144 originally followed on fo. 142, and the insertion of fo. 143 (containing the incident of fetching the ass and the foal) leads to a duplication of the passage beginning *ffrendys be-holde*, which occurs again on fo. 145^v, and to some inconsistency in the movements of Peter and John who 'abyden stylle' and yet advance towards Jerusalem. Again, a cancelled stage direction at the bottom of fo. 148^v, the last page of quire N, shows that quire P originally followed on. The episode of Mary Magdalen in quire O (like that of the ass and the foal on fo. 143) is not mentioned in the proclamation, and the insertion of quire O leads to a duplication of the disciples' questioning of the Saviour as to the traitor (cf. fo. 150 and fo. 155^v). A blank folio follows the last play of the group and the remainder of quire R is filled in with the Doctors' prologue—an entirely disconnected passage. Moreover, besides being of different paper from the rest of the MS. the quires N, P, Q, R show, as will be seen, certain slight differences in writing and rubrication.

The quires S and T offer some similarity in that here again we have traces of a set of plays which were acted separately; and from the appearance of fo. 164, the first of quire S, it would

seem that we have also part of a separate small MS. bound up
to make the series. But the case is more complicated ; the
interpolated folios 184 and 185 could not be removed, for the
last of the Seven Words comes on fo. 184 and the speech of
Anima Christi is continued from fo. 185 to fo. 186 ; and there
is no break at the end of quire T. The writer or compiler of
the set of plays beginning in the S quire clearly had in his
mind at starting a set corresponding with the first *Passion* plays
of the N, P, Q, R quires: 'we intendyn to procede þe matere
þat we lefte þe last ȝere. . . . Now wold we procede how he was
browth þan | be-forn annas *and* cayphas · *and* syth be-forn
pylate | and so forth in his passyon · how mekely he toke it for
man'; but this second group of *Passion* plays does not come to
any definite end, the action being continuous until the end of
the *Appearance to Mary Magdalen* play on fo. 201, the middle of
quire U. The MS., however, affords some traces of the original
plan ; there is evidence of a change of purpose at the close of
the Crucifixion scene, and it can be deduced with some certainty
that the original end of the Passion sequel was discarded to
make room for the material on folios 183^v, 184, 184^v, 185, 185^v,
and 186, and that the Senturyo's speech on fo. 186^v belongs to
a separate *Burial and Resurrection* play now dovetailed on. A
change in the colour of the ink suggests that the scribe paused
at the line: *heloy · heloy · lamaȝabathany* (fo. 183^v). The writing
also becomes smaller with the next line, and, at the bottom of
the folio, is cramped as if to fit in more than was allowed for.
It appears from slight indications in handwriting and ink as if
the scribe went on from the line mentioned to fo. 186^v: *Senturyo.
A now trewly telle weyl I kan* ; and that later he returned and
filled up the remainder of fo. 183^v, continuing on fo. 186 and
fitting in the interpolated folios. That the material on the
interpolated folios and fo. 186, though dovetailed in, is itself an
interpolation is proved by the discrepancies it involves: Mary
having passed into the Temple (fo. 185 interpolated) is yet at
the foot of the Cross to receive her son's body (fo. 188); the
Centurion makes his speech, recognizing the Crucified as the
Son of God, twice (fo. 186 and fo. 186^v), and Nicodemus seems
to make a double entry (fo. 186 and fo. 188). Additional evidence
from the MS. of disturbance at this point is the change in the

pointing; the mid-line point, having been fairly abundant in the preceding folios of the T quire, ceases after fo. 183ᵛ. No instance of borrowing from *The Northern Passion*, otherwise so marked a feature of this group, occurs in this interpolated portion.[1] The MS. throws no additional light on the correspondence of change of paper and change of material at the junction of the T and U quires. Though there is no break in the action here the changes of metre, of the form of Pilate's name from Pylat to Pilatus, and of English for Latin stage directions,[2] the cessation of reminiscences of *The Northern Passion*,[1] as well as the double *exeunt* of the knights (fo. 189ᵛ, last of T, fo. 190, first of V), point to the use of different material, but the scribe appears to have gone on continuously. Quire T is half the size of quire S.

The bulk of the MS. is the work of one scribe, probably of the third quarter of the fifteenth century. The date 1468 is written at the close of the *Purification* play on fo. 100ᵛ, but this play is an interpolation, and it is possible that the date is specially connected with it rather than with the compilation. [Handwriting.]

The writing varies considerably, but the variations are better accounted for as due to parts of the MS. having been written at different times than as due to change of scribe. The system or rather want of system in the use of capital letters remains the same—the preferential use of the capital forms of *t* and *n* at the beginning of lines, the arbitrary use of *ff* and *f*, and of *A* and *a*, and the sporadic use of capital forms of *e* and *r*.

The most noticeable variation is found in the quires N, P, Q, R, where the writing is more angular and altogether rougher and less shapely. Exceptionally large capitals appear throughout these quires, and a few specially tall letters in the top lines. The scribe also more often writes *þe* instead of *þᵉ* than elsewhere in the MS. The writing remains irregular, though less so, in quires S and T—that in the interpolated folios 184, 185, as in the interpolated fo. 143 in quire N, being noticeably firmer.

[1] Cf. *The Northern Passion*, edited F. A. Foster, Ph.D, E.E.T.S. 147, Part II, pp. 90-5.

[2] In quires S and T 56 English stage directions, 12 Latin, two of the latter on the interpolated folios; in quires V and W (excluding *Assumption* play) 2 English, 21 Latin. It may be noted that English and Latin stage directions are mingled in the latter part of the Digby MS. *Mary Magdalen* play.

It recovers its earlier form in quire V. There are minor irregularities in the earlier quires. It is sometimes much more cramped than at others, and once, on the other hand—folios 61-6, the end of quire F—it assumes a larger and more widely spaced form. The genealogies filling the bottoms of some of the folios (ff. 16ᵛ et seq., 21 et seq., 37) and the Psalms and Canticles quoted in the *Mary in the Temple* and *Visit to Elizabeth* plays are in liturgical script; the notes on the five Annas (fo. 37ᵛ) and on the April Calendar (fo. 74ᵛ) are in an ornamental book-hand.

Additions. There are three other handwritings to be found in the MS.: (1) that of the interpolated quire E (ff. 51, 52) containing the conversation between Joseph and his kinsmen in the *Betrothal of Mary* play; (2) that of the interpolated folios 95, 96 in the *Magi* play, and of the interpolated fo. 112 containing the opening of the *Baptism* play; (3) that of the *Assumption* play.[1] The first is a rough cursive hand of the close of the century decidedly later than that of the body of the MS. This writer uses the full *th* form, and neither the þ nor the ȝ appears in this passage. The second is also a cursive hand, but of a less untidy character. Halliwell speaks of it as 'a more recent hand', and this writer was, as will be seen, at work as a corrector of the MS. after its compilation; but the opinion of authorities is that the hand cannot be said to belong to a later type than the original scribe's. This writer uses the þ (in the later y form) and the full *th* form in about the same proportion as the chief scribe; like him he uses the ȝ in the word ȝe, but he uses the forms *you* and *your* where the other invariably uses ȝow (ȝou) and ȝour. This writer uses a curious form of final *s*.

The third hand was also thought by Halliwell to be later, but is now also assigned to the same period as the chief hand. It is of a different character, more cursive and less clear. This writer uses many fewer contractions than the other contributors to the MS. He does not use the ȝ at all; the only time it occurs

---

[1] Miss Swenson (*An Enquiry into the Composition of the Ludus Coventriæ*, University of Minnesota, Studies in Language and Literature) finds two other hands in the Prologue of the Doctors and the Prologue of Contemplacio in the *Passion* play respectively. The slight variation is better accounted for as suggested above.

in the play (fo. 217ᵛ)—*That ȝe schuld ben absent*—it comes in a line which was omitted, and written in the margin apparently in the hand of the chief scribe of the MS. He uses as a rule the full *th* form, but four examples of the þ (written *y*) are found. Fo. 217ᵛ *þu* (*thou*), fo. 219ᵛ *þi*, fo. 220ᵛ *þowth* (*thought*), fo. 221 *þi*. In this play (fo. 218) occurs the only example, in the MS. proper, of the þ in the older form, the two other examples occurring in the notes of some later reviser (fo. 144ᵛ, fo. 145).

The MS. has also been worked over by other hands. Corrections other than those of the original scribe are found scattered in many plays, and range from the alteration of a letter to the rewriting of two or three consecutive lines. The most important are to be found in the *Shepherd*, the *Magi*, the *Harrowing*, and the *Three Maries* plays, on folios 88ᵛ, 89, 89ᵛ, 93, 93ᵛ, 97, 101, and 185ᵛ, 191ᵛ, 192, 197ᵛ, 198. These appear all to be in the hand of the scribe of folios 95, 96, and 112.

<small>Corrections of revisers.</small>

Those in the *Shepherd* and *Magi* plays appear to be changes for the purpose of removing archaic or dialectal (Northern) words, phrases, or forms: *selkowth* > *mervelus*, *shene* > *bryght*, *carpynge* > *spekyng*, *barne* > *child*, *bale* > *sorow*, *buske* > *go*, *tholyn* > *suffyr*, *myrke* > *thyke*; a line is altered (fo. 89) so that þat *hattyht* (*that is called*) may be read as *that hateth*, and *xaln* is altered to *xalle*. The phrases erased and written over in Herod's ranting speech (ff. 93, 93ᵛ) are for the most part indecipherable, but the alterations that can be read in this passage suggest that they would all fall under the same heading: *paphawk* > *paddok*, *shaftys* > *sperys*. The corrections in the *Harrowing* play consist of indications of additions to the text; they take the form of references to fresh characters not found in the play as it stands —*anima* latron*is* (fo. 185ᵛ), þe devyll*e*, *anima* caym (fo. 191ᵛ, fo. 192)—and the manner of the reference—*and þan Cayme xalle sey his spech* (fo. 192)—suggests that the reviser is drawing on some other fuller version of the play known to the users of the MS. The revision in the *Three Maries* play consists in the substitution of eight new lines for four lines in the speech of Mary Magdalen and four in the speech of Mary Jacobi—the whole forming a continuous passage—announcing the resurrection to Peter and John. The purpose of the alteration is to harmonize the

b

account of the resurrection, which in the original follows the narrative of St. Mark (that accepted by Tatian) and represents an angel as announcing the Resurrection to the women, with the version given by St. Matthew, according to which the women see the risen Lord himself before they meet the disciples. A few corrections have been made in the text for the same purpose, as if the marginal rewriting of the lines were a second thought.

A few added marginal stage directions appear also to be in the hand of this reviser: *go homwardys* (fo. 102), various *Incipit hic, nota hic* (ff. 189, 189ᵛ), and it is possible that some of the slighter corrections found singly here and there are due to him. Many of these consist of one word written above the line or over the original word in the text, and some even of the addition or change of a letter. In such cases it is impossible to form a definite opinion as to the handwriting, but a comparison of one or two typical letters leads to the conclusion that most of these corrections are not from the hand of the scribe of folios 95, 96, 112, nor from that of the scribe of fo. 51. They occur singly in the following plays: *Cain and Abel, Mary in the Temple, The Trial of Joseph and Mary, The Birth, The Purification*; three are found in the *Salutation and Conception* play, ten in the *Disputation*, nine in the *Lazarus* play. They consist partly of the insertion of omitted words, partly of changes of apparently archaic or dialectal forms: thus, *beth* (imperative) > *be* (fo. 43ᵛ), *evy* > *hevy* (fo. 135 and fo. 110ᵛ), *dede* > *dyde* (fo. 134), *fende* > *fynde* (fo. 98ᵛ), *glathe* > *gladd* (fo. 134ᵛ), *thei* > *they* (fo. 135), *perysche* (pierce) > *pers* (fo. 107ᵛ), *blysse* > *comfort* (fo. 110ᵛ); *ffor syknes and sorwe* (fo. 58ᵛ) is changed to *bothe eve and morewe* to avoid an identical rhyme. Some of the corrections show a misunderstanding of the original: the line *A mayd milke haue nev*er *man dyde se* is 'corrected' by the deletion of *haue* and the addition of the possessive *is* to *mayd*.

There are also a few added stage directions in different hands, e.g. *here goth he his way*, written against the speech of quartus consolator and nuncius in the *Lazarus* play (fo. 129). Once or twice names of characters about to appear are roughly written in the margin, e.g. *pastores* in the *Barrenness of Anna* play (fo. 39ᵛ). This is perhaps the explanation of the mysterious *Vade Worlych* which is obviously written with some purpose on fo. 206ᵛ and again on fo. 207, where the meeting of Luke and Cleophas with the other disciples occurs.

Of special interest are two notes on fo. 144ᵛ and fo. 145 : *here entrith þe fyrst prophete,* and *here entrith þe parte of þe ij*ᵈᵉ *prophete.* No prophets appear in this play (*The Entry into Jerusalem*), and these notes again suggest a reference to another version of the play, or perhaps rather the use of this play in another combination than that of this compilation. A note on fo. 196 opposite the stage directions at the beginning of the *Three Maries* play. apparently in the hand of the scribe of folios 95, 96, 112—*finem 1ᵃ die · Nota*—seems also to refer to a division of the performance of some group of the plays.

The corrections of the scribe are made generally in black and red ink. Letters or words marked with deleting dot and cancelling strokes are almost always crossed through in red ink. Some of the corrections of the MS. by the original scribe are important for the light they throw on the compilation of the MS. The insertion of a line in the *Assumption* play has already been noticed (p. xvii). If this is accepted as in the hand of the chief scribe it clinches all the other arguments in favour of that play's having formed part of the original compilation, not indeed as conceived by the composer of the prologue but as carried out in the present MS. Mr. Gayley in his *Plays of our Forefathers* (1908) assigns this play to a date later than 1482, on the ground that it is based on Caxton's translation of the *Legenda aurea*—' practically ', he says, ' a transcription from it '. But a comparison of the play with Caxton's translation and the original Latin, so far from giving any evidence in support of this view, proves that the writer of the play used the Latin and not the translation. The only instances of verbal coincidence are found in the translation by both writers of *ammirabile et magnum* by *grete and meruelous,* where both words and order may be fairly said to be inevitable, and of *vasculum vite* by *vessel of lyf,* which again could not easily be avoided. Moreover, in two instances passages of the original Latin not translated by Caxton are used in the play of the *Assumption.*[1]

Particularly significant are the corrections of the numbers

Corrections of the scribe.

1. Of the Assumption play.

[Evidence as to date of the Assumption play.]

2. Of the numbering

[1] See Note A.

of the pageants in the Proclamation.

assigned to the pageants in the Proclamation (ff. 2ᵛ, 3, 3ᵛ, 4). These, being in figures, cannot be positively attributed to a certain hand, but the intimate connexion of the alterations with certain features in the arrangement of the subject-matter, and, in one case, the rubrication of the correction are sufficient evidence that they were made by the original scribe. It is, moreover, to be noted that for the first ten pageants and originally for the one now numbered xii, i.e. *Joseph's Return*, the numbers were given in words; for the pageants now x and xi and from the one now numbered xiv to the end Roman figures were used. This change of system indicates a break in the compilation, and the use of the word 'hellenthe' for the *Joseph's Return* pageant connects it with the original scheme. The corrections affect the plays *Mary's Betrothal, The Salutation and Conception, Joseph's Return, The Trial of Joseph and Mary, The Birth of Christ, The Shepherds*. These were originally numbered: *Mary's Betrothal*, pageants 8 and 9; *The Salutation and Conception*, pageant 10; *Joseph's Return*, pageant 11; *The Trial of Joseph and Mary*, pageant 12; *The Birth*, pageant 13; *The Shepherds*, pageant 14. In the renumbering both the *Betrothal* pageants are numbered 10, *The Salutation and Conception* 11, *Joseph's Return* 12, *The Trial* 14, *The Birth* 15, *The Shepherds* 16. The numbers 8, 9, 13 are thus left for the plays of *The Conception of Mary, Mary in the Temple*, and *The Visit to Elizabeth*, which actually fill those places in the compilation. But the additional stanzas necessary for the Proclamation were never written, and the correction of the numbering of the pageants was not carried beyond the *Shepherds* play, with the result that the numbers 15 and 16 are duplicated.

This renumbering affects the group of plays on the Virgin Mary—the first *Contemplacio* group, which has long been recognized by critics as forming a separate whole; and it suggests what there is much evidence to confirm, that the compiler is here combining two series of plays, one consisting of the *Betrothal, Salutation and Conception*, and *Return of Joseph*, as announced in the Proclamation, the other consisting of the *Conception of Mary, Mary in the Temple, Betrothal, Salutation and Conception*, and *Visit to Elizabeth*, as announced in the prologue of *Contemplacio*.

> This matere here mad · is of þe modyr of mercy
> how be joachym and anne · was here concepcion
> Sythe offred into þe temple · compiled breffly
> than maryed to joseph · and so folwyng þe salutacion
> metyng with Elizabeth · and þer with a conclusyon.

Evidence in support of this is found in certain indications in the arrangement of the plays in the MS., in discrepancies between the *Betrothal* and *Salutation and Conception* plays and the descriptions of them in the Proclamation, and in inconsistencies and incongruities—marks of imperfect amalgamation—in the plays themselves.

That the group of Mary plays from *The Conception of Mary* to *The Trial of Joseph and Mary* is not homogeneous is apparent upon examination. The *Return of Joseph* and the *Trial* plays differ in tone from the others, and they form no part of the Proclamation of *Contemplacio*. A suggestion that the *Return of Joseph* is an interpolation between the *Salutation and Conception* and the *Visit to Elizabeth* plays can be found in the cancelled stage direction at the close of the *Salutation and Conception* play : *And þan Mary seyth*, which would lead on to the *Visit to Elizabeth* play in which Mary is the first speaker, but not to the *Return of Joseph* play, which begins with a dialogue between Joseph and Susanna. In the *Protevangelion of St. James*, and in Tatian's *Diatessaron*, the visit to Elizabeth follows at once on the Salutation and Conception, and Joseph's suspicions are only subsequently aroused, and this order is followed in the *Meditationes* of Bonaventura and in Lydgate's *Lyf of oure Lady*.

Again, though the *Betrothal* play does not appear exactly in the form in which it is described in the Proclamation, where two pageants (and three stanzas) are assigned to it, there are significant discrepancies between it and the *Contemplacio* plays which precede and follow. In the *Mary in the Temple* play the Virgin is represented as left by her parents at the Temple and dwelling there. This also seems to be implied in the speech of *Contemplacio* at the close : *Lo sofreynes here ȝe haue seyn | in þe temple of oure ladyes presentacion | she was nevyr occapyed in thyngys veyn | but Evyr besy in holy ocupacyon*. In Bonaventura's narrative, as in the *Protevangelion* and *Nativity of Mary* gospel, Mary is said to have lived in the Temple from her 3rd to her

14th year. At the opening of the *Betrothal* play, however, Mary is living at home with her parents, and this is the situation conveyed in the words of the Proclamation : *Than Joachym and anne so mylde | þei brynge forthe mary þat blyssyd chylde.* The speech of *Contemplacio* referred to, which appears to introduce the *Betrothal* play, is separated from it in the MS. by a blank folio, and the play begins on a fresh quire (D), which indicates at least a pause—possibly some reconsideration—in the process of compilation. And the name of the bishop Abysakar in the opening stage direction of the *Betrothal* play (fo. 49) represents rather the Abizachar of the *Trial* (fo. 76) and the Abyacar of the Proclamation (fo. 2v) than the Ysakar of the *Conception of Mary* and the *Mary in the Temple* plays.[1]

There are in this play (*The Betrothal*) two interpolated passages in the metre of the preceding *Contemplacio* plays, one of them introducing a character *Minister* not otherwise found in the *Betrothal* play, who appears in the *Mary in the Temple* play. These passages appear to have been fitted in from the rejected *Betrothal* play of the *Contemplacio* series.

On the other hand, the *Salutation and Conception* play is not the play that was in the mind of the writer of the Proclamation, which evidently dealt simply with Gabriel's annunciation. No *parlement of hefne* is mentioned, and special reference is made to the overhearing of Gabriel's speech by the three maidens who dwelt with the Virgin. These maidens are mentioned in the pageant of the *Betrothal* play, and they appear in that and—two of them—in the *Return of Joseph* play, but they do not appear in the *Salutation and Conception* play.

It is not, however, clear that the *Salutation and Conception* play formed part of the original *Contemplacio* series. The *parlement of hefne* is first announced at the end of the second *Contemplacio* play before the *Betrothal* which, as has been seen, does not

[1] Both these names are connected with the Mary story in the pseudo gospels. Abiathar, Abiacar, Isacar, Zacharias appear in different versions of these gospels as the names of the high priest of the Temple during Mary's youth. Isaschar is in one version given as the name of the high priest who rejected Joachym. In some versions a distinction appears to be made between Abiathar sacerdos, who seeks Mary in marriage for his son, and Isacar pontifex, who presides at her Betrothal. This distinction is observed by Lydgate in his *Lyf of oure Lady.*

belong to the series. In *Contemplacio's* first prologue a simple annunciation play only is implied in the line: *Than maryed to joseph • and so folwyng þe salutacion.* This becomes in the later announcement:

> The parlement of hefne sone xal ȝe se
> And how goddys sone come man xal he
> And how þe salutacion aftere xal be.

The *Salutation and Conception* play is again closely connected—in that they are both based on Bonaventura's *Meditatiohes Vitæ Christi*[1]—with the *Purification* play—a play not included in the Proclamation, manifestly an interpolation where it occurs, and the play to which the date 1468 is affixed. Dr. Greg in his *Bibliographical and Textual Problems of the English Miracle Cycles* calls attention to the use in the *Purification* play and in certain passages in the *Return of Joseph* play of a metre not found elsewhere in the collection as evidence of borrowing from an independent source. This source is fairly certainly that from which the *Salutation and Conception*, though not in this metre,[2] is drawn. The *Salutation and Conception* play as we have it would therefore be the result of a further revision of the amalgamated series of Mary

[1] The connexion of the *Ludus Coventriæ* with Bonaventura's *Meditationes* was early pointed out by Hone in his *Ancient Mysteries*, and Halliwell gives the reference to the *Speculum Vitæ Christi* in his note on the *Salutation and Conception* play. The subject has been treated more recently by Mr. S. B. Hemingway in his *English Nativity Plays*, Yale Studies, No. 38. An examination of the borrowing shows that the writer had recourse to Nicholas Love's version of the Meditationes, *The Mirrour of the Blessed Lyf of Jesus Christ*. Not only do words and phrases used by Love in translating occur, but, in one or two cases, phrases or even sentences from Love, for which there is no equivalent in the original. [See Note B.]

[2] Though the different metres in these plays, as has been seen by most critics, undoubtedly give a clue to the stages of revision or sources of compilation, it must be premised that a different metre need not necessarily imply a different stage or source. Some of the writers engaged on these plays were not without power of changing their numbers to suit their theme (cf. the drowsy measures in the *Magi* and the *Resurrection* scenes, the couplet dialogue in the *Trial* scenes), and the changes from one stanza to another (e.g. in the *Magi* and *Shepherd* plays, or in the second *Passion* group, where the change from short to long octaves seems due to a natural movement towards a more emphatic style) may serve an artistic purpose. And on the other hand the 13-lined prologue stanza does not appear to be always of the same type and may not therefore always represent the same stage or source.

plays whereby the *Salutation and Conception* play based on Bonaventura's work was substituted for both alternatives, that of the Proclamation and that of *Contemplacio's* first prologue.

The *Visit to Elizabeth* belongs clearly to the *Contemplacio* series, but there is a curious confusion at the close of this play which suggests revision. As the play ends at present Joseph and Mary are first said to depart, and later Mary is spoken of as remaining with Elizabeth till the birth of John. An alternative ending is, however, suggested. In place of Joseph's speech beginning *Of 30ur dissese thynkys no greff*, a note at the bottom of the folio offers 'si placet' a continuation of Elizabeth's speech : *come I pray 30w specialy | I-wys 3e Are welcome mary | ffor þis comfortabelest comynge good god gramercy*, followed by *Contemplacio* as the name of the next speaker. By the adoption of this version all reference to the departure of Joseph and Mary would disappear. It would seem as if the play originally ended with the first and last stanzas of *Contemplacio's* speech, and that the remainder of this speech (which gives Mary's continued abode with Elizabeth) was intended only to be used as part of the alternative ending. In these additional stanzas the writer appears again to be drawing on Nicholas Love,[1] so that here also we have a trace of this further revision.

On the whole it does not appear rash to conclude that the theory suggested by the renumbering of the pageants is established[2]: namely, that the compiler is here grafting the plays of the 'Contemplacio series'—*The Conception of Mary, Mary in the Temple*, and *The Visit to Elizabeth* with a *Dissponsacion* or *Betrothal* and a *Salutation and Conception* play not used, plays of an ecclesiastical character based at least in part on the *Legenda aurea*,—on to another series of Mary plays described in the Proclamation—*The Betrothal, Return of Joseph, Trial of Joseph and Mary, Birth of Christ*, with a *Salutation and Conception* not used, plays of a simpler and more popular character based directly, as will be shown,[2] on the pseudo gospel of Matthew[3]; and that a later revision or

[1] Cf. Note B.
[2] For additional evidence see p. xxvi and pp. xliv to xlviii.
[3] In both the *Pseudo-Matthaei evangelium de ortu Beatæ Mariæ* and the *Historia de nativitate Mariæ* versions of Mary's story the visit to Elizabeth is omitted.

alteration of purpose led to the substitution of a more elaborate *Salutation and Conception* and to some additions, based on Bonaventura's *Meditationes Vitæ Christi*.

Red ink is used in this MS. (1) for the numbers of the plays and the numbers of the Commandments in the *Moses* play; (2) for paragraph signs marking stanza divisions, looped lines enclosing names of speakers, lines under stage directions, for a few miscellaneous signs—m̃ (ff. 10ᵛ and 16 marking stage directions, fo. 177ᵛ marking a couplet), ⲁ in quires S and T (ff. 168ᵛ–79 and 183ᵛ) and the *Assumption* play to mark couplets, ⱴ on fo. 185ᵛ and in the *Assumption* play to mark Latin versicles ; (3) for the division of two lines written in one to save space here and there when a short perpendicular stroke is made over two similar black strokes or over the point ⁏ (4) for corrections ; (5) for initial capitals for most of the plays, the Commandments, the Psalms in the *Mary in the Temple* play, and a few scattered instances here and there, and for one marginal genealogy (fo. 37ᵛ), and finally red strokes are given to the initial letters of a selection of the first words in the lines of each stanza.¹

Rubrication, &c.

The rubrication is of a rough character ; the initial capitals are ungainly, and the various signs and underlinings, &c., are roughly and carelessly made. An exception is found in the *Doctors' Prologue* standing, as has been shown, by itself, preceded and followed by blank folios. Here each stanza has a shapely red capital, and the lines connecting the rhymes, which elsewhere in the MS. with one exception are black, are here neatly drawn in red. The red initial capitals of the Psalms and versicles in the *Mary in the Temple* and *Visit to Elizabeth* plays (for which liturgical script is used) are also done with more care.

With the exception of the interpolated folios 51, 52, 95, 96, 112, which are unrubricated, all the MS. is rubricated on the same general plan. There are, however, slight variations which seem, like those in the handwriting, to indicate rather that the work was done in different portions at intervals than that another hand was employed. Thus in this, as in the

¹ In this edition all such rubricated letters are printed as capitals as this enables the variation in rubrication, which is more significant than the variation in use of capital forms, to be noted.

writing, quires N, P, Q, R show more carelessness. A more sloping stroke is used for marking the initial letters of the lines, and whereas elsewhere in the MS. three or four of the more important words seem to be picked out, in these quires there are many pages in which every word (*ands* included) has a stroke.

The lines dividing the speeches of the different characters and the lines joining the rhymes (with the exception noted above) are black. In quires N, P, Q, R, and S and T, *x* in black ink stands before stage directions.

Pointing. One or two instances of the use of the point in the line for emphasis or for clearness, to mark an antithesis, or to separate two similar words, or to mark an internal rhyme—(*hevyn and Erth · foulle and best* (fo. 12), *what ȝe xal sey · whan þat ȝe cum* (fo. 223ᵛ), *Ofte ȝoughe is ffowthe with · with sueche vexacion* (fo. 44ᵛ), *þat I haue wretyn · wretyn it is* (fo. 183ᵛ), *And þat he is now · I knowe wel how* (fo. 188)—can be found in most of the plays; but in certain plays or parts of plays the use of the mid-line point becomes more frequent. In a few passages of a specially declamatory character it is used systematically; such passages are the detached *Doctors' Prologue* (ff. 163, 163ᵛ), the *Prologue of the Demon* (which has a specially varied assortment of stops), and that of *John the Baptist*, and the first speech of Annas in the opening *Passion* play (ff. 136-9), Peter's address to the inhabitants of Jerusalem (fo. 143ᵛ), and *Contemplacio's Prologue* (fo. 165). In others, though quite irregular, it still occurs with too great frequency to be considered accidental, and the variations correspond with other differences which indicate the use of material from two sources.

The first passage in the collection in which the pointing becomes noticeable is the episode of Lamech in the *Noah* play (ff. 23, 23ᵛ, 24).[1] This episode is not mentioned in the Proclamation, and may be accepted as an addition to the *Noah* play. The mid-line point next becomes abundant in the group of Mary plays, and here the variations in its use support the division of this group already suggested on other grounds. Pointing occurs to a significant extent in the *Conception of*

[1] The metre of the latter part of the *Noah* play is the same as the metre of this passage, yet the pointing is confined to the Lamech passage.

Introduction xxvii

Mary, the *Mary in the Temple*, the *Salutation and Conception*, and the *Visit to Elizabeth* plays, i. e. the *Contemplacio* group. It is only used in one passage in the *Betrothal* play (fo. 53), and then for the special purpose of marking internal rhymes. It does not occur in the *Return of Joseph* or *Trial of Joseph and Mary* plays. In the *Birth* play again, as in the *Noah* play, the variation in the pointing corresponds with an addition to the original. The only part of the *Birth* play in which pointing occurs with any regularity is the opening episode of the cherry-tree, which is not mentioned in the Proclamation. Of the remaining plays of the first part of the collection the *Purification* alone shows traces of systematic pointing. This play is not included in the Proclamation, and is an obvious interpolation breaking the continuity of the *Magi* and *Massacre of the Innocents* plays. In this play, as verbal reminiscences prove, the writer is again drawing from Bonaventura's *Meditationes* as in parts of the first *Contemplacio* group. It is in the *Passion* plays that most use is made of the mid-line point. Quire N opens, as has been shown, with seven pages (ff. 136-9) in which all the lines are so divided, and though this regularity is not kept up, the point continues to be fairly abundant throughout this quire. The second *Passion* group begins also with systematic pointing in the speech of *Contemplacio*, and the pointing is better maintained than in the former group, continuing to be a marked feature up to the interpolated folio 184, where it ceases abruptly, not being resumed in the remaining folios of quire T, i. e. not in the *Harrowing, Burial*, and *Resurrection* plays. The bearing of this on the question of the composition of the S and T quires has been pointed out.[1]

Of special significance as throwing light on the problem of the connexion of the Proclamation and the collection is the numbering of the plays. This is done in bold red figures in the right-hand margin—in the case of 29 only the number stands at the top of the page—by the original rubricator. The erasure and rewriting of one of the red-lettered names of the apostles at the opening of the *Pentecost* play (fo. 212) in order to leave room for the number disposes of doubt on this point. The numbers 4, 5, and 7 have the earlier Arabic form, which they changed in the course of the

Numbering of the plays.

---

[1] Cf. p. xv.

fifteenth century, and the earlier form is also used in the numbering of the Commandments. The numbers run from 1 to 42 with a duplication of 10, which is written against the closing speech of *Contemplacio* in the *Mary in the Temple* play and against the opening speech of Abysakar in the *Betrothal* play,[1] and with the omission of 17 and 22.

The latter number belongs to the *Baptism* play, and it is not found in the MS. because the original folio containing the opening has been removed at some period subsequent to the compilation to make room for the interpolated folio 112. The omission of 17 is less easy to account for. It occurs in the following sequence: *The Shepherds* play 16, *The Magi* play 18, *The Purification* play 19, *The Massacre of the Innocents* play 20. An erased 1 before the 2 of the 20 in the number of the *Massacre* play is probably connected with the irregularity. It is possible that the compiler's first intention was to put the *Purification* play—which, as has been said, is clearly an interpolation—between the *Shepherds* and the *Magi* plays, the alternative—which he finally chose—being to break up very awkwardly by its insertion the continuous action of the *Magi* and *Massacre* plays. In Tatian's *Diatessaron* the Purification comes between the visit of the Shepherds and of the Magi and in the pseudo gospel, *de Nativitate Mariæ et de infantia salvatoris*, the Purification precedes the visit of the Magi. The order chosen by the compiler is that followed by Bonaventura and Lydgate. There is now only one blank page between the *Shepherds* and *Magi* plays, but there may originally have been two folios corresponding with those removed to make room for the interpolated folios 95 and 96. The correspondence of watermarks in this quire supports this.

The significance of the numbering, however, chiefly appears in the two parts of the *Passion* plays and the *Resurrection* and *Harrowing* plays. Here the rubricator in his numbering, like the writer of the Proclamation in his enumeration of the pageants, is apparently dividing an action, continuous and for practical acting purposes indivisible, into separate sections. It has been held that the Proclamation was written for a pro-

---

[1] This is also evidence that the numbering was done in connexion with the compilation.

Introduction

cessional performance,[1] and the form suggests this, the pageants being described as if they would come forward in order. But a closer examination shows that, in some cases at least, the writer must have had in his mind a group of pageants that could be used simultaneously—that he had in fact a standing and not a processional play in his mind. Pageants 15, 16, and 17 form such a group, and pageants 27 and 29, with the coming and going of Pilate's wife, and pageants 31 and 33 (MS. 23), with the descent and return of *Anima Christi*, could not have been presented successively in a sequence of pageants.[2] But the writer of the Proclamation divides the various episodes between the pageants to form a series more or less corresponding with the usual *Corpus Christi* cycle, and in this the numberer agrees with him. The disregard of the structure of the plays in the numbering is best illustrated by the numbers 28 and 30. No. 28 (fo. 158), dividing the *Betrayal* from the *Last Supper*, is placed after a stage direction which comes in the middle of a speech, so that the speech is actually divided between the two plays. No. 30 (fo. 169ᵛ), dividing the *Trial before Caiaphas* from the *Trial before Pilate*, is placed against Caiaphas's words to the messenger after the episode of the Denial of Peter, and is followed by the coming and going of the messenger and the episode of Judas's return of the money to Caiaphas and Annas, all of which forms part of the action of the preceding play.

That the rubricator had the Proclamation in mind in numbering the plays seems likely in itself. It is supported by the partial correction of the Proclamation, and additional evidence may be found in the numbering of the sections 33–5 (ff. 185, 186, 191). The unequal and arbitrary division of the continuous action of the *Burial, Harrowing*, and *Resurrection* plays as it is divided by these numbers in the MS. (33, the pescent of *Anima Christi*—six stanzas only ; 34, the burial and the setting of the sepulchre guard ; 35, the return of *Anima*

[1] Prof. Hardin Craig's article in the *Athenaeum*, Aug. 16, 1913.
[2] The opening of the *Visit of Elizabeth* with the stage direction *Et sic transient circa placeam* and the speech of *Contemplacio* to occupy the time of the supposed journey of 'myles two and ffyfty' marks a play belonging to a standing group. Note also the phrase *locum interludii* in the stage directions (fo. 23).

Christi with the delivered souls, the appearance to the Virgin, the awakening of the soldiers and the making of the compact between them and Pilate, Annas, and Caiaphas) would seem clearly to be made to indicate as far as possible the correspondence of the plays with the Proclamation.

<small>Evidence of the numbering, &c., as to connexion of Proclamation and compilation.</small>

The close connexion between the numbering and the Proclamation has been hitherto disguised by the alterations made in the former by Halliwell in his edition of the MS. By making the second play begin after instead of before the creation of man, by giving the preliminary matter of the two groups of *Passion* plays, which is not included in the Proclamation, as separate plays—*The Council of the Jews* and *King Herod*, by making a separate play of Pilate's wife's dream, and by making a more convenient division of the *Burial, Resurrection,* and *Harrowing* plays—moving No. 35 so as to include the setting of the guard with the Harrowing instead of with the Burial—by all these rearrangements Halliwell disturbs the agreement of the numbering and the Proclamation; and this has led to the opinion that there is more discrepancy between the Proclamation and the compilation as we have it than is in fact the case.[1]

Most instances of non-correspondence between the two can be shown to be accompanied in the MS. by indications of interpolation or substitution which suggest an explanation. The first instance is the omission in the Proclamation of any mention of the Lamech episode in the description of the fourth pageant, and it has been shown that this bears signs of being an addition to the original *Noah* play. The next discrepancies are found in connexion with the first group of *Contemplacio* plays (Nos. 8–13), and these have been shown to be due to the amalgamation of a new series of Mary plays with those described in the Proclamation and some revision of this amalgamation; and with this is connected the insertion into the middle of the *Magi-Massacre* play of the *Purification* play, which is also (like Nos. viii, ix, xiii) not included in the Proclamation—a play from the same source, Love's version of the *Meditationes Vitæ Christi*, as parts of the *Contemplacio* group.

---

[1] Miss Swenson in her *Enquiry into the Composition of the Ludus Coventriæ* makes a strong case for the close connexion of the Proclamation and the *Ludus Coventriæ* as we have it.

Introduction xxxi

The omission from the Proclamation of the preliminary matter of the two *Passion* groups is explained by the fact that the compiler seems here to have abandoned his plan of arranging a cycle of separate plays and to have incorporated two sets of plays on the Passion which lay to his hand as they had been used for separate yearly performances, using apparently copies already made. With regard to the first *Passion* play the series used corresponds fairly with the Proclamation; the incidents of the ass and the foal and of Mary Magdalen and the box of ointment, which are not mentioned in the Proclamation, are, as has been shown, interpolations in the MS. The Proclamation itself is, however, disturbed at this point, and the use, for the first time, of one stanza for two pageants suggests that the compiler is reconsidering his plan. With the second *Passion* play the divergence is marked. Here there are discrepancies which seem to be fundamental between the Proclamation and the plays.

As regards the preceding portion of the MS., what the MS. shows to be due to interpolation or addition or revision is either absent from the Proclamation or added in manifest revision. The writer had, if not a cycle already in existence, at least a collection of plays or groups of plays ready for compilation in his mind. At this point, however, the plays diverge from the Proclamation; the material of the S and T quires belongs to a different version of the Trial and Death than that described in the Proclamation, and it is material that the MS. shows to be due to interpolation—the descent from the Cross of *Anima Christi*—that corresponds with the Proclamation. The Trial, according to the Proclamation, is before Annas, Caiaphas, and Pilate only; no mention is made of Herod, who does not appear in the *Contemplacio* prologue either. The preliminary scene with Herod can be dispensed with, but the subsequent examination before Herod is an integral part of the action; and the pageants 27 and 28 represent a different treatment of the incidents of the Trial from that of the play. According to the Proclamation the three thieves appear in the first trial before Pilate, at which Pilate's wife is also present, and the remorse and death of Judas has a pageant to itself; in the play the three thieves and Pilate's wife appear in the second scene with Pilate, and the Judas incident occupies only eight lines with two stage directions. There is no mention in the Proclamation of the Veronica incident, and the Proclamation puts the Longeus episode before the descent of *Anima Christi*—i. e. at the close of the Crucifixion as in the York, Towneley, and Chester cycles—instead of after it, at the beginning of the Burial scene, as it stands in the play. On the other hand, the material which the evidence of the MS. shows to be interpolation on folios 183ᵛ–6—the removal of the Virgin by John and the descent of *Anima Christi*—corresponds with the descriptions of pageants 30 and 31 in the Proclamation. With the setting of the sepulchre guards the divergence comes to an end. This coincides with the beginning of a new quire (U), where the metre changes to that used for the interpolated portions on ff. 183ᵛ et seq., and there are, as has been seen, various indications of change of material.

The discrepancies in the case of the *Ascension* and *Pentecost* plays are accompanied by obvious signs of haste and incompleteness in the compilation. The *Ascension* play, to which only the latter part of a stanza is allotted in the Proclamation, breaks off short after the speech of the first of the two angels announced in the Proclamation and given in the stage direction, and the incident of the choice of Matthias, not mentioned in the Proclamation, follows briefly treated. A gap is left in the MS. after the angel's speech, and the speech proposing the choice of Judas's successor is left without any name of speaker. Similarly the *Pentecost* play, a single folio, is manifestly a fragment. The omission of the *Assumption* play from the Proclamation is accounted for by its being, as the MS. shows, an interpolation.

In all cases, therefore, except in the second *Passion* group in the S and T quires, the variation of Proclamation and text is accompanied by some sign of disturbance in the text or, in the case of the first *Passion* group and the *Ascension* play, in the Proclamation.

Conclusions to be drawn from examination of MS. 1. Compilation.

The general evidence of the various features of the MS. that have been examined shows that the collection contains parts or the whole of four separate groups: (1) the composite *Contemplacio* group (viii to xiii); (2) the first *Passion* group (xxvi to xxviii); (3) the second *Passion* group (xxix to xxxii), dovetailed on by means of the *Descent into Hell* (xxxiii), of different style, to

Introduction

a *Burial* play (xxxiv) of similar style, which in its turn is joined (p. 314, beginning of U quire) to a (4) *Resurrection and Harrowing* play connected in style with xxxiii and forming a group with the *Three Maries* and, as it stands in the compilation, with the *Mary Magdalen* play. An examination of the text gives two[1] more groups showing that (5) the first three plays (*Creation, Fall,* and *Cain and Abel*) and (6) the *Visit of the Magi* and *Massacre of the Innocents* and *Death of Herod* (xviii and xx) form respectively continuous sets, diction and metre connecting also the *Shepherds* play with the latter suite.[2] There are also two interpolated separate plays: *The Purification* and *The Assumption*.

The evidence as to the composition of the series to be drawn from the characteristics of the MS. is complicated and often ambiguous, but the following points emerge:

1. MS. Vesp. D. viii is the compiler's book, not a transcript of another MS.
2. It contains a collection of plays made according to a plan which was subject to alteration as it proceeded.
3. Some of the plays and groups of plays had had a separate existence, having been acted as separate plays or groups.
4. One portion of the MS. certainly, and probably two, quires N, P, Q, R, and quires S, T, have also had a separate existence.

In support of 3, besides the evidence already adduced, may be noted the preservation of their distinctive character by the groups in the series, by the *Contemplacio* group, and especially by the two *Passion* groups the stage directions of which are conspicuously different from any others in the collection; the variety of the headings of the single plays—*introitus* (ff. 20^v, 25^v, 31), *modo de* . . . (ff. 106, 212), *hic incipit* . . . (ff. 127^v, 201, 210, 223^v), *hic intrabit pagetu*m *de* . . . (fo. 75); the conclusion of the *Disputation* play addressed to *All þat hath herd þis consummacion of | þis pagent* as to the audience of a separate performance, and to a less extent the sermon-like conclusions of the *Temptation*

---

[1] Or three—the *Baptism* and *Temptation* plays might be taken as continuous.

[2] The general use of stock alliterative phrases makes the evidence to be drawn from diction unreliable, but the following phrases which occur both in the *Shepherds* and *Magi* plays are sufficiently individual to be allowed to count: 'his (þat) bryght blood'; 'in (by) a bestys bynne'; 'Heyl blome on bedde'; and 'To þe blosme upon his bedde'.

and the *Woman taken in Adultery* plays; and the fact that—
as the modernizing revision of certain plays, the added stage
directions (e. g. p. 327), and the worn condition of the *Magi*
portion of the MS. show—certain plays continued to be so
acted.

    5. The compiler had command of other versions of plays or
       groups of plays from which he drew.

The evidence for this is mainly to be deduced from the
variation of the plays from the description of them in the
Proclamation, chiefly in the first *Contemplacio* group and in
the two *Passion* groups, but it is supported by the fact that
the later users of the MS. obviously had access to such other
versions (cf. the references in the later hand in the *Harrowing*
play to *anima latronis*, fo. 185ᵛ, to extra speeches of the devil and
of *anima caym*), and by the presence in the MS. of the detached
Doctors' Prologue written on a blank folio at the end of quire R.
This introduces characters not found together in any play or
group of plays in this collection—the eleven apostles, John the
Baptist, and St. Paul, who only appears in the *Assumption* play.
It would seem to be part of another group in the possession of
the performers of the *Passion* groups, copied for convenience on
a blank sheet of the MS. of the latter. So the allusion to the
1st and 2nd prophets in notes (pp. 240–1) appears to refer to
some other combination of plays than that of the text.

    The evidence of the MS. supports the view that a compiler is
putting together parts to make a whole rather than the view
suggested by Dr. Foster (*Northern Passion*, E.E.T.S. 147, p. 99)
that a reviser has separated a whole into groups, though he drew
apparently on a cycle—or the remains of a cycle—of plays in
thirteen-lined stanzas as well as on single plays and on groups
of plays. To unravel the tissue of compilation and revision
in these plays demands a full study of literary and linguistic
characteristics. The examination of the MS. affords no con-
clusive evidence on such questions, but it gives much to support
the theory that the MS. represents a selection from the repertory
of a body of ecclesiastical actors. As Dr. Smith says: 'Videntur
olim coram populo sive ad instruendum sive ad placendum *a
Fratribus mendicantibus* repraesentata.' Nor does the evidence
from the MS. throw clear light on the question of the identity

of the compiler with any of the writers or revisers of the plays, but some inferences can be drawn.

The relation of the plays to the Proclamation shows, as has been said, that the Proclamation was composed in its present form in close connexion with the compiling of this series. The relation is particularly interesting at the point of divergence in the second *Passion* group. Here the interpolated portion (ff. 184, 185, 186) in the text corresponds with the Proclamation, from which the rest of the text diverges, and that with regard to a feature—the division into two scenes of the Harrowing— that is peculiar to this treatment of the theme. The inference suggests itself that the compiler was himself the writer of this link passage, which again is connected with the latter part of the *Resurrection* play and through that, by the evidence of metre and more elusive evidence of diction, with the *Magi* plays. The conclusion to be drawn from this agrees with that of Mr. Greg (*Problems of English Miracle Cycles*) in that it associates the composition of the *rime couée* portions of the cycle with the process of compilation. Dr. Greg postulates a further overworking by the writer of the *Contemplacio* prologues and possibly of the long octave plays and passages, who would therefore be the actual compiler of the Vesp. D. viii MS. But there is evidence [1] that the long octave *Contemplacio* plays underwent a further revision by a writer drawing on Bonaventura's *Meditationes Vitæ Christi*, to whom, as well as the *Salutation and Conception*, the *Purification* play [2] must be attributed. Now it would appear unlikely, had this reviser been the compiler, that the *Purification* play should not be mentioned in the Proclamation, and that the description of the *Salutation and Conception* in the Proclamation should not have been made to refer to the compiler's own work. The theory that the writer of the *rime couée* parts, with which the corresponding passages in the Proclamation agree closely, is the compiler seems to present the stronger case. This writer draws also on Bonaventura,[3] and both revisions are probably connected with the compilation.

This MS. has known the turn of fortune described by Robert Hegge, its first recorded owner, when after the invention of

2. Compiler.

Marginalia.

---

[1] Cf. pp. xxi to xxv; xlv, xlvii.
[2] It is to this play that the date 1468 is attached.
[3] Cf. p. xlix.

printing 'old MSS. were stright bequeath'd to the Moths: and pigeons and Jack daws became the only students in church libraries'[1]; and many of its blank pages have been used for idle scribbling—attempts at copying the MS. writing, signatures, stray phrases, mostly in sixteenth-century hands. It would appear that it fell early into irreverent schoolboy hands, for some of the scribbled copying of lines of the text seems to be rather rough contemporary work than later imitation of an older script. The first or last line of a folio is often imitated, and in one instance a passage of some length has been carelessly and roughly copied on the blank page opposite (fo. 201). A jumble of ill-formed Greek and fancy letters on fo. 119^v with what appear to be notes on a fraction sum on the opposite folio suggests the hand of a schoolboy. The names that occur among these scribblings are Wylliam Dere (91^v, 136), Polerd (91^v), Hollond, Johan & Hary (151^v, 152^v, 153^v, 155^v), H Kinge the yownger (111^v), John Hasycham (91^v), and on the same page (91^v) 'John Taylphott of parish Bedonson' with the motto 'wee that will not when we paie [sic] when we would we shall find (? or saie) nay'. The oddness of the personal name throws doubt on that of the parish, and no record of a parish of Bedonson has been found.

By a coincidence that is possibly nothing more the names William Kinge the younger his booke 1656, John King his brother and John Holland of Brabant occur among the scribblings in the Chester plays (MS. Add. 10305, ff. 55 and 111); the motto quoted above is also found (fo. 124) in a four-lined form, of which, however, unfortunately the first line is missing.

Of a different character are the two signatures of R. Hegge, Dunelmensis, the title of the plays, and some annotations of the original scribe. The first signature comes on fo. 10 above the opening of the first play. It consisted of the full name Robert Hegge, Dunelmensis, written in large Roman hand, but, having been cut away with the margin, the Christian name is now indecipherable. The second, on fo. 164, the outside sheet of the S and T quires, is more elaborate. At the top of the page in Roman type is written: 'In nomjne Dei. Amen'; in the middle of the page, in a slender cursive hand, 'ego R. H.

[1] *The Legend of St. Cuthbert*, by R. Hegge.

Introduction   xxxvii

Dunelmensis possideo'; and immediately beneath, οὐ κτῆσις [1] ἀλλὰ χρῆσις.

The title 'The plaie called Corpus Christi' in a small Elizabethan hand stands at the top of fo. 1 above the speech of the first vexillator. Mr. Hemingway in his *English Nativity Plays* speaks of this title as being in Hegge's handwriting, but this is not so. It is writing of an earlier and altogether different type.

The notes forming part of the original MS. consist of three genealogies, of Adam to Noah (ff. 16ᵛ et seq.), of Noah to Abraham (ff. 21 et seq.), of the Virgin Mary (fo. 37); the measurements of Noah's ark (fo. 24); a list of the five Annas of scriptural importance (fo. 37ᵛ) and three dates of the ecclesiastical calendar (fo. 74ᵛ). The genealogies are in liturgical script, the note on the dates and on the Annas in smaller bookhand, and the genealogies are handsomely rubricated, one part of that of Mary being entirely in red. These marginal additions give support to the conclusion to be drawn from the general characteristics of the compilation, that it is of ecclesiastical and not of civic origin.

The MS. forms part of the Cottonian collection. It is not included in the 1621 catalogue of the collection (MS. Harl. 6018), and is supposed to have been acquired in 1629 by Sir Robert Bruce Cotton's first librarian, Richard James, on the death in that year at Oxford of Robert Hegge, a member of James's own college, Corpus Christi, in whose possession it had been. Richard James became Sir Robert Cotton's librarian before 1628 [2] and a number of letters exist (MS. Cott. Julius C. III, ff. 212, 214, 217, 219) written by him to his patron from Oxford —unfortunately undated as to the year but clearly belonging to this period—referring to transactions in books, coins, and other rarities. The note on the fly-leaf of the MS. giving the contents is in James's handwriting. The MS. is included in the 1696 catalogue of the Cotton collection drawn up by Dr. Thomas Smith. *History of the MS. and its title.*

The problem of the *Ludus Coventriæ* begins with James's note on the fly-leaf: 'Elenchus contentorum in hoc codice [Vespasian D. viii, added in a later hand] Contenta novi testamenti scenicè expressa et actitata olim per monachos sive *fratres* *R. James.*

---

[1] The word appears to have been first written κησις, and the τ to have been inserted. The οὐ is written ȣ, and the accents on κτησι and χρησις are omitted.

[2] *Dict. of Nat. Biog.*

mendicantes · vulgò dicitur hic liber Ludus Coventriæ · sive ludus corporis Christi · scribitur metris Anglicanis.'

The description of the contents is inaccurate in omitting the Old Testament plays,¹ and the connexion of the collection with Coventry is not supported by any evidence.

T. Smith. In the description of the MS. in the catalogue drawn up by Dr. Thomas Smith in 1696 the reference to Coventry is omitted, and account is taken of the Old Testament material: 'Vespasianus D. viii. A collection of plays in old English metre, i.e. Dramata sacra in quibus exhibentur historiae veteris et N. Testamenti, introductis quasi in scenam personis illic memoratis quas secum invicem colloquentes pro ingenio fingit Poeta. Videntur olim coram populo sive ad instruendum sive ad placendum a Fratribus mendicantibus repraesentata.'

W. Dugdale. James's inscription had, however, in the meantime been accepted by Dugdale, and the passage in his *Antiquities of Warwickshire Illustrated*, in which he describes the Corpus Christi pageants of the Grey Friars of Coventry, referring to this MS. as giving the text of the performance, became the authority for many subsequent writers on the subject of these plays.² It runs as follows: 'Before the suppression of the Monasteries their City [i.e. Coventry] was very famous for the pageants that were play'd therein, upon Corpus Christi day; which occasioning very great confluence of people thither from far and near was of no small benefit thereto; which pageants being acted with mighty state and reverence by the friers of this house [i.e. the Grey Friars] had Theaters for several scenes, very large and high, placed upon wheels and drawn to all the eminent parts of the City for the better advantage of Spectators: And contain'd the story of the New Testament, composed into old English Rithme, as appeareth by an antient MS. ᵖ intituled Ludus Corporis Christi or Ludus Coventriæ [a note in the margin here gives the reference: "ᵖ In bibl. Cotton. sub effigie Vesp.

---

¹ Is it possible that this mistake has any connexion with the fact that R. Hegge's second signature is found on a blank folio preceding the second *Passion* group which shows signs of having been at some time an outside leaf?

² Especially perhaps after the publication in 1722 of Stevens's additional volumes to Dugdale's *Monasticon Anglicanum* in which he prints the first five plays of Vesp. D. viii. Cf. Thomas Pennant, *Journey from Chester to London*, 1782, and 'Q' in *Gentleman's Magazine*, Feb. 1784.

Introduction

D. 9" (so apparently by a slip¹ for viii)]. I have been told by some old people who in their younger days were eye-witnesses of these pageants so acted that the yearly confluence of people to see that show was extraordinary great.'²

The fact that Dugdale follows James in omitting the Old Testament plays shows that he had not examined the MS. Vesp. D. viii for himself. He appears to be accepting the description of the fly-leaf from which the words '*composed into old English Rithme*' seem to be translated. Thomas Sharp in his *Dissertation on the Coventry Mysteries*, 1825, was the first to point out that the Coventry civic cycle was distinct from the *Ludus Coventriæ* preserved in MS. Cott. Vesp. D. viii. The two surviving plays of this cycle and the information concerning it to be derived from Sharp's (and Halliwell's) extracts from MS. sources destroyed in the Birmingham Library fire in 1879, and from other Coventry civic records still preserved, are accessible in the E.E.T.S. edition of Prof. Hardin Craig, and it is now recognized that there is no connexion between *The plaie called Corpus Christi* and the Coventry civic cycle; and the comparison, where it is possible, notably between the two *Shepherds* plays or between the two *Disputation* plays, serves to bring out strongly the more ecclesiastical character of the present collection. [Thomas Sharp. Coventry civic cycle.]

Thomas Sharp accepts Dugdale's attribution of the *Ludus Coventriæ* to the Grey Friars of Coventry on the general ground of his local knowledge. Later scholars have pointed out that the *old people* to whose reminiscences Dugdale refers—speaking evidently of information he has gathered as an adult, not of stories heard as a child—must have been recalling the civic pageants which were only finally 'laid down' in 1580,³ and not the performances of the Grey Friars whose house was dissolved in 1538. Sharp himself speaks somewhat dubiously of the evidence for the acting of the Grey Friars: '*a solitary mention in one MS. not older than the beginning of Charles I's reign of Henry VII's visit to the city in 1492 "to see the Plays acted by the Grey Friers"*'⁴; and this piece of evidence has lately been [Acting 'by the Grey friers,' Coventry.]

---

¹ Unless this is again connected with the separate existence of the *Passion* plays.

² The *Antiquities of Warwickshire Illustrated*, &c., by William Dugdale, 1656.

³ Sharp, *Dissertation*, p. 12.   ⁴ Ibid., p. 5.

weakened if not destroyed by the discovery that in a compilation of the MS. annals of Coventry (Harl. 6388) made by Humphrey Wanley—himself a Coventry man—and said by him to be 'taken out' of a collection of eight MSS., the entry given by Sharp under 1492 is found under 1493 in the following form: '*The King and Queen came to see the playes at ye greyfriers and much commended them*', where the reference is clearly to the place and not the actors. Yet it might perhaps still be questioned whether Dugdale, a careful antiquary, who would have access to the MSS. used by Wanley, having also a personal connexion with Coventry, and living nearly three-quarters of a century nearer the time, could have been misled, by an ambiguous entry and confused local tradition, into the invention of the acting of the Grey Friars.

In any case his association of the MS. Vesp. D. viii with Coventry seems to have been secondhand—a repetition of the statement of James's note.

In the difficulty of explaining the note Halliwell has recourse to the theory of a lost colophon which supplied James with the information. The suggestion might be hazarded that James in the sentence '*vulgò dicitur hic liber Ludus Coventriæ, sive ludus corporis Christi*' is using the words *Ludus Coventriæ* generically, taking the name of the best-known example as the name of a type of dramatic performance. The words *vulgo dicitur* suggest this. It may be gathered from the anecdote in the 'Hundred Merry Tales', quoted by Halliwell, of the Warwickshire village priest who referred his parishioners for confirmation of his doctrine to the Coventry Corpus Christi play, that Coventry was the accepted background for a story concerning the Corpus Christi pageants; Heywood's allusion in the *Four P's* proves that Coventry was recognized as the local habitation of the Mystery devil; and there is evidence that the name Corpus Christi had come to be used for the collective mysteries apart from the occasion of their performance. Weaver, in a passage[1] also quoted by Halliwell, says of a play the subject of which was 'the sacred scriptures from the creation of the world', 'They call this Corpus Christi play in my country'. The title written in a sixteenth-century

---

[1] John Weaver, *Funeral Monuments*, 1631, p. 405.

Introduction

hand on the first page of the present collection, 'The plaie called Corpus Christi', is apparently used in this wider sense, and James in his note on the fly-leaf may be expanding this title: 'vulgò dicitur hic liber Ludus Coventriæ sive ludus corporis Christi', i.e. 'This book is commonly called the Coventry or Corpus Christi play'.

The printing of plays from MS. Vesp. D. viii began in 1722 in John Stevens's additional volumes to Dodsworth's and Dugdale's *Monasticon Anglicanum* (vol. i, pp. 139-53 under Grey Friars of Coventry). Stevens prints the Proclamation and the first five plays, using Gothic type, translating the stage directions, and including the marginal genealogies. Printing of the MS. J. Stevens.

A century later in 1823 Hone in his *Ancient Mysteries described* gives a summary, illustrated by full quotations, of plays viii to xv (incl.) with parallel passages from the Apocryphal Gospels and various illustrations including the 'parliament of heaven' from an English translation (not Love's) of the *Speculum Vitæ Christi*, and the *Cherry-tree* carol. W. Hone.

In 1836 J. Payne Collier included *The Marriage of the Virgin Now first printed from MS. Cott. Vesp. D. viii* in *Five Miracle Plays or Scriptural Dramas* privately printed. J. Payne Collier.

In 1838 William Marriott, Ph.D., in his *A Collection of English Miracle Plays or Mysteries containing ten dramas from the Chester, Coventry, and Towneley series with two of later date*, published at Basle, included two of the *Ludus Coventriæ* plays, *Joseph's Jealousy* and *The Trial of Joseph and Mary*, with one of the Coventry civic cycle, *The Pageant of the Company of Shearmen and Tailors*. W. Marriott.

In 1841 the whole MS. was edited by J. O. Halliwell for the Shakespeare Society, with introduction, notes, and glossary. Various plays have since appeared in representative collections. J. O. Halliwell.

The opening of the *Salutation and Conception* play is included in Mr. A. W. Pollard's *English Miracle Plays* (1890); this play and *Noah and Lamech* are included in Mr. J. M. Manly's *Specimens of Pre-Shakespearian Drama* (1900); and Mr. S. B. Hemingway in *English Nativity Plays* (1909) gives plays 11, 12, 13, 15, and 16 from this series. A. W. Pollard; J. M. Manly; S. B. Hemingway.

In 1915 the *Assumption* play was edited with an exact reproduction of the features of the MS. by Dr. Greg. W. W. Greg.

The present edition.

Such an exact reproduction has not been attempted in the present edition. Stops have been added at the end of stanzas or speeches, though otherwise the text is left unpunctuated as in the MS., the stops which occur, and are reproduced, being metrical except in the case of a few stage directions. A few medial capital letters have been omitted, a few words run together by the scribe have been separated, a few disconnected syllables have been hyphened, where the exact reproduction appeared disproportionately disturbing to the reader. It has not been thought necessary to record all the scribe's slips of the pen. In any special case a note has been appended. *I* and *J*, used indifferently in the MS., have been distinguished according to modern usage and the capital form substituted for the small in the case of the personal pronoun; *z* and *ʒ* have also been distinguished. The lines between the speeches and the lines connecting the rhymes have been omitted. In the MS. the tail-verses of the thirteen-lined stanzas and of the *rime couée* stanzas are sometimes written in the margin, outside the rhyme brackets of the rest of the stanza; such lines are printed as short lines in the stanza. Lines written continuously for reasons of space have been arranged in stanzas, with a note recording the change (cf. p. 120, ll. 125 et seq.); otherwise the irregularities of the MS. have been preserved.

Some of the contraction marks used have clearly lost their significance. This is notably the case with ñ. It is the normal form of *n* for the scribes of fo. 51 and of folios 95, 96, 112. The scribe of the MS. writes *trońe* (fo. 9), *goñe* (fo. 69), *somowñe* (fo. 74ᵛ), and the carelessness with which the stroke is added or not to a series of rhyme-words suggests that it had become a mere flourish. This is also true of *ď*, which occurs here and there as it were accidentally, though the form *viriď* for *viridi* (fo. 25, *cum ramo viriď*) shows that it retained its significance at need. The scribes of fo. 51 and of folios 95, 96, 112 use *ď* for all final *d*'s, and write *de* in full. The case is less clear with regard to *ŧŧ*, though *aŧŧe* and *haŧŧe* are found (fo. 81). On fo. 33 (p. 55, l. 132) a final *e* seems to have been erased in *kyŧŧ*; *knowlacħe* (fo. 110ᵛ) is found, but also *scħ* for *sche* (fo. 78ᵛ and elsewhere). These signs of doubtful significance, *ď, ǵ, ħ, ŧŧ, m̃, ñ*, have been retained in the text.

Introduction

The sign ⁀, not reproduced in the text, is abundantly used in the usual ways: over an *n* after *u* or *w*, e. g. *grou͡ndyd, geaw͡nt*; over *u* or *w* in the combination *ous, ows*, e. g. *gracyous, leprow͡s*; over *u* or *n* occasionally in other cases, e. g. *secundu͡s, opyn͡d*; as sign of a contracted *n* or *m*, e. g. *þy͡g, hy͡*; exceptional instances of this use are *a͡d* (*and*), fo. 193, and on the same folio *a͡resyn* (*am resyn*); for other contractions in *plentevu͡s* (*plentevous*), fo. 137, *co͡nawnt* (*counawnt*), fo. 170, *don͡* (*done*), fo. 69ᵛ, p. 114, l. 165, and *don͡* (*down*), fo. 126. Against *an͡*, fo. 134, *non* has been written as a correction in the margin; *n͡*, fo. 63ᵛ, p. 105, l. 244, stands for *no*. Occasionally this sign serves no purpose, as in fo. 27ᵛ, p. 47, l. 129, *hevy͡ne*; fo. 212ᵛ *crepp͡e*.

The contraction ꝭ for the plural has been expanded as *ys*, as in the fully written plural nouns this form predominates. *Es* and *is* occur, though less frequently. On fo. 202ᵛ *pilgrimes* and *pilgrimys* are found in following lines, elsewhere *sowles* and *sowlys*, &c. On the other hand, on fo. 79, p. 131, l. 234, *legges* has apparently been corrected to *leggys*; and on fo. 209 *eretykis* has been corrected to *eretykys*. There is one instance of the use of the ꝯ for *com* on fo. 112—com*pany*, p. 188, l. 19—and it is possible that the unintelligible *hese juge*, p. 116, l. 34, may be due to the scribe's misreading of this contraction in *considerynge*; the corresponding Latin is: *considerans sui senectutem*. Other contractions occurring once or exceptionally are *mayde̅* (*mayde*n), fo. 90ᵛ, p. 149, l. 102; *hꝭ* (*his*) used by the scribe of folios 95, 96, 112; *wepon̄*, fo. 160 (stage direction) (? *weponys*); *bᵗ* (*but*), fo. 162. The recognized contractions for *er, ur, us, n, m*; *i* in *ion*; final *e* after *r, p, ȝ, t*; *p* + *ra, re*, &c., and those in the Latin passages, have been expanded in italics. Overwritten as well as omitted letters are italized, e. g. *þⁱ* = *þi, þᵘ* = *þou, þᵉ* = *þe*.[1]

The *Ludus Coventriæ* differs from the other Mystery cycles in having, in its New Testament plays at least, a closer dependence on known literary sources. Its borrowings have been

Notes on sources and literary relations.[2]

---

[1] By inadvertence the two forms 'þat' and 'þat' have been used to represent the MS. form 'þᵗ'. In a few instances also 'þer', 'our', and 'ȝour' have been given for 'þer', 'our', and 'ȝour'.

[2] I have not been able to consult *Sprache und Heimath des sogenannten Ludus Coventriæ* by Dr. Max Kramer (1892); nor *Die Quellen des sogenannten Ludus Coventriæ* by E. Falke (1908).

investigated by various writers. Halliwell gives references to the Apocryphal Gospels, to Lydgate's *Lyf of oure Lady*, and his anagrams on the name of the Virgin (cf. *Ludus Coventriæ*, fo. 47, and the *Minor Poems* of J. Lydgate, ed. McCracken, p. 303), to the *Speculum* (or *Meditationes*) *Vitæ Christi* of Bonaventura, and to the thirteenth-century *Harrowing of Hell* (Harl. 2253) (cf. *Ludus Coventriæ*, fo. 192^v, p. 320, ll. 1416 et seq., and lines 43, 44, 31, 55 of the *Harrowing, Altenglische Dichtungen*, ed. Böddeker, p. 271). Hone in *Ancient Mysteries described* gives in more detail the parallels between the Mary plays and the Apocryphal Gospels, and quotes from an English translation of the *Speculum Vitæ Christi* to illustrate the *Parlement of Hefne*. Miss Hope Traver in *The Four Daughters of God* (Bryn Mawr Monographs, No. 6), has discussed the relation of the *Parlement of Hefne* to other English versions of the theme. The use of the *Legenda aurea* for the *Assumption* play has been shown by Mr. Gayley in his *Plays of our Forefathers*, and by Dr. Greg, who in his edition of the *Assumption* prints the chapter from the *Legenda* used by the writer. The most recent discovery has been that of Dr. Foster, who has demonstrated the indebtedness of the writer of the *Passion* plays to the *Northern Passion*. The coincidence of the names of three out of the four Knights of Pilate in the *Ludus Coventriæ*, fo. 189, and the poem on the Resurrection, MS. Ashmole 61 (fo. 138), has been pointed out by E. Falke.

So much of the material used by the writer of Mysteries was common to different possible sources that it is difficult to determine which was the one actually used. Thus the chapter in the *Legenda aurea*: '*De Nativitate Beatæ Mariæ virginis*' (ed. Graesse, 1846, ch. 131) is in great part a transcription of the Apocryphal Gospel *De Nativitate S. Mariæ*, and the material that is versified in the *Conception of Mary* play (the description of Joachim, ll. 25-8; his rejection, ll. 76-80; the angel's speech, ll. 149-74, with the exception of the reference to *joys fyff*) might equally well have been taken from the one or the other. The connexion of the fifteen steps of the Temple with the fifteen Psalms of degree (*Mary in the Temple*, ll. 84 et seq.) is common to the two sources, and found also in both is the original of the interpolated passage in the *Betrothal of*

Mary on fo. 48ᵛ, p. 82, ll. 92 et seq. (cf. *Tunc anxiatus est pontifex eo quod neque contra scripturam quae dicit: vovite et reddite, votum infringendum putaret nec morem genti insuetum introducere auderet*); both give the answer to *Vox*; cf. *The Betrothal*, p. 89, l. 230.

It can, however, be shown that the *Legenda aurea* was the actual authority used for the *Conception of Mary* at least. The marginal genealogies on the folios preceding the *Conception of Mary* are taken from chapter 131 of the *Legenda aurea*; the phrase *regale sacerdocium* (cf. the *Conception of Mary*, l. 15) occurs in it and not in the gospel *De Nativitate*; and the words used of the rejection of Joachim's offering in the *Legenda aurea*, *cum indignatione nimia repulit*, seem rather to be echoed in '*with grett indygnacion þin offeryng I refuse*', than the words in the gospel: *despexit eum et munera eius sprevit*.

The same difficulty occurs in attempting to distinguish the claims of Bonaventura's *Meditationes* and of Lydgate's *Lyf of oure Lady* to be the original of parts of these plays, though the writer's use of Love's translation of the former in certain cases (cf. Note B, p. lviii) gives a clue, and the *Salutation and Conception* and the *Purification* plays may safely be attributed to the influence of the *Meditationes*. To this is due the added passage in a different metre in *Joseph's Return* to which Dr. Greg draws attention (*Bibliographical and Textual Problems of the English Miracle Cycles*, p. 129). The altered tone is consonant with the refined treatment of the theme in the *Meditationes*, and two slight verbal coincidences indicate this source. Cf. '*For on the tone side he sawh hir lyf so holy and no tokene of synne in hir*' and '*I knew never with here so god me spede | tokyn of thynge in word or dede | þat towchyd velany | nevyr þe les*', et seq.; and '*this tribulacion and this desese*' and '*for vnknowlage he is desesyd*' in the same context. From the *Meditationes* comes also the interview between the risen Lord and his Mother. That our Lord appeared first of all to the Virgin is a theory discussed in the *De Resurrectione* chapter of the *Legenda aurea*. Bonaventura describes the appearance: *Sodeynly oure lord Jesu came and aperede to her, and in alther whitest clothes with a glad and louely chere gretynge hir on side half in these wordes: Salue sancta parens that is to say Haile holy moder. . . . My dere moder I am.*

Ego sum? resurrexi et adhuc tecum sum I have uprisen and loo ʒit I am with thee . . . and deth and sorwe and alle peynes and angwische I haue ouercome. . . . And so thei bothe louely and lykyngly talkynge togidre maden a grete joyful feste. This seems to be the original of the passage *Salue sancta parens? my modyr dere,* fo. 192ᵛ, and it is recalled also by the words in the Proclamation, l. 431, *with suche cher and comforth his modyr he doth indew | þat joy it is to here · þer specħ for to devyse.*

Less clear is the source of the material used in the *Mary in the Temple* and *Visit to Elizabeth* plays which is found alike in the *Meditationes,* the *Lyf of oure Lady,* and in part in the *Legenda aurea* and the gospel *De Nativitate*.

The name of Lydgate was early associated with the *Ludus Coventriæ.* In Bishop Tanner's *Bibliotheca Britannico-Hibernica* (1748) these plays are doubtfully identified with the 'pageants' attributed to Lydgate.[1] Halliwell, as has been said, draws attention to the likeness of the passage on the Virgin's name to poems of Lydgate; the hymn sung by the shepherds—*Stella celi extirpauit,* fo. 90—is one that Lydgate translated (*Minor Poems,* ed. McCracken, p. 294). There is also in the *Christ with the Doctors* play an echo of a verse in the *Lyf of oure Lady,* which might however be dismissed as due to the use of a commonplace of theological illustration: cf. *L. C.,* p. 181, ll. 97 et seq., and *Lyf of oure Lady*: ed. 1531, Ch. xix, *Ryght as the sonne perissheth thorowe the glasse | Thorowe the cristall, berall or spectacle | Withoute harme, right so by myracle | Into her closet the father's sapyence | Entred is withouten violence.*

It seems at first sight in favour of the *Lyf of oure Lady* as a source for *Mary in the Temple* that it includes the story of the fifteen steps not found in the *Meditationes, L. C.,* p. 74, and the seven petitions, *L. C.,* p. 79, not found in the *Legenda aurea* or the gospel *De Nativitate.* The imitation of Lydgate's manner in

---

[1] By a fatality attaching to references to Vesp. D. viii his description contains an error. He apparently noticed the change of writing with the *Assumption* play and did not look further:

'A procession of pageants from the creation. In MS. Cotton. Vespas. D. viii. Scripto circa Lydgati tempora sunt collectanea (in 212 paginis in folio) 40 pageants sive ludorum incipiendo a creatione et finiendo tempore descensus Spiritus S. Jamesius existimat hunc fuisse Ludum C.C.C. (sic) sive Ludum Coventriae'.

the anagram (p. 80) favours this view. Against all this there is, however, one piece of evidence that is conclusively in favour of the *Meditationes* as against the *Lyf*. In this work the three precepts observed by Mary are distinct from the seven petitions, though they correspond with the first three. Lydgate gives only the seven petitions. The writer of *Mary in the Temple* follows Bonaventura, and gives the three precepts in the form of the charge of *Episcopus*, p. 77, ll. 152 et seq. Moreover, lines 168, 169, which correspond with the third precept: *Thu xalte hate no thynge but þe devyl* and *synne | god byddyth the lovyn þi bodyly enmy* appear to have some reference to the misleading form in which the third precept is given in the *Meditationes*: *and the thridde is*: *Thou schalt hate thyn enemy* (*Habebis odio inimicum tuum*) only explained later in the text: *his enemyes that is to saie vices* and *synnes*. The order of the seven petitions in *Mary in the Temple* differs from that in the *Meditationes* and the *Lyf*, but the change—putting the most significant last instead of fifth—is one that suggests itself; the phraseology of the passage is closer to Love's than to Lydgate's.

The opening speech of *Contemplacio* in the *Visit to Elizabeth* is taken verbally from the *Legenda aurea*, Ch. 86, ' De Nativitate S. Iohannis baptistae': cf. *L. C.*, p. 116, ll. 23–7, and *David enim rex . . . volens cultum dei ampliare xxiv summos sacerdotes instituit quorum tamen unus maior erat qui princeps sacerdotum dicebatur*. The phrase *The plage of dompnesse*, l. 35, corresponds with *plaga taciturnitatis*. The close of this play has been shown (Note B) to be taken from the *Meditationes*. The matter of the whole play is found there as well as in the *Legenda aurea* and the *Lyf of oure Lady*.

For the *Betrothal of Mary* the *Meditationes* offer no material. The story of the flowering rod and the dove is found in the *Legenda aurea* and in the gospel *De Nativitate S. Mariæ*; in the *Pseudo-Matthew* and the *Lyf of oure Lady* the dove only appears on the rod. In these also the high priest is represented as neglecting Joseph's little rod, whereas in the former as in the play Joseph himself is the delinquent and endeavours to avoid offering the rod. On the other hand, the names of the Virgin's maidens (five in each case, seven unnamed in *De Nativitate*) are given only in the *Pseudo-Matthew* and the *Lyf of oure Lady*, and the forms used for the three in the play, Rebecca, Sephora, Susanna, only in the *Pseudo-Matthew*, the second name in Lydgate being Scaphea. Here, therefore, the *Legenda* and *Pseudo-Matthew* appear to have been used.

The use of the *Pseudo-Matthew* Gospel is clearest in the *Birth* play, where the incident of the midwives follows this original very closely. Cf. *Ludus Coventriæ*, p. 141, stage direction and ll. 181–4 and 265–76 and : *Audiens autem haec Maria subrisit. Cui Ioseph dixit, Noli subridere sed cauta esto ne forte indigeas medicina* ; and *Domine tu nosti quia semper te timui et omnes pauperes sine retributione acceptionis curavi, de vidua et orphano nihil accepi et inopem vacuum a me ire nunquam dimisi. Et ecce misera facta sum propter incredulitatem meam quia ausa fui temptare virginem suam*. The suggestion of the cherry-tree incident comes also from this gospel where a similar story is told of a palm-tree passed on the journey into Egypt. *The Trial of Joseph and Mary* follows less closely this gospel, which does not include the incident of the drinking of the draught by the detractor.

For plays viii to xv there is thus verbal evidence of direct borrowing from three sources : the *Legenda aurea*, the *Meditationes* or *Speculum Vitæ Christi*, and the *Pseudo-Matthew* Gospel, and the mingling of material in the plays corresponds with other evidence (cf. pp. xx to xxiv) to support a theory of dovetailing and revision. The plays of the next group (*Shepherds, Magi, Massacre*, and *Death of Herod*) are of a different character from any of the preceding ones, of higher literary quality, more lyrical, and in parts more dramatic, and in this closer to the *Passion* plays, and no source has been yet indicated for them. By similarities of metre and diction they are connected with the latter part of the *Resurrection* play.

The use made of the *Northern Passion* by the writer of the *Passion* plays differs from the use of the *Legenda aurea* in the plays derived from that source or from the use of the *Meditationes* in the *Salutation and Conception* and *Purification* plays. There is no such slavish transcription of whole passages as we find in the former, nor even such close following as we find in the latter. The parallel passages collected by Dr. Foster do not suggest that the writer set out to dramatize the *Northern*

Passion, but rather that he knew the story well in that form and its phrases recur, as Biblical phrases in a scriptural story would (or should) to-day. There is much in the *Passion* plays (the sermons of Peter and John, our Lord's exposition of the Sacrament, the Virgin's lament on his capture, the sentencing of the two thieves, and the dialogue between the Virgin and John at the foot of the Cross) which is not in the *Northern Passion*; the order of the incidents of the Trial before Pilate is different, and the dramatic handling of the Trial scenes with the accelerated dialogues in couplets is original.

In the *Passion* group, as in the earlier *Contemplacio* group, the layers of superimposed material correspond with the use of different sources. The influence of the *Northern Passion* is crossed by the influence of the *Meditationes*. A possible verbal reminiscence of Love in the scene between the Virgin and her risen Son[1] in the latter part of the *Resurrection* play gives a clue, and though there is no further verbal coincidence it is not rash to assume that the place given to the Virgin at the betrayal, at the death, and at the resurrection, the treatment of the harrowing as well perhaps as the completion of the seven words—matter found at the conclusion of the first *Passion* play, in the interpolated portions of the second, and in the latter part (Quire V) of the *Resurrection* play[2]—is due to the *Meditationes*. The chapter in the *Meditationes* on the taking of our Lord ends with a lament of the Virgin 'in Magdeleyns hous'; cf. *L. C.*, pp. 267, 268; special stress is laid on the 'seuene notable words' in the chapter on the death, and in the account of the descent into hell the actual harrowing following immediately on the death is separated from the appearance of the risen Lord to the rescued souls after the resurrection of his body. This does not exactly tally with the representation in the *Passion* play where the rescued souls are brought forth only on the resurrection morning, but it gives some suggestion for the division of the harrowing into

[1] Cf. p. xlv.
[2] Evidence of metre and diction also, as has been said, connects the interpolated matter on folios 183ᵛ-6 and the end of the *Resurrection* play on quire V. Cf. 'For mannys helpe my body is bred (fo. 185ʳ) and 'For man I haue mad my body in bred' (fo. 192ᵛ), and in similar contexts 'chare awey þe crowe' (fo. 183ᵛ), and 'chare awey þe ravyn' (fo. 194ᵛ).

two parts—the descent at the end of the *Crucifixion* play and the rising with the redeemed souls in the middle of the *Resurrection* play—which is peculiar to the *L. C.* play. The borrowings from the *Northern Passion* cease [1] where the influence of the *Meditationes* comes in—in the interpolated portion of the second *Passion* play whereby it is dovetailed on to the *Resurrection* play and in the conclusion (Quire V) of the *Resurrection* play.[2]

This cycle is later in its compilation than the civic cycles, but it seems to have borrowed little from them beyond the general choice and treatment of material, and even in these respects there are noticeable divergencies. There are more plays peculiar to this cycle than to any other, and the absence of farcical passages in the *Cain, Noah,* and *Shepherds* plays distinguishes it from all the others.

The opening of the speech of God the Father in the first play, *Ego sum a et ω,* is found in all cases, and the further likeness of a few phrases in the Towneley play (cf. 'All maner thyng is in my thoght' and 'all þat evyr xal haue beynge | it is closyd in my mende') would seem due to the same natural coincidence. An interpolated passage in the *Shepherds* play—the solitary attempt at humour—p. 148, ll. 78–89, may be a reminiscence of a similar farcical discussion of the angels' song in the Chester *Shepherds* play; a line in the scourging scene recalls a line in the same context in the Towneley play (cf. p. 277, ll. 168–71, and 'we shall teche hym I wote a new play of yoyll') and there are coincidences in the Towneley and *L. C. Passion* plays due to the use in both of the *Northern Passion,* but otherwise there is nothing that may not be accounted for as due to similarity of theme and form.

Besides the plays which have no counterpart in the other cycles (*The Conception of Mary, Mary in the Temple, The Betrothal*

---

[1] There is perhaps a later reminiscence of the *Northern Passion* (though there are many possible sources) in the line in the *Assumption*: *Sere and ye slept on christis brest · seyng all celestly,* said by Peter of John. Cf. *Northern Passion,* ll. 276 et seq. The corresponding Latin words in the *Legenda aurea* from which this play is directly taken do not give so clearly the reference to the legend of John's vision here implied: *Tu insuper pectus domini recumbere meruisti et exinde sapientiae et gratiae plus ceteris fluenta potasti.*

[2] Cf also p. xiv and p. xxxi for the composition of the second *Passion* play.

of Mary, The Trial of Joseph and Mary) the following episodes are found only in the *L. C.*—the death of Cain at the hands of Lamech (play 4), the testifying of the kings (Matthew i. 6–10) with the prophets (play 7), the 'parliament of heaven' (play 11),[1] the miracle of the cherry-tree (play 15); Veronica appears only in this series, but her action is performed by one of the Maries in the York plays; in no other cycle is the crucifying of our Lord assigned to the highborn Jews ('No man xal towche ȝour kyng | but yf he be knyght · or gentylman born; cf. also fo. 181, stage direction)[2]; characteristic of this collection again are the pithy short sermons put into the mouth of different characters: two from John the Baptist, one at the close of play 22 (p. 192) on repentance, and one as prologue to the *Passion* play (p. 229) on the dangers of overconfidence and of despair; one from Peter in the *Entry into Jerusalem*, scene (p. 238) on the spiritually blind, lame, and dumb; one from Christ expounding the Eucharist (p. 254). To these may be added the ironic discourse of Lucifer, a 'bountevous lord' to 'reward synners', at the opening of the *Passion* play (p. 225), and the exposition of the Commandments in the sixth play, which there is nothing to match in the Chester play in which the Commandments are recited.

In this exposition of the Commandments[3] the writer is drawing on the usual homiletic interpretation of the time. A treatise of the Commandments in a MS. at St. John's Coll., Oxford, ed. by J. F. Royster (University of N. Carolina, Studies in Philology, vol. 6, 1910), who dates it between 1420 and 1434, contains much the same material (cf. *L. C.*, p. 55, ll. 134–7, and 'Bot agaynes þis commaundement doos he þat slaes with hond or with word or will'; and p. 57, ll. 179–82, and 'In the siȝste commaundement is forboden þe dede of lechere ... and in þis commandement þe wille of þe syn is forboden'; cf. also *Court of Sapience*).

[1] This occurs in a different connexion in *The Castle of Perseverance*, cf. p. liv, and in a still more different in the 'Proces de Paradis' of the *Mystère du Vieil Testament*, in which, moreover, only Mercy and Justice are concerned.

[2] In the *Dialogus B. Mariæ et Anselmi de Passione Domini* a distinction is made between the *populus*, who are sympathetic after the Crucifixion, and the *Iudæi*, who continue their persecution of Christ's friends.

[3] For the curious confusion of the burning bush (*rubrus* by mistake for *rubus*) and Mount Sinai, cf. Isidore: 'Interpretatur autem Sinai rubus quod significat ecclesiam.'

The discourse of Lucifer is specially interesting as helping to date the compilation.[1] Attacks on extravagant fashions are common in the fifteenth century; the one in the text seems to be dealing with the same objectionable features in costume as the poem in MS. Harl. 372, printed by J. Payne Collier for the Percy Society, vol. xv, p. 55, and assigned to the mid-century —'not later than 1467' (cf. *L. C.*, p. 227, ll. 69 et seq., and 'Ye prowd galonttys hertlesse | With your high cappis witlesse | And youre schort gownys thriftlesse | Haue brought this lond in gret heuynesse | With your long peked schone ... And your long here in to your eyen ... With your wyde furryd hodes ... leue your short stuffide doublettys and your playtid gownys'). Similar fashions, however, seem to be attacked later in the century in the morality *Nature* (the 'side hair', the 'short gown', and 'wide sleeves' that 'would make a doublet and coat for some lad in this town').

The mingling of the kings of Jesse's race with the prophets in the seventh play is not found in any other play of this type that has come down to us, but Sepet in *Les prophètes du Christ*[2] takes it as the surviving example of a class: 'L'idée de faire paraître à côté des prophètes proprement dits la ligne de Jesse, les rois de Juda, fils de David et ancêtres du Messie, n'est pas particulière au Ludus Coventriæ. La scène a certainement eu ce caractère dans les mystères français.' He quotes in support of this view a description of a similar combination in a Corpus Christi procession still held at Mayenne in the seventeenth century.[3]

The Lamech episode is found in *Le Mystère du Vieil Testament*, but here the boy who guides Lamech and whom he slays is his own son, Tubal-cain. The writer of the *L. C.* interpolation seems to have drawn straight from P. Comestor, cf. pp. 39–41, including stage direction and : 'Lamech vero vir sagittarius diu vivendo caliginem oculorum incurrit, et habens adolescentem

---

[1] Lucifer comes in as a 'prowde galaunt' in *Mind, Will and Understanding*, and there is a reference in *The Castle of Perseverance* to 'longe Crakows on þi schos'.

[2] Bibliothèque de l'École des Chartes, vol. 38, 1877.

[3] Prof. Hardin Craig in a Note appended to Miss Swenson's *Enquiry* points out that apparently a number of kings played a part in the Lincoln pageants, which may, in his opinion, give a clue to the connexion of the *Ludus Coventriæ* with Lincoln.

Introduction

ducem dum exerceret venationem pro delectatione ... casu interfecit Cain inter fructeta aestimans feram quia ad indicium iuvenis dirigens sagittam interfecit. Et cum experiretur quod hominem scilicet Cain interfecisset iratus illic cum arcu ad mortem verberavit eum. Et ideo cum peccatum Cain punitum esset septuplum ut diximus suum punitum est septuagies septies id est septuaginta animae et septem egressae de Lamech in diluvio perierunt.'

The independence of this cycle is perhaps most clearly illustrated in the treatment of a theme common to all the cycles—Christ and the Doctors. Five plays on the subject have come down to us,[1] and, of these, four—those in the York, Towneley, Chester, and Coventry civic cycles—go back to one original ; the fifth, that in the *Ludus Coventriæ*, stands alone ; and the play with the learned brag of the opening speeches and the theological problems and paradoxes introduced into the discussion is characteristic of the more ecclesiastical and perhaps also of the more sophisticated and 'stagy' cast of this compilation. The questions raised—(*a*) the unity of the Trinity, (*b*) the distinction of the three persons, (*c*) the manner of the incarnation, (*d*) the reason for the choice of the second person, (*e*) the 'double birth' of Jesus, (*f*) the reasons for the marriage of Mary—were all theological commonplaces,[2] but

---

[1] Another is mentioned in the Epilogue to the *Candlemas* play: 'And the next yeer as we be purposid in our mynde | The disputacion of the doctours to shew in your presens'.

[2] (*a*) The comparison of the Trinity to *ignis, splendor, calor* is found in V. de Beauvais and goes back to St. Augustine. Cf. also similar comparison in *Candlemas* play of the humanity, soul, and godhead of Christ to the wax, wick, and light of a taper, a comparison used in *Piers Plowman* for the Trinity (C. xx. 168 and Skeat's note on the passage).

(*b*) Cf. *L. C.*, p. 17, &c.

(*c*) Cf. Lydgate, *Lyf of oure Lady*, v. s. p. xlvi.

(*d*) Cf. *L. C.* p 103, and Love: 'Man forfeted by vnwitte and foly'—an addition to Bonaventura; also *Court of Sapience*, where this is the second of three reasons.

(*e*) 'bis genitus dicitur sive quia Pater eum genuit sine Matre in æternitate sive quia Mater sine Patre in tempore'.—Isidore (*Etymologiarum* lib. vii, vol. 2, p. 265 Migne).

(*f*) These are two of the three reasons given by Bonaventura and Comestor. V. de Beauvais gives four, referring to Jerome and Ignatius.

the writer deserves some credit for original ingenuity in collecting them.

A connexion has been often suggested between this cycle and other religious plays of the fifteenth century.[1] There are obvious similarities in the stage directions of the *L. C. Passion* plays and those of such plays as *Mind, Will and Understanding*, the *Candlemas* play, *Mary Magdalene*, *The Croxton Play of the Sacrament*, which suggest similar methods of staging and perhaps similar business on the part of the actors.[2] There are also similarities in diction and in metre and in the handling of metrical form—the varying of the metre to suit the speaker or the situation—and in other characteristics of treatment which deserve further investigation but which lie outside the scope of this introduction. *The Castle of Perseverance* offers one or two special points of contact. It has, like the *L. C.*, a prologue in thirteen-lined stanzas for vexillatores in which the name of the town at which the play is to be performed is left to be filled up. It introduces the debate of the Four Daughters of God (*The Parliament of Heaven*) over the fate of the soul of *Humanum Genus*, though in this case it is the salvation of the individual and not the redemption of the race that is in question, and the debate does not lead up to the Incarnation. It introduces also the figure of Death; the entry of this character has not the quality of dramatic irony[3] that makes the entry in the *Death of Herod* play so impressive (*Humanum Genus*, an old man, has just remarked 'on Coveytyse is al my lay | *And* schal; tyl deth me ouer-throw'— Herod: 'I was nevyr meryer here beforn | Sythe þat I was fyrst born | Than I am now ryght in þis morn')—but the

---

[1] Mr. H. R. Patch ('The *Ludus Coventriæ* and the Digby *Massacre*', Publications of the M. L. A. of America, xxxv), basing his conclusions on an examination of the metres of the plays, makes the suggestion that the Dublin *Abraham and Isaac* and the Digby *Massacre* (or *Candlemas* play) might have formed part of the cycles or groups from which the compiler of *L. C.* made his selection.

[2] The Chester and Towneley cycles have a few, the York cycle a very few, brief directions in Latin; the Coventry civic cycle has fuller directions in English but not of the descriptive kind found in the *L. C. Passion* plays. This similarity as regards *Mind, Will and Understanding* has been noted by Mr. Pollard in his edition of the *Macro Moralities* (E.E.T.S., Extra Series 91).

[3] Cf. also *Everyman*: 'Full lytell he thynketh on my comynge | His mind is on flesshely lustes and his treasure.'

Introduction

general effect is similar as are (naturally) the two speeches of Death. In *The Castle of Perseverance* is found also one of the geographical lists of which there are examples in the Towneley play *Herod the Great* (considered by Mr. Pollard to be one of the group of later plays), in the *L. C. Temptation* play,[1] and in *The Croxton Play of the Sacrament*. The arrangement of names of countries and towns in lists, often alphabetical, in the geographical treatises of the time afforded convenient material for the alliterative artist (cf. *Rel. Ant.*, vol. i, p. 271, and MS. Arundel 123). Though some of the names are common, the lists seem to be independent, but they testify doubtless to some kinship, as do perhaps the attacks in all the *Macro Moralities* and the *Ludus Coventriæ* on the extravagant fashions in clothing of the time.

The *Ludus Coventriæ* has generally been considered inferior in literary merit to the other cycles. It lacks the genuine feeling that dignifies the York cycle (cf. *Birth* and *Flight into Egypt* plays); it has nothing to compare with the pathos of the Chester *Isaac* play or the humour of the Towneley second *Shepherds* play. Though Mary plays such an important part, no speech of hers in the *L. C.* has the poetic quality of the Lament, in the Towneley *Crucifixion* play, of the Mother at the foot of the Cross. Apart from the greetings of the Shepherds, the speech of Death, the lament of Adam and Eve, and a few *rime couée* passages there is very little poetry in the pages of Vesp. D. viii. But the plays of this collection offered their audience compensations for deficiencies—more singing,[2] more 'devices',[3] more processions,[4] more harangues, and, above all,

---

[1] The names are in all cases chosen for reasons other than geographical, but no other writer trades so boldly on the ignorance of his audience as the writer of the *Temptation* play in including *Zebee and Salmana* among the countries to be seen from the Mount (cf. Ps. lxxxiii. 11).

[2] Notably in the *Contemplacio* group (viii to xiii) and the *Assumption*, but also in the *Creation, Noah, Shepherds, Purification*, and *Entry into Jerusalem* plays.

[3] Cf. Proclamation : 'Wit*h* whiche devys as we best may | The holy gost xal ovyr him on'. None of the other cycles attempt the spectacular representation of the Incarnation described in the stage direction, p. 107.

[4] The *Moses* play included apparently a procession of the Commandments. Cf. Proclamation : 'þe ten comaundementys alle be dene | in oure play ȝe xal hem sene'.

in some cases, more acting. The passages of couplet dialogue in the Trial scenes (ff. 168ᵛ, 169, 171ᵛ, 177, 178, 178ᵛ, 179) mark an important advance from recitation and declamation to acting, and in several of the plays the writer or writers show command of stage effect, and understanding of the impressiveness of significant gesture and movement. A comparison of the *Woman taken in Adultery* in this series and in the York and Chester cycles illustrates the superior skill of at least one of the *Ludus Coventriæ* writers in developing a dramatic situation. The preliminary sermon of Christ on forgiveness, the angry comments of the Jews, the arrival of Accusator with his welcome scandal 'a ryght good sporte I kan yow telle', the breaking open of the door of the house, the escape of the man 'If any man my wey doth stoppe ... I xal þis daggere putt in his croppe'; the pleading of the woman 'I pray ȝow kylle me prevely ... lete not þe pepyl up-on me crye'—all these dramatic features are found only in the *Ludus Coventriæ* example. The most dramatic passage in the series is, perhaps, the unnoted entrance of Death in the midst of the revelry of Herod and his knights, where, as has been said, the style rises above the usual level. The effect of silent stealthy movement followed by sudden outcry on the imagination of an audience is again recognized in the stage direction for Pilate's wife's dream: 'her xal þe devyl gon to Pilatys wyf · þe corteyn drawyn as she lyth in bedde *and* he xal no dene make but she xal sone after þat he is come in · makyn a rewly noyse · comyng and rennyng of þe schaffald ... leke a mad woman'; and this direction 'he xal no dene make' is the more significant as Pilate's wife afterwards describes the 'sounds unheard': 'As wylde fyre *and* thondyr blast | he cam cryeng on to me'. Very effective, too, must have been Judas's secret passings to and from the scaffolds in the *Last Supper and Conspiracy* play: 'here Judas rysyth prevely' and 'here Judas goth in sotylly'. The excitement of escape is suggested in the vivid stage direction *et curret* which follows Pilate's dismissal of Barabbas, and again in the description of the flight of the young man in the *Woman taken in Adultery*—'hic iuuenis quidam extra currit in deploydo calligis non ligatis et braccas in manu tenens'. And we can still feel the thrill that stirred

Introduction

the 'gentyllys and ȝemanry' of *N.* town in the fifteenth century when we read the stage direction which opens the first Trial scene: 'here xal a massang*er* com i*n* to þe place ren*n*yng and cr*i*yng Tydyngys tydyngys · and so round abowth þe place · Jh*e*sus of nazareth is take · Jh*e*sus of nazareth is take'. Such an appeal may be, in the words of the mediaeval preacher, more to 'þe siȝt wiþoute forþ' than to 'þe feiþ wiþinne forþ', but there is no doubt of its potency, and in this effective stage representation the *Death of Herod*, the *Woman taken in Adultery*, and the *Passion* plays of the *Ludus Coventriæ* are outstanding.

NOTE A [to page xix].

1. *Legenda sanctorum aurea* (1481).	*The Golden Legend* (Caxton).	*The Play of the Assumption.*
i. |*Tu* in*s*u*p*er *s*u*p*er *p*ectu*s* d*o*mi*n*i recum*b*ere meru*i*ʍti | et exind*e* sap*ient*ie ac gracie plus ceteris fluenta potasti.	And thow oughtest to bere this palme of lyghte atte the xequyes of chastyte and holynes thou that drankest of the fontayn of perdurable clenes.	Sere | *and ye slept on christis brest* · | seyng alle celestly Ye are goddis clene mayde · wyth-outyn ony nay This observaunce is most like · you to do dewly.
ii. Memor e*n*im e*ss*e debes qualiter |*tibi astiti*| et qualiter te accusant*e* ancilla hostiaria te excusavi.	Thow oughtest remember how the chambriere that was Vssher accused thee and [? I] excused thee.	In cayfas halle · when thou were seyne And of thee pet*er* a mayde acusid there | *I halpe the tho* | · now helpe me ageyne.

iii. Caxton translates all the versicles which are given in the original Latin in the play, and in one instance the writer of the play has a variant reading.

| habebit fructum i*n* re-spectio*n*e[1] animarum sanctarum. | She shalle haue fruyte i*n* refection of noble sowles. | Habebit requiem i*n* respec*t*u a*n*imarum sa*n*ctaru*m.* |

---

[1] Graesse's edition *refeccione.*

NOTE B [to pages xxiii, xxiv, and xlv].

The following passages show the correspondence of the text of the plays and passages dependent on the *Meditationes* with the translation rather than with the original work.

Meditationes Vitæ Christi.	*The Mirrour of the Blessed Lyf of Jesus Christ,* p. 29.	'*Salutation and Conception*' *Play.*
Bonaventura, *Opera*, 1588.	Nicholas Love, ed. L. F. Powell.	
i. Intuere hic pro deo et meditare qualiter tota Trinitas est ibi expectans responsionem et consensum huius suae filiae singularis amanter et delectabiliter aspiciens verecundiam eius et mores et verba; et etiam qualiter angelus diligenter et sapienter inducit eam et ordinat verba sua stans inclinatus et reverens coram domina sua . . .	Now take good heed and haue in mynde how fyrst all the holy trinyte is there abidynge a fynal answer and assent of his blessid douȝter Marye takynge hede and byholdynge lykyngliche her schamfast semblaunt, her sad maneres and her wise wordes,	*and furthermore howe alle the blessid spirites of heuene, and alle the ryȝtwis lyuynge men in erthe and alle the chosen soules that weren that tyme in helle as adam, abraham, david and alle othere desireden hir assent in the whiche stood the sauacioun of all mankynde,*	and also how the aungel gabriel stondynge with reverence byfore his lady enclynynge . . .	Mary come of and haste the and take hede in thyn entent whow þe holy gost blyssyd he be abydyth þin answere and þin assent	*fferthermore take hede þis space whow alle þe blyssyd spyrytys of vertu þat are in hefne by ffore goddys face and alle þe gode levers and trew That are here in þis erthely place thyn owyn kynrede · þe sothe ho knew And þe chosyn sowlys þis tyme of grace þat are in helle and byde rescu as Adam, abraham and david in fere and many othere of good reputacion þat þin answere desyre to here and þin assent to þe incarnacion in whiche þou standyst as persevere of alle man-kende savacion.*	

Introduction

Meditationes Vitæ Christi.	The Mirrour of the Blessed Lyf of Jesus Christ, p. 39.	'Visit to Elizabeth' Play.
O qualis domus, qualis camera, qualis lectus, in qua et quo pariter commorantur et requiescunt tales matres talibus filiis foecundatae, Maria et Elisabeth, Jesus et Joannes. Sunt et ibi magnifici senes, scilicet Zacharias et Joseph	| *A lord god what house was that* | or what chamber and what bedde in the whiche dwelleden to gidre and resteden so worthi moderes with so noble sones that is to saie Marie and Elizabeth Jesu and John And also with them dwellynge tho worschipful olde men zacharie and Joseph.	| *A lord god what hous was þis on* | þat þese childeryn and here moderys to as mary and elizabeth jhesus and john and joseph and zakarye also.
Et tunc apertum est os Zachariae, et prophetavit, dicens: *Benedictus Dominus Deus Israel*, etc. · Et sic in domo illa haec duo cantica pulcherrima, scilicet *Magnificat* et *Benedictus*, facta fuerunt	And the mowthe and the tunge of the fader zacharye *by-fore closed for vntrowynge* was than opened: and so he prophecied seienge: . . . And so in that house thise two noble and worthy canticles that is to seyen Magnificat and benedictus weren first spoken and made.	And þan zakarye spak i-wus þat had be dowm and his speche lorn he and Elizabeth prophesyed as þus they mad BENEDICTUS . them be-forn and so · MAGNIFICAT · and BENEDICTUS · ffyrst in þat place þer made worn.
Tandem valefaciens Elisabeth et Zachariae, ac benedicens Joanni rediit ad domum suae habitationis in Nazareth.	And at the laste | *whan al this was done* | sche toke hir leue at Elizabeth and Zacharye and blessid the child John and so went home et seq.	| *Whan all was don* | oure lady fre toke here leve than aftere this at Elizabeth and at Zakarie and kyssyd johan and gan hym blys.
	Ch. ix, *preliminary paragraph not in original.*	'Purification' Play.
	After the kynges had performed her offerynges . . . ȝit stode that worthy lady of all the world in that symple herborwe . . . pacieatly abidyng unto the fourty day ordeyned	To be purefyed · haue ȝe no nede Ne þis on · to be offeryd · so god me spede ffor fyrst þou art ful clene *Vndefowlyd in thought* and *dede*

Meditationes Vitæ Christi.	Ch. ix. *preliminary paragraph not in original.*	'*The Purification*' *Play.*
	by the lawe to hir purificatioun as sche were another womman of the peple defoyled thoruȝ synne, and as the childe Jesu were a pure man and not god nedy to kepe the observaunce of the lawe.	and *anothyr* þi *son withowtyn drede is god* and *man to mene Wherefore it nedyd not to bene* but to kepe þe lawe · on Moyses wyse.
	The Mirrour of the Blessed Lyf of Jesus Christ, pp. 62–3.	
ii. Accipite pater excellentissime unigenitum vestrum quem secundum mandatum vestrae legis vobis offero quia primogenitus matris est.	Taketh now *hiȝest fader ȝoure owne dere sone* whom I offre here to ȝow after the biddynge of ȝoure lawe: for he is the firste born of his moder.	hyest ffadyr god of powere ȝour owyn dere son I offre ȝow here as I to ȝour lawe am sworn receyve þi childe in glad manere for he is þe fyrst þis childe so dere þat of his modyr is born.
iii. Accipite clementissime pater hanc oblationem et munusculum et primum donum quod parvulus vester hodie de sua paupertate praesentat.	*Al miȝty and merciful fader of heuene* vnderfonge ȝe this litel ȝifte and offerynge and the firste ȝifte that ȝoure litel child this day presenteth *vnto ȝoure hiȝe maieste of his symple pouerte.*	alle myghty ffful fadyr · mercyful kynge receyvyth now þis lytyl offerynge ffor it is þe fyrst in degre þat ȝour lytyl childe so ȝynge presentyth to day be my shewyng to ȝour hyȝ mageste of his sympyl poverte.

The plaie called Corpus Christi

¶ Now gracyous god groundyd of all goodnesse
as þi grete glorie nevyr þe gynnyng had
So þou socor & save all þo þat sytt & sese
& lystenyth to oure talkyng wt sylens stylle & sad
ffor our purpose is purpose is pleyn stylle in mo[o]d
ȝe pepyl to plese wt pleyes ful glad
now lystenyth us lovely bothe more & lesse
gentyllys & ȝemanry of goodly lyff lad
þis tyde
þe whiche of oure lady
of oure ladyes bok
good wyll þi speche a-byde
þat god wyll a-byde
þat god made bothe molde & man

¶ In þe fyrst pagent we purpose to play
how god dede make þorwe his owyn myght
hevyn so cler upon þe fyrst day
& þerin he sett aungell ful bryth
Than aungell wt songe þis is no nay
shall worchep god as it is ryth
but lucyfer þat aungell so gay
in suche pompe þan is he pyth
& goddys sete he gynnyth to take
hese lordys pes hy selfe to make
but þan he fallyth a fende ful blake
ffrom hefne to helle to go
þis voyelle

¶ Than þe secunde pagent by godys myth
we purpose to shewe & play be-dene
in þe other syx dayes by opyn syth
what þynge þat is wrowyght þer shall be sene
How best was made & foule of flyth
and last was man made as I wene
of mannys ryb as I ȝow plyth
þe woman wrowyth mannys make to bene
& put in paradyse

B.

LUDUS COVENTRIÆ.

Cotton MS. Vespasian D. viii.

The Proclamation.

The Plaie called Corpus christi.[1]

          1us vexillat[or

¶ Now gracyous god groundyd of aħ goodnesse
as þi grete glorie nevyr be-gynning had
So þou socour and saue aħ þo þat sytt and sese
and lystenyth to oure talkyng with sylens stylle and sad
Ffor we purpose us pertly stylle in þis prese   5
þe pepyl to plese with pleys ful glad
now lystenyth us louely bothe more and lesse
Gentyllys and ȝemanry of goodly lyff lad
   þis tyde
we xal ȝou shewe · as þat we kan   10
how þat þis werd ffyrst be-gan
and how god made bothe molde and man
   Iff þat ȝe wyl a[byde.
         ¶ 2us vexilla[tor

¶ In þe ffyrst pagent we þenke to play
how god dede make þurowe his owyn myth   15
hevyn so clere upon þe fyrst day
and þer in he sett Angeħ fful bryth
Than Angeħ with songe þis is no nay
xal worchep god as it is ryth
but lucyfer þat Angeħ so gay   20
in suche pompe þan is he pyth
   and set in so gret pride
þat goddys sete he gynnyth to take
hese lordys pere hym self to make
but þan he ffallyth a ffend ful blake   25
   ffrom hevyn in helle to a[byde.

---

[1] A note at the top of the folio in small Elizabethan handwriting.

3ⁿˢ vexill[ator

¶ In þe secunde pagent by godys myth
we þenke to shewe and pley be-dene
in þe other sex days by opyn syth
what þenge was wrought þer xal be sene 30
How best was made and foule of flyth
and last was man made as I wene
Of mannys o ryb as I ȝow plyth
was woman wrougth mannys make to bene
 and put in paradyse 35
Fo. 1ᵛ Ther were flourys bothe blewe and blake
of all frutys þei myth þer take
Saff frute of cunnyng þei xulde for-sake
 And towche it no wyse.

¶ The Serpent toke Eve an Appyl to byte 40
and Eve toke Adam a mursel of þe same
whan þei had do þus a-ȝens þe rewle of ryte
than was oure lord wroth and grevyd al with grame
Oure lord gan appose þem of þer gret debyte¹
both to Askuse hem of þat synful blame 45
and þan almythy god ffor þat gret dyspite
Assygned hem grevous peyn · as ȝe xal se in game
 In dede
Seraphyn An Angell gay
with brennyng swerd þis is verray 50
From paradise bete hem a-way
 in bybyl as we rede.

1ᵘˢ vexillator

¶ We purpose to shewe in þe thryd pagent
the story of Caym · and of hese brother Abelle
of here tythyngys now be we bent 55
In þis pagent þe trewth to telle
How þe tythyng of Abel with feyr was brent
and accept to god yf ȝe wyl dwelle
We purpose to shewe as we haue ment
and how he was kyllyd of his brother so felle 60
 And than
how Caym was cursyd in al degre

¹ Halliwell reads *delyte*. The MS. is not clear.

  of godys owyn mowthe þer xal ȝe se
  of trewe tythyng þis may wel be
    exawple¹ to every man.   65
           2us *vexilator*

¶ The iij^{de} pagent is now ȝow tolde
 þe ffourte pagent of Noe xal be
 how god was wroth · with man on molde
 because fro synne man dede not fle

Fo. 2 He sent to Noe An Angel bolde   70
 A shyp ffor to makyn and swymmen on þe se
 vpon þe water both wood and coolde
 And viij sowles þer savyd xulde be
 And i peyre of everich bestys in brynge
 whan xl^{ti} days þe flode had fflowe   75
 þan sente Noe out a crowe
 and After hym he sent a dowe
 þat brouth ryth good tydyng.
           3^{xs} *vexil[lator*

¶ Of Abraham is þe fyfte pagent
 and of ysaac his sone so fre   80
 how þat he xulde² with fere be brent
 and slayn with swerd as ȝe xal se
 Abraham toke with good A-tent
 his sone ysaac and knelyd on kne
 his suerd was than ful redy bent   85
 and thouth his chylde þer offered xuld be
 Vpon An hyll full Ryff
 than god toke tent to his good wyl
 and sent An Angel ryth sone hym tyl
 and bad Abraham a shep to kyl   90
 And sauyd his chyldys lyff.
           1us *vexil[lator*

¶ The sexte pagent is of Moyses
 and of tweyn tabelys þat god hym took
 in þe which were wrete with-out les
 þe lawes of god to lerne and lok   95
 and how god charged hym be wordys these
 þe lawes to lerne al of þat book
 Moyses than doth nevyr more sese

  ¹ So in MS.   ² A letter or letters obliterated between *xulde* and *with*.

4 *The Proclamation*

        but prechyth duly · bothe ȝere and woke
Fo. 2ᵛ The lawes as I ȝow telle   100
        þe ten comaundementys alle be-dene
        In oure play ȝe xal hem sene
        to alle þo þat þere wyl bene
        If þat ȝe thenke to duelle.

                                      2ᵘˢ *vexillator*
    ¶ Off þe gentyl Jesse rote   105
        þe sefnt pagent for sothe xal ben
        out of þe which doth sprynge oure bote
        as in prophecye we redyn and sen
        Kyngys and prophetys with wordys fful sote
        Schull prophesye al of a qwen   110
        þe which xal staunch oure stryff and moote
        And wynnen us welthe with-outyn wen
        In hevyn to Abyde
        they xal prophecye of a mayde
        All ffendys of here xal be Affrayde   115
        here sone xal saue us be not dismayde
        With hese woundys wyde.

                                      3ⁿˢ *vexillator*
    ¶ Of þe grete bushop Abyacar
        þe tende [1] pagent xal be with-out lesyng
        þe which comaundyth men to be war   120
        and brynge here douterys to dew weddyng
        All þat ben xiiij ȝere and more
        to Maryage he byddyth hem bryng
        wher evyr þei be he chargyth sore
        þat þei not ffayle for no lettyng   125
        þe lawe byddyth so than [2]
        Than Joachym and anne so mylde
        þei brynge forthe mary þat blyssyd chylde
        but she wold not be de-fylyde
        with spot nor wem of man.   130

Fo. 3 ¶ In chastyte þat blysful mayde
        A-vowyd there here lyff to lede

      [1] A later correction. The original word is erased.
      [2] The word *sertayn* is written before *than* and crossed out. Halliwell retains it.

þan is þe busshop sore dysmayde
And wonderyth sore al of þis dede
he knelyd to god as it is sayde 135
and prayth than for help and rede
þan seyth an Angel be not a-frayde
of þis dowte take þou no drede
but for þe kynrede of dauyd þou sende
lete hem come with here offryng 140
And in here handys white ȝerdys brynge
loke whose ȝerde doth ffloure and sprynge
And he xal wedde þat mayden hende.

         1us *vexillator*

¶ In þe x^{te} [1] pagent sothe to say
A masangere fforthe is sent 145
Dauyd is kynrede with-out de-lay
they come fful sone with good Entent
Whan joseph offeryd his ȝerde þat day
Anon ryth fforth in present
þe ded styk do floure fful gay 150
and þan joseph to wedlok went
Ryth as þe Angel bad
Than he plyth to his wyff
In chastyte to ledyn here lyff
þe busshop toke here iij · maydonys ryff 155
Som comforte þere she had [2].

         2us *vexillator*

¶ In þe xi^{de} [3] pagent goth Gabryell
And doth salute oure lady ffre
Than grett with chylde as I ȝow tell
þat blyssyd mayde for sothe is she 160
þo iij maydenys þat with here dwelle
here gret specħ but noon þei se
than they suppose þat sum Angell
goddys masangere þat it xuld be
And thus 165
þe holy gost in here is lyth

[1] This is corrected from an original ix^{te}.
[2] First written *hadde* and corrected.
[3] This is corrected from an original x^{de}; the writing becomes slightly smaller and there is a faint difference in the colour of the ink with this stanza.

and goddys sone in her*e* is pygth
þe Aungell doth telle what he xal hyght
And namyth þe chylde jhesus.
                                                              3us *vexillator*

¶ In þe xij[1] pagent as I ȝow telle   170
Joseph comyth hom fro fer countre
Our*e* ladyes wombe w*ith* chylde doth swelle
and þan joseph ful hevy is he
He doth forsake her*e* w*ith* hert ful felle
out of countre he gynnyth to fle   175
he nevyr mor*e* thenkyth w*ith* her*e* to dwelle
and than our*e* lady ryth sor*e* wepyth she
An Angell seyd hym ryf
god is w*ith* þi wyff sertayn[2]
þerfore Joseph turne hom a-gayn[2]   180
þan is Joseph in herte ful fayn
And goth a-geyn onto his wyff.
                                                              1us *vexillator*

¶ The xiiij^{te}[3] pagent I sey ȝow be-dene
xal be of joseph *and* mylde mary
how they were sclawndryd w*ith* trey *and* tene   185
and to her*e* purgacion þei must hem hy.[4]
                                                              2us *vexillator*

Fo. 4

¶ In þe xv[5] pagent shewe we xal
how joseph went w*ith*-oute varyauns
for mydwyuys to helpe our*e* lady at all
of childe that she had delyuerau*n*s.   190
                                                               3us *vexillator*

¶ In þe xvj[6] pagent Cryst xal be born
of þat joy Aungelys xul synge
and telle þe shepherdys in þat morn
the blysseful byrth of þat kyng

---

[1] Here the original word *hellenthe* has been crossed out in red ink, and xii written in fresher black ink above.

[2] *Serteyn* and *ageyn* appear to have been first written and corrected by scribe.

[3] This is corrected from an original xii^{te}.

[4] A space is left at the bottom of Fo. 3^v, beneath this speech, and at the top and in the middle of Fo. 4 above and beneath the next speech, as if to be filled in later.

[5] This is corrected from an original xiii.

[6] This is corrected from an original xiv.

The shepherdys xal come hym be-fforn 195
with reuerens and with worchepyng
ffor he xal sauyn þat was for-lorn
and graunt us lyf evyr more lestyng
   i-wys
þis gle in gryth 200
is mater of myrth
now crystys byrth
  Bryng¹ us to his blys.

           1^{ns} vexillator

¶ The xv^{te} pagent come kyngys iii
Fo. 4^v with gold myrre and ffrankynsens 205
kyng herowdys styward hem doth se
And bryngyth all to his presens
The kyngys of Coleyn with hert ful ffre
tolde kyng herownde here dylygens
that þei south in þat countre 210
A kyng of kyngys ffrom fere thens
A sterre led hem þe way
The Chylde is ȝoung and lyth in stall
he xal be kyng of kyngys all
beffore hym we thynk on kne to fall 215
And worchep hym þis day.

          2^{ns} vexillator

¶ In þe xvi pagent as wroth as wynde
is kyng herownde þe soth to say
And cruel knytys and vn-kende
to sle male chylderyn he sendyth þat day 220
But cryst jhesu þei may not ffynde
Ffor joseph hath led þat childe away
Vnto Egypth as we haue mende
As Angel to joseph dyd byd and say
  In hyȝht 225
þo chylderyn þat syt in here moderys lap
to sowkyn ful swetly here moderys pap
þe knythtys do sle hem euyn at A swap
þis is a rewly syth.

¹ Some word beginning with *s* (or f) and ending with *l* is crossed out before *Bryng*.

The Proclamation

               3ᵘˢ *vexillator*

¶ In þe xvii pagent þe knyhtys be-dene   230
Shull brynge dede childeryn be-for þe kyng
whan kyng herownde þat syth hath sene
fful glad he is of here kyllyng
Than kyng herownde with-owtyn wene
is sett to mete at his lykyng   235
in his most pride xal come gret tene
As 3e xal se at oure pleyng

Fo. 5   his sorwe xal a-wake
whan he is sett at hese most pryde
Sodeyn deth xal thrylle his syde   240
and kylle his knyttys þat with hym byde
þe devyl þer soulys xal take.

               1ᵘˢ *vexillator*

¶ In þe xviii pagent we must purpose
to shewe whan cryst was xij 3er of Age
how in þe temple he dede appose   245
and answerd doctoris ryth wyse and sage
The blyssyd babe with-owte glose
ouer cam olde clerkys with suych langage
þat þei meveylyd · 3e xal suppose
how þat he cam to suche knowlage   250
    and in þis whyle
thre days he was oute
ffro his modyr with-out doute
wepyng she sowth hym rownde aboute
    jheruselem many a myle.   255

               2ⁿˢ *vexillator*

¶ in þe xix pagent xal seynt jhon
baptyse cryst as I 3ow say
in þe watyr of flom jordon
with which devys as we best may
The holy gost xal ouyr hym on   260
þe ffaderys voys xal be herd þat day
out of hevyn þat blisful tron
þe fadyr xal be herd þis is no nay
    and forth with pleyn
þe holy gost xal be his gyde   265
in to desert þer-in to. A-byde

xlti days A terme ful wyde
and xlti nygthtys to faste sarteyn̄ [1].

3us vexillator

¶ in þe xxti pagent all þe deuelys of helle
they gadere a parlement as ȝe xal se   270
they haue gret doute þe trewth to telle
Of cryst jhesu whath he xulde be
Fo. 5v They Sende Ssathan þat Ffynde so Ffelle
Cryst for to tempte in fele degre
We xal ȝow shewe if ȝe wyl dwelle   275
How cryst was temptyd in synnys thre
Of þe devyl sathan̄
And how cryst Answeryd on to Alle
and made þe fende awey to falle
As we best may þis shewe we xalle   280
Thorwe grace of god and man

1ns vexillator

¶ The xxiti pagent of a woman xal be
þe which was take in Adultrye
The pharysewys ffalsed þer ȝe xal se
Cryst to convycte how they were slye   285
they conseyvyd þis sotylte
yf cryst þis woman dede dampne trewly
A-geyn his prechyng than dede he
Which was of pete and of mercy
And yf he dede here save   290
þan were he A-ȝens moyses lawe
þat byddyth with stonys she xulde be slawe
þus they thowth vndyr þer Awe
Cryst jhesu for to haue.

2us vexillator

¶ The grettest meracle þat evyr jhesus   295
In erthe wrouth be-forn his passyon̄
In xxiiti pagent we purpose vs
to shewe in dede þe declaracion
þat pagent xal be of lazarus
In whos place and habytacion   300
Cryst was logyd þe gospel seyth thus

---

[1] An original *e* in the first syllable appears to have been corrected by the scribe to an *a*.

        and ofte[1] tyme toke *þer* consolacion
              but ȝyt

Fo. 6   Lazarus As I ȝow say
        was iiij days ded *and* beryed in clay   305
        ffrom deth to lyve þe iiij^{te} day
        Cryst reysed hym ffrom *þat* pyt.
                                      3us *vexill*ator

¶ In þe xxiij^{ti} pagent palme sunday
    in pley we purpose ffor to shewe
    how chylderyn of Ebrew w*ith* flowr*ys* ful gay   310
    þe wey *þat* cryst went þei gu*n* to strewe
                                  1us *vexill*ator

    In þe xxiiij^{ti} pagent as *þat* we may
    cryst *and* his apostelys alle on rewe
    the mawnde of god *þer* xal they play
    and sone declare it w*ith* wordys ffewe   315
        And tha*n*
    Judas *þat* fals trayto*ur*
    ffor xxx^{ti} platys of werdly treso*ur*
    xal be-tray ou*r*e savyo*ur*
        to þe jewys certa*n* [2].   320
                                  2us *vexill*ator

¶ ffor grevous peyn þis is no les
    in þe xxv^{ti} pagent cryst xal pray
    to þe fadyr of hevyn *þat* peyn for to ses
    his shamful deth to put Away
    Judas *þat* trayto*ur* be-for gret pres   325
    xal kys his mouth *and* hy*m* be-tray
    All his dyscyples than do dyscres
    and forsake cryst þe soth to say
    ffor doute þei do hem hede
    hese dyscyplys all every-cho*n*   330
    do renne awey *and* leve hy*m* alo*n*
    they lete hym stondyn amonge his ffon
    And ronne Away ffor drede.
                                  3us *vex*[*illa*tor

Fo. 6^{v} Than in þe xxvj^{ti} pagent
       to Cayphas cryst xal be brouth   325

---

[1] The *o* is blotted and may have been corrected from another letter.
[2] The *e* is blotted and not clearly decipherable.

The Proclamation

þo jewys fful redy þer xul be bent
Cryst to Acuse with worde and thouth
Seynt petyr doth folwe with good intent
to se with cryst what xuld be wrouth
Ffor crystys dyscyple whan he is hent   340
thryes he doth swere he knew hym nowth
A kok xal crowe and crye
Than doth petyr gret sorwe make
Ffor he his lord þus dede for-sake
But god to grace hym sone doth take   345
Whan he doth aske mercye.
1us vexillator

¶ In þe xxvij pagent sere Pylat
Is sett in sete as hy justyce
Whan he is set in his astat
thre thevys be brout of synful gyse   350
And cryst þat louyd nevyr stryff nor bat
but trewth and goodnesse on every wyse
as for a thef with ryth gret hat
is browth to stondyn at þat same syse
And þan as I ȝow say   355
b coveryd with clothis al of þe best¹
a the Wyff of Pylat goth to rest¹
Than for to slepe she is ful prest
All þis we thenke to play.
2us vexillator

¶ in þe xxviij^{ti} pagent xal judas   360
þat was to cryst a ffals traytour
with wepyng sore evyr crye Alas
þat evyr he solde oure savyour
He xal be sory ffor his trespas
And brynge a-ȝen all his tresour   365
all xxx pens to sere Cayphas
he xal them brynge with gret dolowre
Ffor þe which cryst was bowth
Ffor gret whanhope as ȝe xal se
he hangyth hym self vpon a tre   370
Ffor he noth trostyth in godys pete
to helle his sowle is browth.

¹ So, in wrong order with indication of the mistake, in MS.

                                                        3ᵘˢ *vexillator*

¶ In þe xxix pagent to Pylatus Wyff
  in slepe aperyth þe devyl of helle
  ffor to savyn crystys lyff   375
  The devyl here temptyth as I ȝow telle
  Sche sendyth to pylat a-non ful ryff
  And prayth þat cryst he xuld not qwelle
  þan pylat is besy *and* ryth blyff
  cryst for to savyn he ȝevyth councelle   380
  ffor he dede neuyr trespas
  The jewys do crye fast ffor to kylle[1]
  the[2] rythful man þei aske to spylle
  A thef þei saue with herty wylle
  þat Callyd is barrabas.   385

                                                        1ᵘˢ *vexillator*

¶ In þe xxxᵗⁱ pagent þei bete out crystys blood
  And nayle hym al nakyd upon a rode tre
  betwen ij thevys · i-wys they were to wood
  they hyng cryst jhesu · gret shame it is[3] to seˑ
  vii wurdyes cryst spekyth hangyng upon þe rode   390
  þe weche ȝe xal here all þo þat wyl þer be
  þan doth he dye ffor oure allether good
  his modyr doth se þat syth gret mornyng makyth she
  Ffor sorwe she gynneth to swowne
  Seynt johan Evyn þer as I ȝow plyth   395
  doth chere oure lady with al his myth
  And to þe temple anon forth ryth
  he ledyth here in þat stownde.

                                                        2ᵘˢ *vexillator*

¶ we purpose to shewe in oure pleyn place[4]
  in þe xxxjᵗⁱ pagent þorwe godys myth
  how to crystys herte a spere gan pace   400
  and rent oure lordys bryst in ruly plyth
  Ffor longeus þat olde knyth blynd as he was
  A ryth sharpe spere to cristys herte xal pyth

---

[1] The writing becomes slightly less regular with this line to the bottom of the folio.
[2] *ryff* written before the next word and crossed through by scribe.
[3] *was* first written and corrected by scribe to *is*.
[4] The writing in this speech is slightly more sprawled than heretofore.

The Proclamation

    þe blod of his wounde to his eyn xal tras   405
    and þorwe gret meracle þer hath he syth
    Than in þat morn
    crystys soule goth down to helle
    and þer ovyr comyth þe fende so felle
    comfortyth þe soulys þat þer in dwelle   410
    and savyth þat was fforlorn.

                                                              3us vexillator

¶ Joseph and nycodemus to cryst trew servaunt [1]
    in þe xxxij page [2] þe body þei Aske to haue
    Pylat ful redyly þe body doth hem graunt
    þan þei with reverens do put it in grave   415
    þe jewys more wyckyd þan ony geawnt
    Ffor crystys ded body kepers do þei craue
    pylat sendyth iiij knytys þat be ryth hardaunt
    to kepe þe blody body in his dede conclaue

Fo. 8   And ȝit be his owyn myth   420
    The body þat was hevy as led
    be þe jewys nevyr so qwed
    A-ryseth from grave þat þer lay ded
    And ffrayth than every knyth.

                                                              1us vexillator

¶ in þe xxiij [3] pagent þe soule of cryst jhesu   425
    xal brynge all his ffrendys ffrom helle to paradyse
    þe soule goth þan to þe graue · and be ryth [4] gret vertu
    þat body · þat longe ded hath loyn · to lyf aȝen doth ryse
    Than doth cryst jhesu on to his modyr sew
    and comfortyth all here care in temple þer she lyse   430
    with suche cher and comforth his modyr he doth indew
    þat joy it is to here · þer spech for to devyse
            and than
    oure lady of hefne so cler
    In herte sche hath ryth glad chere   435
    whan here sone þus doth apere
    here care awey is tan.

                                                              2us vexillator

¶ in þe xxxiiij^{ti} pagent xal maryes thre

---

[1] The writing for the first five lines of this speech is smaller.
[2] So in MS. for *pagent*.   [3] So in MS. for xxxiii.
[4] This word omitted and written over the line by the scribe.

seke cryst jhesu in his grave so coolde
An Aungel hem tellyth þat aresyn is he 440
and whan þat þis tale to them is tolde
To crystys dyscyplis with wurdys fful fre
they telle these tydyngys with brest ful bolde
than petyr and johan as ʒe xal se
down rennyn in hast ouer lond and wolde 445
The trewth of þis to haue
whan þei þer comyn as I ʒow say
he is gon ffrom vndyr clay
þan þai wytnesse a-noon þat day
he lyth not in his grave. 450

          3us *vexill*ator

Fo. 8^v ¶ On to Mary Mawdelyn as we haue bent
Cryst jhesu xal than Apere
In þe xxxv^{ti} pagent
and she wenyth he be A gardenere
Mary be name verament 455
Whan cryst here callyth with speċh ful clere
She ffallyth to ground with good entent
to kys his fete with gladsom̅ chere
but cryst byddyth here do way
he byddyth his feet þat sche not kys 460
tyl he haue styed to hefne blys
to crystys dyscyplys mary i-wys
than goth · þe trewtħ to say.

          1us *vexill*ator

¶ In þe xxxvj^{ti} pagent xal cleophas
And sent luke · to a castel¹ go 465
Of crystys deth as þei fforth pas
they make gret mornyng and be ful wo
Than cryst þem ovyr-tok as his wyl was
and walkyd in felachep fforth with hem too
to them he doth expowne bothe more and las 470
aɫɫ þat prophetys spak ad ² of hym self Also
that nyth in fay
whan þei be set · with-in þe casteɫɫ

   ¹ The initial *c* is corrected from some other letter.
   ² So in MS.

In brekyng of bred · þei know[1] cryst weŀŀ
Than sodeynly as I ȝow teŀŀ   475
Cryste is gon his way.

       2ᵘˢ *vexillator*

¶ In þe xxxvij^(ti) pagent þan purpos we
to Thomas of ynde cryst xal apere
And thomas euyn þer as ȝe xal se
xal put his hand in his woundys dere.   480

       3ᵘˢ *vexillator*

Fo. 9 ¶ In þe xxxviij^(ti) pagent up stye xal he
in to hefne þat is so clere
Aŀŀ hese Apostele þer xul be
And woundere sore and haue gret dwere
Of þat fferly syth   485
Þer xal come Aungeŀŀ tweyn
and commforte hem þis is certeyn
and tellyn þat he xal comyn ageyn
Evyn by his owyn myth.[2]

       1ᵘˢ *vexillator*

¶ Than ffolwyth next sekyrly   490
of Wyttsunday þat solempne ffest
whych pagent xal be ix and thretty
to þe apostelys to apere be crystys hest
In hierusalem were gaderyd xij opynly
to þe Cenacle comyng ffrom west and Est   495
þe holy gost apperyd fful veruently
with brennyng ffere thyrlyng here brest
procedyng from hevyn trone
Aŀŀ maner langage hem spak with tung
latyn grek and Ebrew among   500
And Affter þei departyd and taryed not long
here deth · to take ful sone.

       2ᵘˢ *vexillator*

¶ The xl^(ti) pagent xal be þe last
and domysday þat pagent xal hyth
who se þat pagent may be agast   505
to grevyn his lord god eyther day or nyth
The erth xal qwake bothe breke and brast

---

[1] Or *knew*; the letter is partially blotted.
[2] This speech forms one stanza with the last speech, but it is marked as a separate stanza in the MS.

                    beryelys *and* gravys xul op*e* ful tyth
                    ded men xul rysyn *and* þat þer in hast
                    And ffast to here ansuer*e* þei xul hem dyth   510
Fo. 9ᵛ¹   Beffore godys fface
                    but prente wyl þis in ȝo*ur* mende
                    Who so to god · hath be vnkende
                    Ffrenchep þer xal he non ffynde
                    ne þer get he no grace.   515
                                                    ¶ 3ᵘˢ *vexillator*

    ¶ Now haue we told ȝow all be-dene
        the hool mater þat we thynke to play
        whan þ*at* ȝe come þer xal ȝe sene
        this game wel pleyd in good a-ray
        Of holy wrytte þis game xal bene   520
        and of no fablys be no way
        now god þem save from trey *and* tene
        ffor us þat prayth upon þat day
        And qwyte them wel þ*er* mede
        A sunday next yf þat we may   525
        At vj of þe belle we gynne our*e* play
        In N. town̄ wherfore we pray
        That god now be ȝou*re* Spede.   Amen.

            Remainder of Fo. 9ᵛ—3½ inches—left blank.

Fo. 10 ²   *Ego sum alpha et* oo · *principiu*m *et finis.*   Deus

    ¶ My name is knowyn god *and* kynge   **1**
        My werk for to make · now wyl I wende
        in my self restyth my reynenge
        it hath no gynnyng ne non ende
        And all þ*at* evyr xal haue beynge   5
        it is closyd in my mende
        whan it is made at my lykynge
        I may it saue I may it shende

---

¹ Corrected to 10 and then crossed through.

² At the top of this page in the MS., the bottom part of the letters of the name of the owner Hegge can be seen. The remains of the letters of the Christian name are indistinguishable, but *Hegge Dunelmensis* can still be deciphered.

³ This line is written in larger form and underlined in red.

The Creation of Heaven and the Angels 17

After my plesawns
So gret of myth is my pouste
Aħ thyng xal be wrowth be me
I am oo god · in personys thre
knyt in oo substawns.

¶ I am þe trewe trenyte
here walkyng in þis wone
thre personys myself I se
lokyn in me god Alone
I am þe ffadyr of powste
my sone with me gynnyth goñ
my gost is grace in mageste
weldyth welthe up in hevyn tron
O god thre · I calle
I am fadyr of myth
my sone kepyth ryth
my gost hath lyth
and grace with-alle.

¶ My-self begynnyng nevyr dyd take
And endeles I am thorw myn owyn myth
now wole I be-gynne my werke to make
Ffyrst I make hevyn with sterrys of lyth
In myrth and joy euermore to wake
In hevyn I bylde Angeħ fful bryth
my servauntys to be and for my sake
with merth and melody worchepe my myth
I belde them in my blysse
Aungeħ in hevyn evyr more xal be
In lyth ful clere bryth as ble
With myrth and song to worchip me
Of joye þei may not mys.

hic cantent angeli in celo. ¶ Tibi omnes angeli tibi celi et vniuerse potestates · Tibi cherubyn et seraphyn incessabili voce proclamant · Sanctus · Sanctus · Sanctus · Dominus deus sabaoth.

lucifere

¶ To whos wurchipe synge ȝe þis songe
to wurchip god or reverens me
but ȝe me wurchipe ȝe do me wronge
ffor I am þe wurthyest þat evyr may be.

                                    Angeli boni
      We wurchipe God of myth most stronge
      whiche hath fformyd · bothe vs *and* the¹   45
      we may nevyr wurchyp hym to longe
      Ffor he is most worthy of mageste
      On knes to god we ffalle
      oure lorde god wurchyp we
      And in no wyse honowre we² the   50
      A gretter lord may nevyr non be
      than he þat made us alle.
                                    lucifere
      ¶ A wurthyer lord forsothe am I
      and worthyer than he · Euyr wyl I be
Fo. 11 In evydens þat I am more wurthy   55
      I wyl go syttyn in goddys se
      Above sunne *and* mone *and* sterrys on sky
      I am now set as ȝe may se
      now wurchyp me ffor most mythy³
      and for ȝour lord honowre now me   60
      Syttyng in my sete.
                                    angeli mali
      Goddys myth we for-sake
      and for more wurthy we þe take
      þe to wurchep honowre we make
      and ffalle down at þi ffete.   65
                                    Deus
      ¶ Thu lucyfere ffor þi mekyl pryde
      I bydde þe ffalle from hefne to helle
      And all þo þat holdyn on þi syde
      in my blysse nevyr more to dwelle
      At my comawndement anoon down þou slyde   70
      with merth *and* joye nevyr more to melle
      In myschyf *and* manas evyr xalt þou abyde
      in byttyr brennyng *and* fyer so felle
      in peyn evyr to be pyht⁴.

---

¹ The *e* here is thickly written as if a correction, and a faint second *e* is written above.
² This word omitted and written over the line by the scribe.
³ *wurthy* first written and crossed through.
⁴ *pyth* first written and crossed through.

Fall of Lucifer

                                              lucyfere

At thy byddyng þi wyl I werke
and pas fro joy to peyne smerte   75
now I am a devyl ful derke
þat was An Aungell bryht.

¶ Now to helle þe wey I take
in endeles peyn þer to be pyht   80
Ffor fere of fyre a fart I crake
In helle donjooñ · myn dene is dyth.

                                              Deus 2

Fo. 11ᵛ Now hevyn is made ffor Aungell sake
þe fyrst day and þe fyrst nyth
The secunde day watyr I make   85
The walkyn also ful fayr and b[r]yth¹
The iij^{de} day · I parte watyr from erthe
tre and every growyng thyng
both erbe and floure of Suete smellyng
the iij^{de} day is made be my werkyng   90
now make I þe day þat xal be þe fferthe.

¶ Sunne and mone and sterrys Also
þe forthe day I make in same
þe v^{te} day · wermñ and ffysch þat swymme and go
byrdys and bestys · bothe wylde and tame   95
the sexte day my werk I do
And make þe man Adam be name
In erthelech paradys with-owtyn wo
I graunt þe bydyng lasse þou do blame.

¶ Fflesch of þi fflesch · And bon of þi boñ   100
Adam here is þi wyf and make
both ffysche and foulys þat swymmyn and goñ
to everych of hem a name þou take
Bothe tre and frute and bestys echoñ
red and qwyte bothe blew and blake   105
þou ȝeve hem name be þi self aloñ
Erbys and gresse both beetys and brake
þi wyff þou ȝeve name Also

---

¹ This word has been altered by erasure to *lyth*.

loke þat ȝe not ses
ȝowre ffrute to encres 110
þat þer may be pres
  me worchipe for to do.

Fo. 12 ¶ Now come Fforth Adam to paradys
ther xalt þou haue all maner thynge
bothe flesch and ffysch and frute of prys 115
all xal be buxum at þi byddyng
Here is pepyr pyan and swete lycorys
take hem all at þi lykyng
both appel and pere and gentyl rys
but towche nowth þis tre þat is of Cunnyng 120
All thynge saff þis ffor þe is wrought
here is all þinge þat þe xulde plese
All redy made on to þin ese
Ete not þis frute ne me dysplese
ffor þan þou deyst þou skapyst nowth. 125

¶ Now haue I made all thynge of nowth
hevyn and Erth · foull and best
to all thynge þat myn hand hath wrowth
I graunt myn blyssyng þat evyr xal lest [1]
My wey to hefne is redy sowth 130
of werkyng I wole þe [2] vij^{te} day rest
And all my creaturys þat be a-bowth
my blyssyng ȝe haue both Est and west
of werkyng þe vij^{te} day ȝe sees
and all þo þat sees of laboryng here 135
þe vij^{te} day with-outyn dwere
and wurchyp me in good manere
þei xal in hefne haue endles pes.

¶ Adam go forth and be prynce in place
Fo. 12^v ffor to hefne I sped my way 140
þi wyttys wel loke þou chase
And gostly gouerne þe as I say.
       Adam

¶ Holy ffadyr blyssyd þou be

[1] *last* first written, corrected to *lest*.
[2] A word, possibly *sefte*, written before vij^{te} and crossed through.

Creation of the World and Man 21

Ffor I may walke in welthe anow [1]
I ffynde datys gret plente 145
and many ffele frut*ys* ful every bow
All þis wele is ʒevyn to me
and to my wyf þat on me lowh
I haue no nede to towche ʒon tre
Aʒens my lordys wyl to werke now 150
I am a good gardene*re*
Euery frute of rythe name
I may gaderyn w*ith* gle *and* game
to breke þat bond I were to blame
þat my lord bad me kepyn here. 155

         Eua

¶ We may both be blyth *and* glad
oure lordys comaundement to fulfyll
w*ith* ffele frutys be we ffayr ffad
wou*n*dyr dowcet *and* nevyr on ill
Euery tre w*ith* frute is sprad 160
of them to take as plesyth us tyll
oure witte were rakyl *and* ovyr don bad
to fforfete Ageyns oure lordys wyll
   in ony wyse
in þis gardeyn I wyl go se 165
all þe fflour*ys* of fayr bewte
and tastyn þe frut*ys* of gret ple*n*te
   þ*at* be i*n* paradyse.

         Serpens

¶ Heyl Ffayr Wyff *and* comely dame
þis ffrute to Ete I þe cownselle 170
take þis Appyl *and* Ete þis ssame
þis frute is best as I þe telle.

         Eua

That appyl to Ete I were to blame
ffrom joy oure lorde wold us expelle
we xuld dye *and* be put out w*ith* schame 175
in joye of paradyse nevyr more to duelle

---

[1] The final *e* of *welthe* and the *a* (?) of *anow* are obscured by a large blot in the MS. which may also hide some letter or mark above *be* in the first line of the stanza.

God hym self þus sayde¹
what day of þat frute we Ete
wi*th* þese wurdys god dyd us threte
þat we xuld dye o*ur* lyff to lete   180
þ*er*ffore I am affrayde.

                Serpens

¶ Of þis Appyl yf ȝe wyl byte
Evyn as god is so xal ȝe be
wys of Connyng as I ȝow plyte
lyke on to god in al degre   185
Su*n*ne *and* mone *and* sterrys bryth
ffyscħ *and* foule boþe sond *and* se
at ȝo*ur* byddyng bothe day *and* nyth
alł thynge xal be in ȝowre powste
ȝe xal be goddys pere   190
Take þis appyl in þin hond²
and to byte þer of þo*u* ffond
take a nother to þin husbond
þer of haue þo*u* no dwer*e*.

                Eua

¶ So wys as god is in his gret mayñ   195
and ffelaw in ku*n*nyng ffayn wold I be.

                Serpens

Fo. 13ᵛ   Ete þis Appyl *and* in certeyñ
þ*at* I am trewe sone xalt þo*u* se.

                Eva

¶ To my*n* husbond wi*th* herte fful fayñ
þis appyl I ber*e* as þo*u* byddyst me   200
þis frute to Ete I xal asayñ
so wys as god is yf we may be
And goddys per*e* of myth
To my*n* husbond I walke my way
and of þis appyl I xal a-say   205
to make hym to Ete yf þat I may
and of þis ffrewte to byth.

¶ *hic eua reueniet ade viro suo et dicet ei.*
¶ My semely spowse *and* good husbond

---

¹ The *a* in *sayde* has been corrected from an *e* by the scribe.
² Corrected by the scribe from *hand*, the *a* marked with deleting dot beneath and *o* written above the line.

Fall of Man 23

lysteneth to me sere I ȝow pray
take þis ffayr appyl all in ȝour hond 210
þer of a mursel byte *and* a-say
To Ete þis appyl loke þ*at* ȝe fonde
Goddys ffelaw¹ to be al-way
All his wysd*am* to vndyrstonde
and goddys pere to be ffor Ay 215
all thyng for to make
both ffysch *and* foule se *and* sond
byrd *and* best watyr *and* lond
þis appyl þou take out of my*n* hond
a bete þerof þou take. 220

Adam

¶ I dare not towch þi*n* hand ffor dred
of our*e* lord god omnypotent
if I xuld werke after þi reed
of god our*e* makere I xuld be shent
Fo. 14 If þ*at* we do þis synful dede 225
we xal be ded by goddys jugement
out of þi*n* hand w*ith* hasty spede
cast out þ*at* appyl a-non present
ffor fer of goddys threte.

Eva

Of þis appyl yf þ*ou* wylt byte 230
goddys per*e* þ*ou* xalt be pyht
So wys of kun*n*yng I þe plyht
þis frute yf þ*ou* wylt Ete².

Adam

¶ If we it Ete our*e* self we kylle
as god us told we xuld be ded 235
to ete þ*at* frute · *and* my lyf to spylle
I dar not do aftyr þ*i* reed.

Eua

A ffayr Aungell þus · seyd me tylle
to Ete þ*at* appyl take nevyr no dred
so ku*n*nyng as god in hevyn hille 240
þ*ou* xalt sone be w*ith*-inne a sted
þerfore þis frute þ*ou* Ete.

¹ The second *f* is doubtful here. It seems to have been faintly inserted.
² Final *e* blotted.

  Adam

  Off goddys wysdam for to lere
  and in kunnyng to be his pere
  of thyn hand I take it here   245
  and xal sone tast þis mete.
                                            Adam dicit sic
  Alas Alas ffor þis fals dede
  my flesly frend · my fo I fynde
  Schameful synne doth us vn-hede¹
  I se vs nakyd be-fore and be-hynde   250
  oure lordys wurd wold we not drede
Fo. 14ᵛ þerfore we be now caytyvys vn-kynde
  oure pore preuytes ffor to hede
  Summe ffygge levys fayn wolde I fynde
  Ffor to hyde oure schame   255
  Womman ley þis leff on þi pryvyte
  And with þis leff I xal hyde me
  Gret schame it is vs nakyd to se
  Oure lord god þus to grame.
                                                    Eva
  ¶ Alas þat evyr þat speche was spokyn   260
  þat þe fals Aungel seyd on to me
  Alas oure makers byddyng is brokyn
  Ffor I haue towchyd his owyn dere tre
  oure fflescly eyn² byn al vnlokyn
  nakyd for synne oure sylf we se   265
  þat sory Appyl þat we han sokyn
  to deth hath brouth my spouse and me
  Ryth grevous is oure synne
  of mekyl shame now do we knowe
  Alas þat evyr þis Appyl was growe   270
  to dredful deth now be we throwe
  in peyne vs evyr to pynne.
                                                    Deus
  ¶ Adam þat with myn handys I made
  where art þou now · what hast þou wrought.
                                                    Adam
  A lord for synne oure flourys do ffade   275
  I here þi voys · but I se þe nought.

          ¹ *vn* corrected by scribe from *un*.
          ² An *h* crossed out between *eyn* and *byn*.

Fall of Man

                                                                 Deus

Adam why hast þou synnyd so sone
þus hastyly to breke my bone
And I made þe mayster vndyr mone
    Trewly of Euery tre   280
O tre I kept for my owe
Lyff and deth þer-in I knowe
þi synne fro lyf now þe hath throwe
    Ffrom deth þou mayst not fle.

                                                Adam

¶ Lord I haue wrought aȝens þi wyll   285
I sparyd nat my sylf to spylle
þe woman þat þou toke me tylle
    sche brougth me þer to
It was here counsell and here reed
sche bad me do þe same deed   290
I walke as werm with-outyn wede
    A-wey is schrowde and sho.

                                                Deus

¶ Womman þat arte þis mannys wyffe
why hast þou steryd ȝour bothers stryffe
now ȝe be¹ ffrom ȝour ffayr lyffe   295
    and are demyd for to deye
Vnwys womman sey me why
þat þou hast don þis fowle foly
and I made þe a gret lady
    in paradys for to pleye.   300

                                                Eua

¶ Lord whan þou wentyst from þis place
A werm with An Aungelys face
he hyth vs to be ful of grace
    þe frute yf þat we Ete
I dyd his byddyng Alas Alas   305
now we be bowndyn in dethis las
I suppose it was sathanas
    to peyne he gan vs pete.

                                                Deus

¶ Thou werm with þi wylys wyk
þi fals fablis þei be ful thyk   310

---

¹ *be* omitted and written above the line.

Fall of Man

why hast þou put dethis pryk
in Adam *and* his wyff
thow þei bothyn my byddyng haue brokyn
out of whoo ȝet art not wrokyn
in helle logge þou xalt be loky [1] 315
And nevyr mo lacche lyff.

Diabolus

¶ I xal þe sey where ffore *and* why
I dede hem all þis velony
ffor I am ful of gret envy
Of wreth *and* wyckyd hate 320
That man xulde leve above þe sky
where as su*m* tyme dwellyd I
and now I am cast to helle sty
streyte out at hevyn gate.

Deus [2]

¶ Adam ffor þou þat appyl boot 325
A-ȝens my byddyng well I woot
Go teyl þi mete wit*h* swynk *and* swoot
in to þi lyvys ende
Goo nakyd vngry *and* bare ffoot
Ete bot*h* erbys gres *and* root 330
thy bale hath non other boot
as wrecch in werlde þou wende.

¶ wo*m*man þou sowtyst þis synnyng
And bad hym breke my*n* byddy*n*g
þerfore þou xalt ben vndyrlyng 335
to mannys byddyng bend
what he byddyth þe · do þou þat thynge
and bere þi chyldere · wit*h* gret gronynge
Fo. 16 In daungere *and* in deth dredynge
in to þi lyvys ende. 340

¶ Thou wyckyd worm ffull of pryde
ffowle envye syt be þi syde
Vpon þi gutt þou xalt glyde
As werm wyckyd in kende
tyl a mayden in medyl-erth be born 345
þou ffende I warn þe be-forn

---

[1] So in MS. for *lokyn*.   [2] *Ad* first written and crossed through.

thorwe here þi hed xal be to-torn
On wombe a-wey þou wende.

Diabolus

¶ At þi byddyng ffowle I falle
I krepe hom to my stynkyng stalle 350
helle pyt and hevyn halle
xul do þi byddyng bone
I ffalle down here a ffowle freke
ffor þis ffalle I gynne to qweke
with a ffart my brech I breke 355
my sorwe comyth ful sone.

Deus

¶ Ffor ȝour synne þat ȝe haue do
out of þis blysse sone xal ȝe go
in erthly labour to levyn in wo
and sorwe þe¹ xal a-tast 360
Ffor ȝour synne and mys-doyng
An Angell with a swerd brennyng
out of þis joye he xal ȝow dyng
ȝour welth awey is past

Hic recedit deus et angelus seraphicus cum gladio fflammea verberat adam et Euam extra paradisum.

Seraphim

¶ ȝe wrecchis vnkend and ryht vnwyse 365
out of þis joye hyȝ ȝow in hast
with flammyng swerd ffrom paradyse
to peyn I bete ȝow of care to tast
ȝour myrth is turnyd to carfull syse
ȝour welth with synne a-wey is wast 370
ffor ȝour ffalse dede of synful gyse
þis blysse I spere ffrom ȝow ryth fast
here-in come ȝe no more
Tyl a chylde of a mayd be born
and vpon þe rode rent and torn 375
to saue all þat ȝe haue forlorn
ȝour welth for to restore.

Eva

¶ Alas alas and wele away
þat evyr towchyd I þe tre
I wende as wrecch in welsom way 380

¹ So in MS. *ye* for *ȝe*.

in blake busshys my boure xal be
In paradys is plente of pleye
Ffayr frutys ryth gret plente
þe ȝatys be schet with godys keye
my husbond is lost be-cause of me 385
leve spowse now þou fonde
Now stomble we on stalk and ston
my wyt a-wey is fro me gon
wrythe on to my necke bon
with hardnesse of þin honde. 390

          Adam

 Wyff þi wytt is not wurth a rosch
leve woman turne þi thought

Fo. 17 I wyl not sle fflescly of my fflesch
Ffor of my flesch · þi fflesch was wrought
Oure hap was hard · oure wytt was nesch 395
to paradys whan we were brought
my wepyng xal be longe ffresch
schort lykyng xal be longe bought
no more telle þou þat tale
Ffor yf I xulde sle my wyff 400
I sclow my self with-owtyn knyff
in helle logge[1] to lede my lyff
with woo in wepyng dale.

¶ But lete vs walke forth in to þe londe
with ryth gret labour oure fode to fynde 405
with delvyng and dyggyng with myn hond
oure blysse to bale and care to-pynde
And wyff to spynne now must þou ffonde
oure nakyd bodyes in cloth to wynde
tyll sum comforth of godys sonde 410
with grace releve oure careful mynde
Now come go we hens wyff.

          Eva

Alas þat ever we wrought þis synne
oure bodely sustenauns for to wynne

---

[1] Or longge. There is a stroke over the *og*, but it seems to be in a different ink, higher and not quite like the ordinary mark of the contracted *n*. The words *in helle logge* occur on Fo. 15ᵛ.

Fall of Man

ȝe must delve *and* I xal spynne 415
in care to ledyn our*e* lyff.

                                                   Abeł

I wolde ffayn knowe · how I xuld do   **3**
  to serue my lord god to his plesyng
þ*er* fore Caym brother lete us now go
vn-to our*e* ffadyr wit*h*-owte lettyng *
Fo. 17ᵛ Suenge hym in vertu and in nortur*e*   5
to com to þe hyȝe joy celestyałł
remembryng to be clene *and* pur*e*
for in mys-rewle we myth lythly fałł
A-ȝens hevyn kynge
lete us now don our*e* dyligens   10
to come to our*e* fader*ys* pr*es*ens
Good brother passe we hens
to knowe ffor our*e* levynge.

                                                  Caym

¶ As to my fadyr lete us now tee
to knowe what xal be his talkyng   15
and þat[1] I holde it but vanyte
to go to hy*m* ffor Any spekyng
to ler*e* of his lawe
Ffor if I haue good[2] a-now plente
I kan be mery so moty the   20
thow my fadyr I nevyr se
I ȝyf not þer of An hawe.

                                                 Abel

¶ Ryth sovereyn fadyr semely sad *and* sure
eu*er* we thank ȝow in hert body and thowth

---

* Ff. 16–17 at the foot of both pages crosswise in large liturgical script with rubrication :

Adam. genuit { Caym / Abel / Seth } Caym. gen*uit*. Enoch. gen*uit*. Iradh. gen*uit* Maynael. gen*uit* Matussahel. gen*uit* lamech. þat slow Caym. þis lamech had 2 wyffys. Ada and[3] Sella. Of Ada com Jabel. fadere of tent*ys* and of herdme*n*. [The rest of this note on Lamech's offspring is illegible, the letters having been cut away with the margin. It is in the handwriting of the scribe of the MS.]

    [1] MS. y*t* for ȝet ?
    [2] The *d* is written over some other letter.
    [3] Some miswritten letters (? Se) crossed out after *and*.

And alwey shull whyll oure lyf may indure 25
as inwardly in hert it kan be sought
bothe my brother and I[1]
Ffadyr I ffalle on-to ȝour kne
to knowe how we xul rewlyd be
ffor godys þat fallyth bothe hym and me 30
I wolde ffayn wete trewly.

               Adam

¶ Sonys ȝe arn to spekyn naturaly
 The ffyrstffrute · of kendely engendrure
 Be-fforn whom saff ȝour modyr and I
Fo.18 were nevyr non of mannys nature 35
 And ȝit were we al of a nother portature[2]
 As ȝe haue me oftyn herd seyd sothly
 Wherfore sonys yf ȝe wyl lyff sad and sure
 Ffyrst I ȝow counseyll most syngulerly
 God ffor to loue and drede 40
 And suche good as god hath ȝow sent
 the fyrst frute offyr to hym in sacryfice brent
 hym evyr be-sechyng with meke entent
 In all ȝour werkys to save and spede.

               Abeell

¶ Gramercy ffadyr ffor ȝour good doctrine 45
 Ffor as ȝe vs techyn so xal we do
 And as ffor me þorwe[3] goddys grace[4] dyvyne
 I wyl fforth with applye me þerto.

               Cayme

And þow me be loth I wyl now also
On to ȝour counsell ffadyr me inclyne 50
and ȝitt I say now to ȝow both too
I had levyr gon hom well ffor to dyne.

               Adam

Now god graunt good sacryfice to ȝow both too
he vowche-saff to acceptyn ȝow and all myne

---

[1] The writing becomes noticeably smaller from this line to the bottom of the folio.

[2] *As ȝe haue me oftyn seyd sothly*, with *haue* written above the line, is written as the next line and crossed through.

[3] MS. þõ.   [4] A letter y or þ obliterated after *grace*.

and ȝeve ȝow now grace to plesyn hym soo 55
þat ȝe may come to þat blysse þat hym self is inne
With gostly grace
þat all ȝour here levyng¹
may be to his plesyng
and at ȝour hens partyng 60
to com to good place.*

            Abell dicit

fo.18ᵛ ¶ Al-myhtty god and god ful of myth
be whom all þing is made of nowth
to þe myn hert is redy dyht
for upon þe is all my thought 65
O souereyn lord reygnyng in eternyte
with all þe mekenesse þat I kan or may
This lombe xal I offre it up to the
accept it blyssyd lord I þe pray
my ȝyft is but sympyl þis is no nay 70
but my wyl is good and evyr xal be
þe to servyn and worchepyn both nyht and day
and þer to þi grace grawnt þou me
throwh þi gret mercy
which in a lombys lyknes 75
þou xalt for mannys wyckydnes
Onys ben offeryd in peynfulnes
and deyn ful dolfoly.

¶ Ffor trewly lord þou art most worthy
þe best to haue in eche degre 80
both beste and werst ful certeynly
all is had þurowe grace of þe
The best schep full hertyly
amonges my flok þat I kan se
I tythe it to god of gret mercy 85
And bettyr wold if bettyr myht be
Evyn here is myn offryng
I tythe to þe with ryht good wylle

¹ Lines 58-61 are written in two lines divided by two black strokes.
* Ff. 17ᵛ-18. At the foot of the pages as before : Seth genuit Enos. genuit Caynan. genuit Malachel. genuit Jared genuit Enok genuit Matussalem. genuit Lamech genuit Noe.

of þe best þou sentyst me tylle
now gracyous god on hevyn hille 90
Accept now my tythyng.

Caym

Fo. 19 ¶ Amonges all folys þat gon on grownd
I holde þat þou be on of þe most
to tythe þe best þat is not sownd
and kepe þe werst þat is nere lost 95
But I more wysly xal werke þis stownde
to tythe þe werst and make no bost
Off all my cornys þat may be fownde
In all my ffeldys both croft and cost
I xal lokyn on every syde 100
here I tythe þis vnthende sheff
lete god take it · or ellys lef
þow it be to me gret repreff
I 3eve no ffors þis tyde.

Abell

¶ Now Caym brother þou dost ful ill 105
Ffor god þe sent both best and werst
þerfore þou shewe to hym good wyll
and tythe to god evyr of þe best.

Caym

In feyth þou shewyst now a febyll skyll
it wolde me hyndyr and do me greff 110
what were god þe bettyr þou sey me tyll
to 3evyn hym awey my best sheff
and kepe my self þe wers
he wyll neyther ete nor drynke
Ffor he doth neyther swete nor swynke 115
þou shewyst a ffebyl reson me thynke
what þou fonnyst as a best I gesse.

Abell

3it me thynkyth my wyt is good
to god euer more sum loue to shewe[1]
Off whom we haue oure dayly food 120
and ellys we had but lytyl drewe.

Caym

3itt me thynkeht þi wytt is wood

---

[1] The writing in this and the following three lines at the top of the next folio is smaller.

Ffor of þi lore I ffynde but ffewe
I wylt neuer þe more chawnge my mood
Ffor no wordys þat þou dost shewe 125
I sey I wylt tythe þe werst.

Abell

Now god þat syt in hefne aboue
On whom is sett all myn hool loue
þis wyckyd wylt from þe he showe[1]
As it plesyth hym best. 130

¶ *Hic ardent decimum Abel · et Caym quo facto dicit*   *Caym*

¶ herke abel brother what a-ray is þis
thy tythyng brennyth as ffyre fful bryght
it is to me gret wondyr i-wys
I trow þis is now a straunge syght.

Abell

Goddys wylt fforsothe it is 135
þat my tythyng with fyre is lyth
Ffor of þe best were my tythis
and of þe werst þou dedyst hym dyght
bad thyng þou hym bede
of þe best was my tythyng 140
and of þe werst was þin offryng
þerfor god almyghty hevyn kyng
Alowyht ryht nowth þi dede.

Caym

¶ What þou stynkyng losel and is it so
doth god þe love and hatyht me 145
þou xalt be ded I xal þe slo
þi lord þi god þou xalt nevyr se
Tythyng more xalt þou nevyr do
With þis chavyl bon I xal sle þe
þi deth is dyht þi days be go 150
out of myn handys xalt þou not fle
With þis strok I þe kylle
Now þis boy is slayn and dede
Of hym I xal nevyr more han drede
He xal here after nevyr ete brede 155
With þis gresse I xal hym hylle.

[1] So in MS. for *shoue*.

Deus

    Caym come fforth *and* answer*e* me
a-soyle my qwestyon anon ryght
thy brother Abel wher is now he
ha don and answer*e* me as tyght.   160

Caym

My brothers keper*e* ho made me
Syn whan was I his kepyng knyght
I kan not telle wher þat he be
to kepe hym was I nevyr dyght
I know not wher he is.   165

Deus

A cursyd Caym þ*ou* art vntrewe
and for þ*i* dede þ*ou* xalt sore rewe
þ*i* brothers blood þ*at* þ*ou* slewe
Askyht vengeauns of þ*i* mys.

¶ Thu xalt be cursyd on þe grounde   170
Vn-pr*o*phitable wher*e* so þ*ou* wende
both veyn *and* nowthty *and* no thyng sounde
wi*th* what þing þ*ou* medele þ*ou* xalt it shende.

Caym

Alas in whoo now am I wounde
a-cursyd of god as man vn-kende   175
of any man yf I be founde
He xal me slo I haue no ffrende
       Alas and wele Away.

Deus

Fo. 20ᵛ   Of what man þat þou be sclayn̄
he xal haue vij folde more payn   180
hym wer*e* bettyr to be sayn̄ [1]
      on lyve be nyth ne day.

Caym

¶ Alas alas whedyr may I go
I dar*e* nevyr se man in þe vesage
I am woundyn as a wrecch in wo   185
A*nd* cursyd of god ffor my ffalfage [2]

---

[1] The word *nevyr* is written above the line after *bettyr* in blacker ink and a different hand. The original construction may be compared with the following on Fo. 22: *I pray to god . . . that he me kepe in such a plyght . . . I to affendyn hym day nor nyght.*

[2] So in MS. for *ffalsage.*

Vn-profytabyl and vayn also
In felde *and* tow͡n in strete and stage
I may nevyr make merthis mo
I wot nevyr whedyr to take passage 190
I dare not here abyde
Now wyl I go wende my way
With sore syeng *and* wel away   Introitus Noe
to loke where þat I best may
Ffrom mannys ssyht me hyde. 195

        Remainder of Fo. 20ᵛ—4 inches—left blank.

                                       ¶ Noe

Fo. 21   G Od of his goodnesse · and of grace grounde   **4**
quire)   By whoys gloryous power all thyng is wrought
in whom all vertu plentevously is ffounde
with-owtyn whos wyl may be ryth nought
Thy seruauntys saue lord fro synful sownde 5
In wyl in werk in dede *and* in thouht
oure welth in woo lete nevyr be fownde
Vs help lord from synne þat we be in brought
Lord god fful of myght
Noe serys my name is knowe 10
my wyff *and* my chyldere here on rowe
to god we pray with hert ful lowe
to plese hym in his syght.

¶ In me Noe þe secunde age
in dede be-gynnyth as I ȝow say 15
afftyr Adam with-outyn langage
þe secunde fadyr am I in fay
But men of levyng be so owt-rage
bothe be nyght and eke be day
þat lesse þan synne þe soner swage 20
god wyl be vengyd on vs sum way
       In dede
Ther may no man go þer owte
but synne regnyth in every rowte
In every place rownde a-bowte 25
Cursydnes doth sprynge *and* sprede.

                                                            Vxor Noe

    ¶ All myghty god of his gret grace
       enspyre men w*ith* hertely wyll
       for to sese of here trespace
       Ffor synfull levyng oure sowle xal spyll *   30
Fo. 21^v Synne offendyth god in his face
       and a-grevyth oure lorde ffull ylle
       It causyth to man ryght grett manace
       and scrapyth hym out of lyvys bylle
       þ*at* blyssyd book   35
       What man in synne doth all wey scl*e*pp*e*
       He xal gon to helle ful d*e*pp*e*
       than xal he nevyr aft*er* cr*e*pp*e*
       out of þ*at* brennyng brook.

    ¶ I am ȝo*ur* wyff[1] ȝo*ur* childeryn þese[2] be   40
       On to us tweyn it doth longe
       hem to teche in all degre
       Synne to for-sakyn *and* werkys wronge
       Therfor*e* fere[3] for loue of me
       Enforme hem wele evyr amonge   45
       Synne to for-sake *and* vanyte
       and vertu to ffolwe þ*at* þei ffonge
               oure lord god to plese.

                                     Noe
       I warne ȝow childeryn on *and* all
       Drede oure lord god in hevy[4] hall   50
       and in no forfete þat we[5] ne fall
               oure lord for to dysplese.
                                  Shem

    ¶ A dere ffadyr god for-bede
       þat we xulde do in ony wyse
       Ony werke of synful dede   55
       oure lord god þat xulde a-gryse

---

* In the bottom margin of Fo. 21 as before:—Noe genuit { Sem / Cham / Japhet }

[1] þese *ch* written after *wyff* and crossed through.
[2] þese here appears to be the correction of another word.
[3] So in MS. for *sere*.   [4] So in MS. for *hevyn*.   [5] So in MS. for ȝe.

my name is Shem[1] ȝour son of prise
I xal werke aftere ȝour rede
and also wyff þe weyll a-wyse
Wykkyd werkys þat þou non brede 60
    never in no degre.

                                  Vxor Seem

For-sothe sere be goddys grace *
I xal me kepe from all trespace
þat xulde offende goddys face
    be help of þe trynyte. 65
                                  Cham

¶ I am cham ȝour secunde son
and purpose me be goddys myght
nevyr suche a dede for to don
þat xuld a-greve god in syght.
                                  Vxor Cham

I pray to god me grawnt þis bone 70
that he me kepe in such A plyght
mornynge hevenynge mydday and none
I to affendyn hym day nor nyght
lord god I þe pray
Bothe wakynge and eke in slepe 75
gracyous god þou me keppe
þat I nevyr in daunger crepe
On dredfull domys day.
                                  Japhet

¶ Japhet þi iij[de] sone is my name
I pray to god wher so we be 80
þat he vs borwe fro synfull shame
and in vertuous levynge evyr more kepe me.
                                  Vxor Japhet

I am ȝour wyff and pray þe same
þat god vs saue on sonde and se

---

* At the foot of Ff. 21ᵛ-2 : Sem genuit Arfaxat. genuit Sale. genuit Heber. genuit Phaleg. genuit Reu. genuit Sarug. genuit Nachor. genuit Thare.
    Abraham⎫
genuit. Nacor ⎬
    Aran   ⎭

[1] Manly gives *Chem*, but the *S* is clear in the MS. Some miswritten letter between *Shem* and ȝour has been crossed out by the scribe.

With no grevauns þat we hym grame 85
he grawnt vs grace synne to fle
  lord god now here oure bone.
           Noe

Gracyous god þat best may
with herty wyl to the we pray
þou save us sekyr bothe nyght and day 90
Synne þat we noon done.
           Deus

Fo. 22ᵛ ¶ Ow what menyht this mys-levyng man
whiche myn hand made and byldyd in blysse
Synne so sore grevyht me ȝa in certayn [1]
I wol [2] be vengyd of þis grett mysse 95
Myn aungel dere þou xalt gan [3]
to noe þat my servaunt is
A shypp to make on hond to tan̄
þou byd hym swyth ffor hym and his
Ffrom drynchyng hem to save 100
Ffor as I am god off myght
I xal dystroye þis werd down̄ ryght
Here synne so sore grevyht me in syght
þei xal no mercy haue.

¶ Ffecisse hominem nunc penitet me 105
þat I made man sore doth me rewe
myn handwerk to sle sore grevyth me
but þat here synne here deth doth brewe
Go sey to noe as I bydde þe
hym self his wyf his chylderyn trewe 110
tho viij sowlys in shyp to be
thei xul not drede þe flodys fflowe
þe fflod xal harme them nowht
of all ffowlys and bestys thei take a peyre
In shypp to saue both ffoule and ffayere [4] 115
Ffrom all dowtys and gret dyspeyre
This vengeauns or it be wrought.

---

[1] First written *tayn* and corrected to *teyn* or vice versa.
[2] *he* written after this word and crossed through.
[3] *gon* first written and corrected by scribe to *gan*.
[4] In this line and the remaining six to the bottom of the folio the writing is thicker and slightly more cramped.

Noah   39

                                             Angelus ad noe

¶ Noe · Noe · A shypp loke þou make·[1]
  and many a chaumbyr þou xalt haue þerinne
  Of euery kyndys best a cowpyl þou take·   120
  with-in þe shypp bord · here lyvys to wynne·*

Fo. 23   Ffor god is sore grevyd with man for his synne
  þat all þis wyde werd xal be dreynt with flood
  saff þou and þi wyff xal be kept from þis gynne
  and also þi chylderyn with here vertuys good.   125
                                                 Noe

¶ How xuld I haue wytt a shypp for to make
  I am of ryght grett Age V. C. ȝere olde
  it is not for me þis werk to vndyr-take
  Ffor ffeynnesse[2] of Age my leggys gyn ffolde.
                                               Angelus

  This dede ffor to do be bothe blythe and bolde   130
  God xal enforme þe and rewle þe ful ryght
  of byrd and of beste take as I þe tolde
  A peyr in to þe shypp and god xal þe qwyght.
                                               Noe

¶ I am ful redy as god doth me bydde
  A shypp for to make be myght of his grace   135
  Alas þat ffor synne it xal be so be-tydde
  þat vengeauns of flood xal werke þis manase
  god is sore grevyd with oure grett tresspas
  þat with wylde watyr þe werd xal be dreynt
  A shyppe for to make now lete us hens pas   140
  þat god a-ȝens us of synne haue no compleynt

Hic transit noe cum familia sua pro naui quo exeunte locum interludij sub intret statim lameth conductus ab adolescente et di[cit
                                                 lameth

¶ Gret mornyng I make and gret cause I haue
  Alas now I se not · for age I am blynde
  blyndenes doth make me of wytt for to rave
  Whantynge of eye syght · in peyn doth me bynde   145
  whyl I had syht þer myht nevyr man fynde
  my pere of Archerye · in all þis werd A-boute

---

* At the foot of Fo. 22ᵛ. Aran. gen*uit*. loth.
[1] The stops at *make, take,* and *wynne* are in fresher coloured ink.
[2] MS. *ffeyynnesse* with a dot under the second *y*. The letters for *þ* and for *y* in this MS. are indistinguishable; Halliwell reads *ffeyþnnesse.*

Fo. 23ᵛ   Ffor ȝitt schet I nevyr at hert Are · nere hynde
           but yf þat he deyd · of þis no man haue doute.

     ¶ Lameth þe good archere my name was ovyr aH   150
       Ffor þe best Archere myn name dede ever sprede
       Record of my boy here wytnes þis he xal
       What merk þat were set me · to deth it xuld blede.
                                                                Adolescens
           It is trewe mayster · þat ȝe seyn in dede
           Ffor þat tyme ȝe had ȝoure bowe bent in honde   155
           If þat ȝour prycke had be half a myle in brede¹
           ȝe wolde þe pryk han hitte if ȝe ny had stonde.
                                                                lameth
     ¶ I xuld nevyr affaylid · what marke þat ever were sett
       whyl þat I myght loke · and had my clere syght
       and ȝit as me thynkyht no man xuld shete bett   160
       Than I xuld do now · if myn hand were sett a-ryght
       A-spye som marke boy · my bow xal I bende wyght
       and sett myn hand euyn to shete at som best
       and I dar ley a wagour his deth for to dyght
       þe marke xal I hitt · my lyf do I hest.   165
                                                                Adolescens
     ¶ Vndyr ȝon grett busche mayster · A best do I se
       take me þin hand swyth and holde it ful stylle
       now is þin hand evyn as euyr it may be
       Drawe up þin takyH ȝon best for to kylle.
                                                                lameth
           My bowe xal I drawe ryght with herty wylle   170
           this brod arwe I shete þat best ffor to sayH
           now haue at þat busch · ȝon best for to spylle
           A sharppe schote I shote þer of I xal not fayH.
                                                                Caym
     ¶ Out out and alas · myn hert is on sondyr
       With a brod arwe I am ded and sclayn   175
Fo. 24 I dye here on grounde · myn hert is aH to tundyr
       With þis brod arwe it is clovyn on twayn,
                                                                lameth
           Herke boy cum telle me þe trewth in certeyn
           what man is he þat · þis cry doth þus make.
                                                                Adolescens
           Caym þou hast kyllyd I telle þe ful pleyn   180

       ¹ *dede* first written and crossed through.

with þi sharp shetyng · his deth hath he take.

lameth

¶ Haue I slayn cayme · alas what haue I done
þou stynkynge lurdeyn · what hast þou wrought
þou art þe¹ why I scle hym so sone
þer fore xal I kyll þe here · þou skapyst nowght   185

Hic lameth cum arcu suo verberat Adolescentem ad mortem, dicente Adolescente

Adolescens

Out out I deye here · my deth is now sought
þis theffe with his bowe hath broke my brayn
þer may non helpe be · my dethe is me brought
Ded here I synke down · as man þat is sclayn.

lameth

¶ Alas what xal I do wrecch wykkyd on woolde   190
God wyl be vengyd ful sadly on me
Ffor deth of Caym I xal haue vii folde
more peyn þan he had · þat abell dede sle.
These to mennys deth full sore bought xal be
Vpon all my blood god wyll venge þis dede   195
where fore sore wepyng hens wyl I fle
and loke where I may best my hede sone heyde.

*Hic recedat lameth et statim intrat noe cum naui cantantes.**

Noe

Fo. 24ᵛ ¶ With doolful hert syenge sad and sore
Grett mornyng I make ffor this dredful flood
of man and of best is dreynte many a skore   200
All þis werd to spyll þese flodys be ful wood
and all is for synne of mannys wylde mood
þat god hath ordeyned þis dredfull vengeaunce
In þis flood spylt is many a mannys blood'
Ffor synfull levynge of man we haue gret grevauns.   205

¶ All þis hundryd ȝere ryght here haue I wrought
this schypp for to make as god dede byd me

---

¹ So in MS.
* Note at bottom of Fo. 24 in what might be the less careful writing of the scribe of the MS. but might also be a different hand:
    Noe Schyp was in lenght CCC. Cubytes. ⎫ þe fflod 15 Above
    In brede ffyfty. And þe heyth. thretty. ⎭ hyest montayn.

of aɫɫ maner bestys a copyɫɫ is in brought
With-in my Shypp borde · on lyve for to be
Ryght longe god hath¹ soferyd · amendyng to se 210
Aɫɫ þis hundyrd ȝere god hath shewyd grace
Alas fro gret syn man wyl not fle
God doth þis vengeauns for oure gret trespase.

                Vxor noe

¶ Alas for gret ruthe of þis gret vengeaunce
 gret doyl it is to se · þis watyr so wyde 215
 but ȝit thankyd be god of þis ordenaunce
 þat we be now savyd on lyve to abyde.

                Seem̄

Ffor grett synne of lechory aɫɫ þis doth betyde
Alas þat evyr such synne xulde be wrought
þis fflood is so gret on every asyde 220
þat aɫɫ þis wyde werd to care is now brought.

                Vxor Seem̄

¶ Becawse of chylderyn of God · þat weryn good²
 dede forfete ryght sore · what tyme þat þei were
 synfully compellyd to caymys blood
 Ther-fore be we now cast in ryght grett care. 225

                Cham

Ffor synful levynge þis werde doth for-fare
So grevous vengeauns myght nevyr man se
ouyr aɫɫ þis werd wyde þer is no plot bare
with watyr and with flood god vengyd wyɫɫ be.

                Vxor Cham

¶ Rustynes of Synne is cawse of þese wawys 230
 Alas in þis fflood þis werd xal be lorn
 ffor offens to god · brekyng his lawys
 On Rokkys ryght sharp is many a man torn.

                Japhet

So grevous fflodys were nevyr ȝett be-forn̄
Alas þat lechory þis vengeauns doth gynne 235
it were weɫɫ bettyr euer to be vnborn̄
than ffor to forfetyn evyr more in þat synne.

                Vxor Japhet

¶ Oure lord god I thanke of his gret grace
 þat he doth us saue from þis dredful payn̄

---

¹ An *f* between *hath* and *soferyd* obliterated.
² *gode* first written and crossed out.

Noah

hym for to wurchipe in euery stede *and* place   240
we beth gretly bownde w*ith* myght *and* w*ith* mayn.

Noe xl^{ti} days and nyght*ys* hath lasted þis rayn
And xl^{ti} days þis grett flood be-gynnyth to slake
this Crowe xal I sende out to seke sum playn
Good tydyng*ys* to brynge þis massage I make   245

*hic emittat coru*u*m et paru*m *expectans iteru*m *dicat*

¶ This crowe on su*m* careyn is fall for to ete
þer fore a Newe masangere I wyll fforth now sende
ffly fforth þou fayr dove ovyr þese wate*rys* wete
and aspye afftere sum drye lond oure mornyng to A-mend

hic euolet columba qua redeinte [1] *cum ramo virid*' *oliue.*

Joye now may we make of myrth þat þat were frende   250
A grett olyve bush þis dowe doth us brynge
Ffor joye of þis token ryght hertyly we tende
oure lord god to worchep a songe lete vs synge

Hic decantent hos versus · ¶ *Mare vidit et fugit ·*/ *jordanis conuersus
est retrorsum · Non nob*is *dom*i*ne non nob*is ·/ *sed nom*i*ni tuo da
gloriam · Et sic recedant cum naui.*

Introitus abrahe &c.

5

Most myghty make*re* of Sunne and of mone
Kyng of kyng*ys and* lord oue*r* all
All myghty god in hevyn trone
I þe honow*re and* evyr more xal
My lord my god to þe I kall   5
w*ith* herty wyll lord I þe pray
In synfull lyff lete me nevyr fall
but lete me leve evyr to þ*i* pay.

¶ Abraham my name is kydde
and patryarke of Age ful olde   10
A*nd* ȝit be þe grace of god is bredde
In my*n* olde age a chylde full bolde
Ysaac lo here his name is tolde
My swete sone þat stondyth me by

---

[1] So in MS.

| | Amonges all chylderyn þat walkyn on wolde | 15 |
| | A louelyer chylde is non trewly. | |

¶ I thanke god · with hert well mylde
of his gret mercy and of his hey grace
and pryncepaly ffor my suete chylde
þat xal to me do gret solace¹   20

Fo. 26   Now suete sone fayre fare þi fface
Fful hertyly do I love the
Ffor trewe herty love now in this place
my swete chylde com kysse now me.

                                  Ysaac

¶ At ȝoure byddynge ȝour mouthe I kys   25
with lowly hert I ȝow pray
ȝoure fadyrly love lete me nevyr mysse
but blysse me ȝour chylde both nyght and day.

                                 Abraham

Al-myghty god þat best may
his dere blyssyng he graunt þe   30
And my blyssyng þou haue all way
in what place þat evyr þou be.

¶ Now Isaac my sone so suete
Al-myghty god loke þou honoure
wich þat made both drye and wete   35
Shynyng sunne and scharpe schoure
thu art my suete childe and paramoure
Fful wele in herte do I þe loue
loke þat þin herte in hevyn toure
be sett to serve oure lord god a-bove.   40

¶ In þi ȝonge ² lerne god to plese
and god xal quyte þe weyl þi mede
now suete sone of wordys these
with all þin hert þou take good hede
Now fare weyl sone god be þin spede   45
Evyn here at hom þou me a-byde
I must go walkyn ffor I haue nede
I come a-ȝen with-inne a tyde.

¹ The initial s (ſ) is thickened and may be intended to be double (ſſ).
² So for ȝongþe (?).

Ysaac

¶ I pray to God ffadyr of myght
þat he ȝow spede in all ȝour waye
From shame and shenshipp daye and nyght
God mote ȝow kepe in ȝour jornay.

Abraham

Now fare weyll sone I þe pray
Evyr in þin hert loke god þou wynde
hym to serue bothe nyght and day
I pray to god send þe good mynde.

¶ Ther may no man love bettyr his childe
þan Isaac is lovyd of me
Almyghty god mercyful and mylde
Ffor my swete sone I wurchyp þe
I thank þe lord with hert ful fre
Ffor þis fayr frute þou hast me sent
now gracyous god wher so he be
to saue my sone evyr more be bent.

¶ Dere lord I pray to þe Also
me to saue for þi seruuaunte
and sende me grace nevyr for to do
thyng þat xulde be to þi displesaunte
Bothe ffor me and for myn infaunte
I pray þe lord god vs to help
thy gracyous goodnes þou us grawnt
and saue þi serwaunt from helle qwelp.

Angelus

¶ Abraham how abraham
lyst and herke weyll on to me.

Abraham

Al redy sere here I am
tell me ȝour wyll what þat it be.

Angelus

Almyghty god þus doth bydde þe
ysaac þi sone Anon þou take
and loke hym þou slee a-noon lete se
and sacrafice to god hym make.

¶ Thy welbelouyd childe þou must now kylle
to god þou offyr hym as I say

Evyn vpon ȝon hey hylle
þat I þe shewe here in þe way
Tarye not be nyght nor day 85
but smertly þi gate þou goo
Vpon ȝon hille þou knele and pray
to god · and kylle þe childe þer and scloo.

           Abraham

¶ Now goddys comaundement must nedys be done
Alł his[1] wyl is wourthy to be wrought 90
but ȝitt þe fadyr to scle þe sone
grett care it causyth in my thought
In byttyr bale now am I brought
my swete childe with knyf to kylle
but ȝit my sorwe avaylith ryght nowth 95
for nedys I must werke goddys wylle.

¶ With evy hert I walke and wende
my childys deth now for to be
now must þe fadyr his suete son schende
Alas for ruthe it is pete 100
my swete sone come hedyr to me
how Isaac my sone dere
com to þi ffadyr my childe so fre
ffor we must wende to-gedyr in fere.

           Isaac

¶ Alł redy fadyr Evyn at ȝour wylł 105
and at ȝour byddyng I am ȝow by
Fo. 27ᵛ With ȝow to walk ovyr dale and hilł[2]
At ȝoure callyng I am redy
To þe fadyr evyr most comly
It ovyth þe childe evyr buxom to be 110
I wyl obey ful hertyly
to alł thyng þat ȝe bydde me.

           Abraham

¶ Now son in þi necke þis fagot þou take
and þis fyre bere in þinne honde
Ffor we must now sacrefyse go make 115
evyn aftyr þe wylł of goddys sonde

---

[1] The *h* is partly effaced through stain on MS.
[2] This line is preceded by part of the last line—*And at ȝour byddyng I am*—written again by mistake and crossed out.

Take þis brennyng bronde
my swete childe and lete us go
ther may no man þat levyth in londe
haue more sorwe than I haue wo. 120

Ysaac

¶ Ffayre fadyr ȝe go ryght stylle
I pray ȝow fadyr speke on to me.

Abraham

Mi gode childe what is þi wylle
telle me thyn hert I pray to the.

Ysaac

Ffadyr fyre and wood here is plente 125
but I kan se no sacryfice
what ȝe xulde offre fayn wold I se
þat it were don at þe best avyse.

Abraham

¶ God xal þat ordeyn þat sytt in hevynne
my swete sone ffor þis offryng 130
A derere[1] sacryfice may no man nempne
þan þis xal be my dere derlyng.

Ysaac

Lat be good fadyr ȝour sad wepynge
ȝour hevy cher agrevyth me sore
teñ me fadyr ȝour grett mornyng 135
and I xal seke sum help þer-fore.

Abraham

Fo. 28 ¶ Alas dere sone for nedys must me
Evyn here þe kylle as god hath sent
thyn owyn fadyr þi deth must be
Alas þat evyr þis bowe was bent 140
With þis fyre bryght þou must be brent
An Aungelle seyd to me ryght so
Alas my chylde þou xalt[2] be shent
þi careful fadyr must be þi ffo.

Ysaac

¶ Al-myghty god of his grett mercye 145
Fful hertyly I thanke þe sertayne
At goddys byddyng here for to dye
I obeye me here for to be sclayne

---

[1] Double r corrected in the middle of this word.
[2] u apparently first written and altered to a.

I pray ȝow fadyr be glad and fayne
trewly to werke goddys wyll 150
take good comforte to ȝow agayne
and haue no dowte ȝo*ur* childe to kyll.

¶ ffor godys byddyng for sothe it¹ is
þat I of ȝow my deth schulde take
Aȝens² god ȝe don amys 155
his byddyng yf ȝe xuld for-sake
ȝowr*e* owyn dampnac*i*on xulde ȝe bake
if ȝe me kepe from þis reed
wi*th* ȝo*ur* swe*r*d my deth ȝe make
and werk evy*r* mor*e* þe wyll of god. 160

Abraham

¶ The wyll of god must nedys be done
to werke his wyll I seyd nevyr nay
but ȝit³ þe ffadyr to⁴ sle þe sone
my hert doth clynge and cleue as clay.

Ysaac

Fo. 28ᵛ ȝitt werk goddys wyll fadyr I ȝow pray 165
and sle me her*e* anoon forth ryght
and turne fro me ȝo*ur* face a-way
myne heed whan þat ȝe xul of smyght.

Abraham

¶ Alas der*e* childe I may not chese
I must nedys my swete sone kylle 170
my der*e* derlyng now must me lese
my*n* owyn sybb blood now xal I spylle
ȝitt þis dede or I fulfylle
my swete sone þ*i* mouth I kys.

Ysaac

Al redy fadyr evyn at ȝo*ur* wyll 175
I do ȝo*ur* byddyng as reson is.

Abraham

¶ Alas der*e* sone her*e* is no grace
but nedis ded now must þo*u* be

---

[1] *it* omitted and written above the line in slightly darker ink.
[2] *Aȝeng* first written and crossed through.
[3] An *e* stands above the *it* of *ȝit*.
[4] The letters *sch* between *to* and *sle* are crossed through.

Abraham and Isaac

with þis kerchere I kure þi face
In þe tyme þat I sle the 180
Thy lovely vesage wold I not se
not for all þis werdlys good
with þis swerd þat sore grevyht me
my childe I sle and spylle his blood.

Angelus

¶ Abraham Abraham þou fadyr fre. 185

Abraham

I am here redy what is ȝour wylle.

Angelus

Extende þin hand in no degre
I bydde þou hym not kylle
here do I se by ryght good skylle
All myghty god þat þou dost drede 190
For þou sparyst nat þi sone to spylle
God wyll aqwhyte þe well þi mede.

Abraham

¶ I thank my god[1] in hevyn a-bove
and hym honowre for þis grett grace
and þat my lord me þus doth prove 195
I wyll hym wurchep in every place
my childys lyff is my solace
I thank myn god evyr for his lyff
In sacrifice here or I hens pace
I sle þis shepe with þis same knyff. 200

¶ Now þis shepe is deed and slayn
with þis fyre it xal be brent
Of Isaac my sone I am ful fayn
þat my swete childe xal not be shent
This place I name with good entent 205
þe hill of[2] godys vesytacion
Ffor hedyr god hath to us sent
his comforte aftyr grett trybulacion.

Angelus

¶ Herke Abraham and take good heyd
by hym self god hath þus sworn 210

[1] Some miswritten word (? me) after *god* crossed through.
[2] *of* omitted and written above the line.

Ffor þat þou woldyst a done þis dede
he wyll þe blysse both evyn and morn
Ffor þi dere childe þou woldyst haue lorn
at goddys byddyng as I the telle
god hath sent þe word be-forn   215
þi seed xal multyplye wher so þou duelle.

¶ As sterrys in hevyn byn many and fele
so xal þi seed encrese and growe
þou xalt ovyr-come in welth and wele
all þi fomen reknyd be rowe   220
As sond in þe se doth ebbe and flowe
Fo. 29ᵛ
(marked 30)   Hath cheselys many vnnumerabyll
so xal þi sede þou mayst me trowe
encres and be evyr prophytabyll.

¶ Ffor to my spech þou dedyst obeye   225
thyn enmyes portys þu shalt possede
and all men on erthe as I þe seye
thei xal be blyssed in þi sede
Al-myghty god þus þe wyll mede
Ffor þat good wyll þat þou ast done   230
þerfore thank god in word and dede
Both þou þi self and ysaac þi sone.

         Abraham

¶ A my lord god to wurchep on kne now I fall
I thank þe lord of þi mercy
now my swete childe to god þou kall   235
and thank we þat lord now hertyly.

         Isaac

with lowly hert to god I crye
I am his seruuant both day and nyght
I thank þe lord in hevyn so hyȝe
with hert with thought with mayn with myght.   240

         Abraham

¶ Gramercy lord and kyng of grace
gramercy lord ouer lordys all
now my joye returnyth his trace
I thank þe lorde in hevyn þin halle.

Abraham and Isaac

              Isaac
  Ovyr all kyngys crownyd kyng I þe kalle   245
  at þi byddyng to dye with knyff
  I was fful buxvm evyn as þi thralle
  lord now I thank the þou grauntyst me lyff.
            Abraham

Fo. 30 ¶ Now we haue wurchepyd oure blyssyd lorde
  on grounde knelyng upon oure kne   250
  now lete us tweyn sone ben of on A-corde
  and goo walke hom in to oure countre.
            Ysaac

  Ffadyr as ȝe wyll so xal it be
  I am redy with ȝow to gon
  I xal ȝow folwe with hert full fre   255
  all þat ȝe bydde me sone xal be don.
            Abraham

 ¶ Now god all thyng of nowth þat made
  evyr wurcheppyd he be on watyr and londe
  his grett honowre may nevyr more fade
  in feld nor town se nor on sonde   260
  As althyng lord þou hast in honde
  so saue us all wher so we be
  whethyr we syttyn walk or stonde
  Evyr on þin handwerke þou haue pyte.
            explicit.

  Remainder of Fo. 30—3⅜ inches—and Fo. 30ᵛ left blank.

(Fo. 31
written
he side
crossed
out)   ¶ Introitus Moyses
  HE þat made all thynge of nought   **6**
   Hevyn and erth both sunne and mone
  Saue all þat his hand hath wrought
  All myghty god in hevyn trone
  I am Moyses þat make þis bone   5
  I pray þe lord god with all my mende
  to us incline þi mercy sone
  þi gracyous lordchep lete us fynde.

 ¶ The to plesyn in all degre
  Gracyous god and lord ovyr all   10

þou graunte us grace wher so we be
and saue us sownd fro synfull fall
thy wyll to werke to us þi thrall
enforme and teche us all þi plesans
in purenesse put us þat nevyr not fall 15
and grounde us in grace ffrom all grevauns.

Hic moyses videns rubrum ardentem admirande dicit[1]

¶ A mercy god what menyth 3on syte
A grene busch as fyre doth flame
and kepyth his colowre fayr and bryghte
Ffresch and grene with-owtyn blame 20
It fyguryth sum thynge of ryght gret fame
I kan not seyn what it may be
I wyll go nere in goddys name
and wysely loke þis busch to se.

Deus

¶ Moyses how Moyses 25
herke to me a-non þis stounde.

Moyses

Fo. 31ᵛ I am here lorde with-owtyn les
3owre gracyous wyll to do I am bounde.

Deus

Thu take þi schon anon ful rownde
of þi fete in hast lete se 30
fful holy is þat place and grownde
þer þou dost stonde I sey to the.

Moyses

¶ Barfoot now I do me make
and pull of my schon fro my fete
now haue I my schon of take 35
what is 3our wyll lord · fayn wold I wete.

Deus

Com nere moyses with me to mete
these tabell-is I take þe in þin honde
with my ffynger in hem is wrete
all my lawys þou vndyrstonde. 40

¶ Loke þat þou preche all abowte[2]
hoo so wyll haue frenshipp of me

---

[1] This is written in larger form than other stage directions.
[2] The colour of the ink changes slightly with this line.

to my lawys loke þei lowte
þat þei be kept in all degre
Go forth and preche a-non let se 45
loke þou not ses nyght nor day.

Moyses

ȝour byddyng lord all wrought xal be
ȝour wyll to werk I walk my way.

Custodi precepta domini dei tui · deutronomini · vj^{to}.

¶ The comaundement of þi lord god man loke þou kepe
where þat þou walk · wake or slepe 50
Euery man take good hede
and to my techynge take good intent
for god hath sent me now in dede
ȝow for to enforme his comaundment
ȝow to teche god hath me sent 55
his lawys of lyff þat arn ful wyse
them to lerne be dyligent
ȝour soulys may þei saue at þe last Asyse.

¶ The preceptys þat taught xal be
be wretyn in þese tablys tweyn 60
In þe fyrst ben wretyn thre
that towch to god þis is serteyn
In þe secund tabyl be wretyn ful pleyn
þe tother vij þat towch mankende
herk now well man what I xal seyn 65
and prent þise lawys well in þi mende.

1us Primum mandatum non habebis deos alienos.

¶ The fyrst comaundement of god as I ȝow say
of þe fyrst tabyl for sothe is this
þou xalt haue neythyr nyght nore day
noon other god but þe kyng of blysse 70
Vndyrstonde wele what menyth this
Euery man in hys degre
And sett neuyr ȝour hert a-mys
Vpon þis werdlys vanyte.

Moses

¶ Ffor if þou sett þi loue so sore   75
   Vpon ryches and werdly good
   þi wurdly rycches þou takyst ever more
   evyn ffor þi god · as man ovyr wood
   Amend þe man and chaunge þi mood
   lese not þi sowle for werdlys welth   80
   Only hym loue · which bodyly ffood
   doth ȝeve all day and gostly helth.

Fo. 32ᵛ 2. Secundum mandatum · non assumens nomen dei tui in vanum.

¶ The secund precept of þe fyrst tabyll
   þe name of god take nevyr in vayne
   Swere none othis be noon fals fabyll   85
   þe name of god þou nevyr dysteyn
   Be whare of othis for dowte of peyn
   Amonges ffelacheppe whan þou dost sytt
   A lytyl othe þis is serteyn
   may dampne thy sowle to helle pytt.   90

¶ Man whan þou art sett at þe nale
   and hast þi langage as plesyth þe
   loke þin othis be non or smale
   and ȝett alwey loke trewe thei be
   But swere not oftyn by rede of me   95
   ffor yf þou vse oftyn tyme to swere
   it may gendyr custom in the
   by-ware of custom ffor he wyl dere.

3. Tercium mandatum · memento vt sabbatum sanctificet.

¶ The iijᵈᵉ comaundment of god as I rede
   doth bydde the · halwe well þin haly day   100
   kepe þe well ffro synfull dede
   and care not gretly ffor rych a-ray
   A ryght pore man þis is non nay
   of sympyl astat in clothis rent
   may-be bettyr than rych with garmentys gay   105
   Oftyn tyme doth kepe þis comaundment.

¶ Ffor rych men do showe oftyn tyme pompe and pride
   On halydays as oftyn is sene

Yn halyday þ^e bodyst not clene
yn glotony to lede þⁱ lyff
yn goddys howes 3e yulde be done
honowr to god both mayden & wyff
Quartū mandatū honora p^{at}rem tuū & m^{at}rem tuam.

Of þ^e secunde tabyll þ^e fyrst comaundmet
And in þ^e order þ^e nyt^e I sey yn fay
he byddyth þ^e eu^{er}mo lyst kept bowt
both fadyr & modyr to wurchep alway
Thow þⁱ þⁱ fadyr be por of aray
& yow neu^{er} so gret of goldes & good
3itt loke þ^u wurchep hy myght & day
of whom þ^u hast both flesch & blode

In þis comaundmente includyd is
thi bodyli fadyr & modyr also
Includyd also I fynde in þis
thi gostly fadyr & modyr y^e to
To þⁱ gostly fadyr Obyr pen3ens do
þⁱ gostly modyr is holy chyrch
these aleyn bene þⁱ sibbe fro Woo
eu them to wurchep loke þ^u þ^u wyrch

Quintū mandatū non occides

The ffyrst comaundement byddyth all us
Sle no man. no wight þ^t þ^u kyll
undyr stonde þis p^{re}cept þus
Sle no wyght with wurd nor will
wykkyd worde wykkyst of all tyme spell
be way þ^t fey of wykkyd langage
wykkyd speche many on doth spyll
therfor of speche hold not ouer large

  Whan pore men passe *and* go be-syde
  At wurthy festys riche men woll bene   110
Fo. 33 Thyn halyday þou kepyst not clene
  In gloteny to lede þi lyff
  In goddys hous ȝe xulde be-dene
  honoure ȝour god · both mayden *and* wyff.

 4. Quartu*m* mandatu*m* hono*r*a p*at*rem tuu*m* et m*at*rem tuam.

 ¶ Off þe secunde tabyll þe fyrst comaundement   115
  And in þe ordyr þe iiij^te I sey in fay
  he byddyth þe eue*r* more wit*h* hert bent
  both ffadyr *and* modyr to wurchep Alway
  Thow þ*at* þi fadyr be pore · of array
  and þow neuer so rych of golde *and* good   120
  ȝitt loke þou wurchep hy*m* nyght *and* day
  Of whom þou hast both fflesch and blood.

 ¶ In þis comaundmente includyd is
  thi bodyli fadyr *and* modyr also
  Includyd also I fynde in þis   125
  thi gostly fadyr *and* modyr þer to
  To þi gostly ffadyr evyr reuerens do
  þi gostly modyr is holy cherch
  these tweyn saue þi sowle fro woo
  euer them to wurchep loke þ*at* þou werch.   130

 5. Quintum mandatum non occides.

 ¶ The ffyfft comaundement byddyth all us
  Scle no man · no whight þ*at* þou kyll¹
  Vndyr stonde þis p*re*cept þus
  Scle no wyght wit*h* wurd nor wyll
  Wykkyd worde werkyht ofty*n* tyme grett ill   135
  be war þer fore of wykkyd langage
  wyckyd spech many on doth spyll
  therfore of spech beth not owt-rage.

Fo. 33^v 6. Sextu*m* mandatu*m* non makaberis.

 ¶ The sexte comaundement byddith every man
  þat no wyght lede no lecherous lay   140

   ¹ A final *e* has perhaps been erased here.

fforfet neue*r* be no woman
lesse þan þe lawe a-lowe þ*i* play
Trespas nevyr wi*th* wyff ne may
wi*th* wedow nor wi*th* non othyr wyght
kepe þe clene as I þe say 145
to whom þou hast þ*i* trowth plyght.

7. Septim*um* mandat*um* non furtum facies.

¶ Do no thefte no thynge þou stele
þe vij^te precept byddyth þe ful sor*e*
whyłł þou arte in weltħ and wele
euyłł gett good loke þou restor*e* 150
Off handys *and* dede be trewe euyr mor*e*
ffor yf þ*in* handys lymyd be
þou art but shent þ*i* name is lor*e*
in ffelde *and* tow*n and* in ałł cou*n*tre.

8. Octau*um* mandat*um* · no*n* loquer*is* co*n*tra p*ro*xim*um* tuu*m* fals*um* testimoni*um*.

¶ The viij^te p*re*cept þus doth þe bydde 155
Ffals wyttnes loke non þou bere
þe trowtħ nevyr more loke þ*at* þou hyde
wi*th* ffals wyttnes no ma*n* þou dere
Nowther ffor love ne dred ne fer*e*
Sey non other than trowtħ is 160
Ffals wytnes yf þ*at* þou rere
Aȝens god þou dost grettly amys.

9. Nonu*m* mandat*um* non desiderabis vxorem proximi tui et cetera.

Fo. 34 ¶ The ix^te precept of lawe of lyff
evyn þus doth bydde every ma*n*
desyre not þ*i* neybor*ys* wyff 165
þow she be fayr and whyte as swa*n*
And þ*i* wyff brown ȝitt natt for tha*n*
þ*i* neybor*ys* wyff þou nevyr rejoyse
kepe þe clene as evyr þou ca*n*
to þ*in* owyn wyff *and* þ*in* owyn choyse. 170

10. Decimum mandatum non concupisces domum proximi tui non seruum non ancillam non bos non asinum nec omnia que illius sunt et cetera.

¶ The x^de comaundement of god and last is þis
thi neyborys hous desyre þou nowth
Maydon nor servaunt nor nowth of his
desyre hem nevyr in wyll nor thowth
Oxe nere Asse þat he hath bought   175
nere no thynge þat longyht hym to
Godys lawe must nedys be wrought
desyre no thynge þin neybore ffro.

¶ The vj^te comaundement of lechory
doth exclude þe synfull dede   180
but theys tweyn last most streytly
both dede and thought þei do for-bede
In wyll nere thought no lechory þou lede
þi thought and wyll þou must refreyn
All þi desyre as I þe rede   185
in clennes of lyff þi self restreyn.

¶ Ffrendys þese be þe lawys þat ȝe must kepe
therfore every man sett well in mende
Fo. 34^v Wethyr þat þou do wake or slepe
these lawys to lerne þou herke ful hynde   190
And godys grace xal be þi ffrende
he socowre and saue ȝow in welth fro woo
Fare well gode frendys for hens wyll I wende
my tale I haue taught ȝow my wey now I goo.
                              Explicit moyses.

Remainder of Fo. 34^v—6½ inches left blank (except for a scribbled *wethyr þat þou do.* in another hand).

Fo. 35   I Am þe prophete callyd Isaye   Ysaias
         Replett with godys grett influens
         and sey pleynly be spyryte of prophecie   7
         þat a clene mayde thourgh meke obedyens

Shall bere a childe which xal do resystens 5
Ageyn foule Zabulon þe devyl of helle
mannys soule ageyn hym to defens
Opyn in þe felde þe fend he xal felle.

¶ Where fore I seye ‖ quod virgo concipiet
et pariet filium ‖ nomen emanuel 10
Oure lyf for to saue he xal suffyr deth
and bye us to his blysse · in hevyn for to dwell
Of Sacerdotale lynage þe trewth I ȝow tell
Fflesch and blood to take god wyll be born
Joye to man in erth · and in hevyn Aungell 15
At þe chyldys byrth · joye xal make þat morn.

         Radix Jesse

¶ Egredietur virga de radice jesse
Et flos de radice eius ascendet
A blyssyd braunch xal sprynge of me
That xal be swettere þan bawmys breth
Out of þat braunch in nazareth
A flowre xal blome of me jesse rote 20
The which by grace xal dystroye deth
and brynge mankende to blysse most sote.

         Dauyd' Rex

¶ I am david of jesse rote
the fresch kyng by naturall successyon
and of my blood xal¹ sprynge oure bote 25
As god hym self hath mad promyssyon

Fo. 35ᵛ Of Regall lyff xal come suche foyson
þat a clene mayde modyr xal be
Ageyns þe devellys fals illusyon
with regall power to make man fre. 30

        Jeremias propheta

¶ I am þe prophete Jeremye
And fullich a-corde in all sentence
with kyng dauid and with ysaie
Affermynge pleynly be-forn þis Audyens
That god of his high benyvolens 35
of prest and kynge wyll take lynage
And bye us all ffrom oure offens
in hevyn to haue his herytage.

  ¹ *spyr* miswritten before *sprynge* and crossed through.

The Prophets

            Salamon rex

¶ I am Salamon þe secunde kynge
And þat wurthy temple for sothe made I   40
which þat is fygure of þat mayde ȝynge
þat xal be modyr of [1] grett messy.

         Ezechiel propheta

A vysion of þis fful veryly
I Ezechiel haue had also
Of a gate þat sperd was trewly   45
and no man but a prince myght þer-in go.

           Roboas rex

¶ The iij^de kynge of þe jentyll [2] jesse
my name is knowe kyng Roboas
of oure kynrede ȝitt men xul se
A clene mayde trede down foule sathanas.   50

         Micheas propheta

And I am a Prophete calde mycheas
I telle ȝow pleynly þat þus it is
Evyn lyke as Eve modyr of wo was
So xal a maydyn be modyr of blyss.

           Abias Rex

¶ I that am calde kynge Abias   55
conferme for trewe þat ȝe han seyd
and sey also as in þis cas
þat all oure myrth comyth of a mayd.

         Danyel propheta

I prophete Danyel am well apayed
in fygure of þis I saw a tre   60
all þe fendys of hell xall ben affrayd
whan maydenys ffrute þer on þei se.

           Asa Rex

¶ I kynge Asa be-leve all þis
þat god wyll of a maydyn be born
And vs to bryngyn to endles blys   65
Ruly on rode be rent and torn.

          Jonas propheta

I jonas sey þat on þe iij^de morn
ffro deth he xal ryse þis is a trew [3] tall [4]

---

[1] *of* omitted and written above the line in darker ink.
[2] *jeng* first written and crossed through.
[3] Or *trow*—the vowel is blotted.
[4] There has been some subsequent attempt to correct *tall* and *qwall* to *tale*

fyguryd in me þe which longe beforn
lay iij days beryed with in þe qwall.¹   70
                                                    Josophat rex

¶ And I josophat þe vj^te kynge serteyn
of jesse rote in þe lenyall successyon
All þat my progenitouris hath be-for me seyn
Ffeythfully be-leve with-owtyn all dubytacion.
                                                    Abdias propheta
I abdias prophete make þis protestacion   75
þat aftyr he is resyn to lyve onys a-ȝen
deth xal be drevyn to endles dampnacion
And lyff xal be grawntyd of paradys ful pleyn.
                                                    Joras Rex
¶ And I Joras also in þe numbre of sefne
of jesse rote kynge · knowlych þat he   80
aftyr his resureccion returne xal to hefne
both god and verry man ther endles to be.
                                                    Abacuch propheta
I Abacuch prophete holde wele with the
whan he is resyn he xal up stye
in hevyn as juge sitt in his se   85
Vs for to deme whan we xal dye.
                                                    Ozias Rex
Fo. 36ᵛ ¶ And I Ozyas Kynge of hygh degre
Spronge of jesse rote dar² well sey this
whan he is gon to his dygnyte
he xal send þe sprytt to his discyplis.   90
                                                    Joell propheta
And I Joel knowe full trewe þat is
god bad me wryte in prophesye
he wolde sende down his sprytt i-wys
On ȝonge and olde ful sekyrlye.
                                                    Joathas rex
¶ My name is knowe kyng Joathan   95
the ix^e kynge spronge of jesse
Of my kynrede god³ wol be man
mankend to saue and þat joyth me.

and *qwale*. There are faint traces of an *e* in darker ink over the second *l* in each case.
¹ See note 4 on p. 59.   ² Or *dare*.   ³ *gold* first written.

A pphete callyd Sophonye
of pis matyr do bere wytnes
ffor tholpth to testyfie
ȝ maydens byrth ouȝ lesse þat blesse

Of þe nobyll tlengthy genelacow
thes xij· kyng am j manasses
Wytnessynges hes be thes testyficacow
ȝ maydens chylde þal be truce of pees

Manasses yc

And j baynk phete confyrmes þis dyo thes
lord & prince of pees ȝolt þ chylde be
At his somon A soyn kyng þis pees
By hit A gyuen syȝ at domys day þal he be

Baynk pphē

Amon kynges ffor þ last conclusyow
All thynges be form sayd ffor tholpth do testyfie
prayynge þ lord of our synne remyssyon
At þ dredful day he us grannt mercye
Thus be all of pis genealogye
Acordynge in on hed in pis place
By þ heyȝ lord when ȝ les þal dye
Of his gret goodnesse to grawnt us his grace

Amon Rex

Explicat Jesse

Carpauter
Asiuaria
Bisakar
Nasaphat
Joachim
Anna
Cleophas
Anna
Salome
Anna

genuit Joachim
gen Anna
Spons Joseph falso
gen Maria mat ihū ꝫ
Spons Alpheo
gen ij Maria uxr Symone ⁊ juda jacobū minore ꝫ
Spons Zebedeo
gen iij Mari uxr Johān euāg Jacobū maiore

Emeria fuit soror āne
q̄ heluc quoud tiham
elizabeth q̄ nupta fuit
zakarie de q̄ genit johē
baptam princep dō

Eliud. Emīt filia hs
Geruasius epūs.

Joseph uyst

The Prophets

                                          Aggeus propheta

With ȝow I do holde þat am prophete Aggee
Com of þe same hygh and holy stok   100
god of oure kynrede in dede born wyl be
From þe wulf to saue al shepe of his flok.

                                          Achas rex

¶ Off jesse kyng Achas is my name
þat falsly wurchepyd ydolatrye
tyl Ysaie putt me in blame   105
And seyd a mayd xulde bere messye.

                                          Ozyas propheta

Off þat byrthe wyttnes bere I
A prophete Osyas men me calle
And aftyr þat tale of Isaye
þat mayd xal bere Emanuelle.   110
                                          Ezechias rex

¶ My name is knowyn kyng Ezechyas
þe xj^{te} kyng of þis geneologye
And say ffor sothe as in þis cas
A mayde be mekenes xal brynge mercye.

                                          Sophosas propheta

Fo. 37   I A prophete callyd Sophonye   115
of þis matyr do bere wyttnes
and for trowth to sertyfie
þat maydens byrth oure welth xal dresse.

                                          Manasses rex

¶ Of þis nobyll and wurthy generacion
the xij^e kyng am I manasses   120
wyttnessynge here be trew testyficacion
þat maydenys childe xal be prince of pes.

                                          Baruk propheta

And I baruk prophete conferme wurdys thes
lord and prince of pes þow þat chylde be
Al his fomen ageyn hym þat pres   125
Ryght a grym syre at domys day xal he be.

                                          Amon Rex

¶ Amon kynge ffor þe last conclusyon
Al thynge be-forn seyd ffor trowth do testyfie
Praynge þat lord of oure synne Remyssyon
At þat dredful day he us graunt mercye   130

Thus we all of þis genealogye
Acordynge in on here in þis place
Pray þat hey3 lorde whan þat we xal dye
Of his gret goodnesse to grawnt us his grace.

The remainder of Fo. 37 is filled with the following genealogical table and note written like those on Ff. 16, 17 and 21, 22:

Explicit Jesse.

Barpanter Asmaria } genuit Joachym

Ysakar Nasaphat } gen. Anna

Joachym Anna } gen. sponsa Joseph fabro Maria mater ihesu Christi

Cleophas et Anna } gen. ij^a sponsa Alpheo Maria mater Symonem et Judam Jacobum minorem et Joseph just[um

Salome et Anna } gen. iij^a sponsa Zebedeo. Maria mater Johannem euangelistam et Jacobum majorem.

Emeria fuit soror Anne que habebat quondam filiam Elizabeth que nupta fui[t Zakarie de quo peperit Johanne[m baptistem precursorem domini Elyud. Eminẽ filia. beatus Geruasius episcopus

Fo. 37^v   Contemplacio

¶ Cryst conserve þis congregacion   8
    fro perellys past · present and future
    and þe personys here pleand · þat þe pronunciacion
    of here sentens to be seyd · mote be sad and sure
    And þat non oblocucyon · make þis matere obscure   5
    but it may profite and plese eche persone present
    ffrom þe gynnynge to þe endynge so to endure
    þat cryst and every creature · with þe conceyte be content.

¶ This matere here mad · is of þe modyr of mercy
    how be joachym And Anne · was here concepcion   10
    Sythe offred into þe temple · compiled breffly
    than maryed to joseph · and so folwyng þe salutacion
    Metyng with Elyzabeth · and þer with a conclusyon
    in fewe wordys talkyd þat it xulde nat be tedyous
    to lernyd nyn to lewd · nyn to no man of reson   15
    þis is þe processe [1] · Now preserve 3ow jhesus.

[1] of oure first written after processe, and crossed through.

*þerfore of pes I ȝow pray aⱡⱡ þat ben here present
and tak hed to oure talkyn · what we xal say
I be-teche ȝow þat lorde þat is evyr omnypotent
to governe ȝow in goodnes as he best may   20
In hevyn we may hym se
Now god þat is hevyn kynge
sende us aⱡⱡ hese dere blyssynge
and to his towre he mote vs brynge
Amen ffor Charyte.   25

The bottom of Fo. 37ᵛ is filled with the following two genealogical notes :

Quinque sunt Anne ⎰ mater Samue ⎱ lis
                  ⎪ Vxor Rague ⎰
                  ⎨ Vxor Tob   ⎱ ie
                  ⎪ Mater beate Mar ⎰
                  ⎩ Anna¹ prophetissa

Est Ysakar Anne pater : Melophat sic quoque mater vel Nasaphat²

Fo. 38   ¶ Ysakar³

THe prestys of god offre sote Ensens
Vn-to here god *and* þer fore they be holy
we þat mynistere here in goddys presens
in vs xuld be fownd no maner of foly
Ysakar prynce of prestys am I   5
þat þis holyest day here haue mynystracion
Certyfyenge aⱡⱡ tribus in my cure specyaly
þat þis is þe hyest fest of oure solemnyzacion.

¶ This we clepe festum Encenniorum
þe newe ffest · of which iij · in þe ȝere we exercyse   10
now aⱡⱡ þe kynredys to jerusalem must cum
In to þe temple of god · here to do sacryfyse
Tho þat be cursyd my dygnyte is to dysspyse
and þo þat be blyssyd here holy sacrefyse to take
We be regal sacerdocium · it perteyneth vs to be wysse 15
be fastyng · be prayng · be almes · *and* at du tyme to wake.
                                                Joachym
¶ Now aⱡⱡ þis countre of Galyle

¹ This name appears to have been added, though possibly by the MS. scribe. It is not in more ornate book-hand like the others.
² This note is in red ink and in large liturgical script. The bottom half of the letters of the words *vel Nasaphat* has been cut away with the margin.
³ A small reference sign in ink of the same colour as the text stands to the right of this name.

with þis cetye of Nazareth specyal
þis ffest to jerusalem must go we
to make sacrefyce to god eternal 20
My name is joachym a man in godys substancyall
Joachym is to say he þat to god is redy
so haue I be *and* evyr more xal
Ffor þe dredful domys of god sore dred I.

¶ I am clepyd Ryghtful why wole ȝe se 25
Ffor my godys in to thre partys I devyde
On to þe temple · *and* to hem þat þer servyng be
A nodyr to þe pylgrimys *and* pore men · þe iij^{de}[1] ffor hem with
    me abyde

Fo. 38^v So xulde euery curat in þis werde wyde
ȝeve a part to his chauncel i-wys 30
A part to his parochonerys þat to povert slyde
the thryd part to kepe for hym *and* his.

¶ But blyssyd wyff anne sore I drede
In þe temple þis tyme to make sacryfice
be-cawse þat no frute of vs doth procede 35
I fere me grettly þe prest wole me dysspice
Than grett slawndyr in þe tribus of vs xulde aryse
but þis I Avow to god with all þe mekenes I can
ȝyff of his mercy he wole a childe us devyse
we xal offre it up in to þe temple to be goddys man. 40
                          Anna

ȝour swemful wurdys make terys trekyl down be my face
i-wys swete husbond þe fawte is in me
my name is Anne þat is to sey grace
we wete not how gracyous god wyl to us be
A woman xulde bere cryst þese profecyes haue we 45
if[2] god send frute · *and* it be a mayd childe
with all reuerens I vow to his mageste
sche xal be here foot mayd to mynyster here most mylde.
                          Joachym

¶ Now lete be it as god wole þer is no more
tweyn turtelys ffor my sacryfice with me I take 50

---

[1] This ^{de} is written in red ink.
[2] This is written in the margin, some miswritten letters standing at the beginning of the line.

and I be-seche wyff and evyr we mete more
þat hese grett mercy · vs meryer mut make.
                                              Anna
For dred and ffor swem of ȝour wourdys I qwake
thryes I kysse ȝow with syghys ful sad
and to þe mercy of god · mekely[1] I ȝow be-take   55
and þo þat departe in sorwe god make þer metyng glad.
                                              Senior tribus
¶ Worchepful sere joachym be ȝe redy now
    all ȝour kynrede is come ȝow to exorte
Fo. 39 þat þei may[2] do sacrifice at þe temple with ȝow
    Ffor ȝow be of grett wurchep as men ȝow report.   60
                                              Joachym

All synfull seke and sory · god mote comforte
I wolde I were as men me name
Thedyr in goddys name · now late us all resorte
A Anne · Anne · Anne · god scheeld us fro shame.
                                              Anne
¶ Now am I left alone · sore may I wepe   65
A husbond a-geyn god wel mote ȝow brynge
And fro shame and sorwe he mote ȝow kepe
tyl I se ȝow a-geyn · I kan not sees of wepynge.
                                              Senior
Prynce of oure prestys · if it be ȝour plesynge
we be com mekely to make our sacrefice.   70
                                              Ysakar
God do ȝow mede bothe elde and ȝynge
than devowtly we wyl begynne servyse.

m̄ *There they xal synge þis sequens · Benedicta sit beata trinitas*
And in þat tyme Ysakar with his ministerys ensensyth þe Autere
and þan þei make her offryng and Isaker seyth

¶ Comyth up serys and offeryth all now
ȝe þat to do sacryfice worthy are
A-byde a qwyle sere · Whedyr wytte[3] þou   75
þou and þi Wyff arn barrany and bare
neyther of ȝow · ffruteful nevyr ȝett ware
Whow durste þou a-monge fruteful presume and Abuse

---

[1] Some letter (? I) erased before *mekely*.
[2] *may* omitted and written above the line.   [3] So in MS.

It is a tokyn þou art cursyd þare
Whereffore with grett indygnacion · þin offeryng I refuse. 80
et refudit sacrificium Joachim.

¶ amonge all þis pepyl barreyn be no mo
therefore comyth up and offeryth here alle
þou joachym I charge þe · fast out þe temple þou go
et redit flendo.
than with goddys holy[1] wourde blysse ʒow I shalle
ministra[2] *cantando*

Fo. 39ᵛ Adiutorium nostrum in nomine domini 85
(marked Qui fecit celum et terram Chorus
40) Sit nomen domini benedictum minister
Ex hoc nunc et usque in seculum chorus
Episcopus
Benedicat vos diuina[3] maiestas et vna deitas
✠Pater✠et filius✠et spiritus sanctus chorus · Amen. 90
Signando manu cum cruce solenniter[4] *et recedant tribus extra templum.*

Now of god and man blyssyd be ʒe alle
homward a-ʒen now returne ʒe
And in þis temple A-byde we xalle
to servyn god in trinyte.

Joachym
¶ A mercyfful lord what is this lyff 95
What haue I do lorde to haue þis blame
Ffor hevynes[5] I dare not go hom to my wyff
And amonge my neyborys · I dare not abyde ffor shame
A Anne · Anne · Anne · Al our joye is turnyd to grame
Ffrom ʒour blyssyd ffelacheppe I am now exilyd 100
and ʒe here onys of þis ffowle fame
sorwe wyl sle ʒow · to se me thus revylyd.

¶ but son[6] god soferyth thys · vs must sofron[7] nede
now wyl I go to my sherherdys[8] · and with hem abyde

---

[1] *holy* omitted and written above the line.
[2] This word is rubbed—the end letters barely visible.
[3] Some word or words (? *et vna*) after *diuina* obliterated. [4] So in MS.
[5] *dare* written before *hevynes* and crossed through.
[6] So in MS. for Sen?
[7] The *r* in *sofron* seems to be the correction of some other letter.
[8] So in MS.

and þer evyr more levyn in sorwe and in drede 105
shame makyth many man his hed for to hyde
ha how do ȝe felas in ȝow is lytel pryde [1]
how fare ȝe *and* my bestys þis wete wolde I veryly.
                              Primus pastor (a) [2]
A welcom͡ hedyr blyssyd mayster · we pasture he*m* ful wyde
they be lusty *and* fayr · *and* grettly multyply 110
how de ȝe mayster ȝe loke al hevyly
how doth oure dame at hom sytt she *and* sowyht.
                              Joachym [3] (b)
to here þe speke of here it sleyth my*n* hert veryly
how I *and* sche doth god hym self knowyth [4]
The meke god lyftyth up þe proude over throwyht (b)
Go do what ȝe lyst se [5] ȝo*ur* bestys not stray. 116
                              ii^us pastor (a)
Aftere grett sorwe mayster · evyr gret grace growyht
Sympyl as we kan · we xal for ȝow pray.
                              iii^us pastor (d)
¶ ȝa to pray ffor careful · it is grett nede
we all wul prey ffor ȝow knelende 120
god of his goodnes send ȝow good spede
and of ȝo*ur* sorwe ȝow sone amende.
                              Joachym [6]
¶ I am nott wurthy lord · to loke up to hefne
my synful steppys Anvemynyd þe grounde
I loth-folest [7] þ*at* levyth · þ*ou* lord hyest in þ*i* setys sefne 125
What art þ*ou* lord · What am I wrecche · werse þan an hownde
þ*ou* hast sent me shame · which my*n* hert doth wounde
I thank þe more here fore · þan for all my prosperite
þis is a tokyn þ*ou* lovyst me · now to the I am bounde
þ*ou* seyst þ*ou* art w*ith* hem · þ*at* in tribulacion be. 130

---

[1] *Pastores* is scribbled underneath the words *for to hyde* and a rough line extending into the margin is drawn under this line on the other side.

[2] These letters seem to indicate some rearrangement, possibly to reduce the shepherds' parts to two.

[3] *ii^us pastor* first written and crossed through by the red loop of name of speaker; *Joachym* carelessly written, but probably by scribe of MS.

[4] The y seems to be written over an h.   [5] Or *so*.

[6] Some indistinct marginal sign opposite Joachym's speech (? nota).

[7] MS. fo lest; cf. hy est in same line.

The Conception of Mary

¶ And ho so haue þe · he nedyth not care thanne
 my sorwe is feryng¹ I haue do sum offens
 punchyth me lorde · and spare my blyssyd wyff anne
 þat syttyth and sorwyth ful sore of myn Absens
 Ther is not may profyte but prayour to ȝour presens   135
 With prayorys prostrat · by-fore þi person I wepe
 haue mende on oure a-vow · for ȝour mech magnyficens
 and my lovyngest wyff Anne · lord for þi mercy kepe.
                              Anna

¶ A mercy lord mercy mercy mercy
 we are synfolest it shewyth þat ȝe send us all þis sorwe   140
 Why do ȝe thus to myn husbond lord why · why · why ·
 for my barynes · he may amend þis · þi self and þou lyst to
     morwe
 And it plese so þi mercy · þe my lord · I take to borwe
 I xal kepe myn a-vow · qwhyl I leve and leste
 I fere me · I haue offendyd þe · myn hert is ful of sorwe   145
Fo. 40ᵛ most mekely I pray þi pety · þat þis bale þou wyl breste.
                              Joachym

here þe Aungel descendith · þe hefne syngyng Exultet celum laudi-
bus · resultet terra gaudijs Archangelorum gloria sacra canunt
solemnia.

¶ Qwhat art þou in goddys name · þat makyst me a-drad
 It is as lyth a-bowt me as al þe werd were fere.
                              Angelus
 I am an Aungel of god · com to make þe glad
 God is plesyd with þin helmes · and hath herd þi prayere   150
 He seyth þi shame · þi repreff · and þi terys cler
 God is a vengere of synne · and not nature doth lothe
 Whos wombe þat he sparyth · and maketh barreyn her
 he doth to shewe his myth · and his mercy bothe.

¶ Thu seest þat Sara was nynty ȝere bareyn   155
 sche had a son Ysaac · to whom god ȝaff his blyssynge
 Rachel also had þe same peyn
 She had a son joseph · þat of Egypt was kynge
 A strongere þan sampson · nevyr was be wrytynge
 nor an holyere þan samuel it is seyd thus   160
 ȝett here moderys were bareyn bothe in þe gynnynge
 þe concepcion of all swych · it is ful mervelyous.

¹ The r seems to have been inserted—perhaps over another letter.

The Conception of Mary

¶ And in þe lyke wyse · Anne þi blyssyd wyff
   sche xal bere a childe · xal hygth mary
   which xal be blyssyd · in here body · *and* haue joys ffyff   165
   and ful of þe holy goost · inspyred syngulyrly
   Sche xal be offryd in to þe temple · solemply
   þ*at* of here[1] non evyl ffame · xuld sprynge thus
   And as sche xal be bore · of a barrany body
   So of here xal be bore · w*ith*-out nature Jhesus.   170
                       þ*at* xal be savyour[2]

(fo. 41 ¶ That xal be savyour · vnto al man-kende
quire)   In tokyn whan þo*u* come · to jherusalem to þe gyldyn gate
   þo*u* xalt mete anne þ*i* wyff · haue þis in þ*i* mende
   I xal sey here þe same here sorwys to rebate.
                             Joachym

Of þis imc*om*parabyl c*om*fort · I xal nevyr for-gete þe date   175
my sorwe was nevyr so grett · but now my joy is more
I xal hom in hast be it nevyr so late
A Anne blyssyd be þ*at* body · of þe xal be bore.

¶ Now fare wel my*n* shepherdys · governe ȝow now wysly.
                                  1us pastor
Haue ȝe good tydyng*ys* mays*ter* · þan be we glad.   180
                                  Joachym
Prayse god for me · for I am not wourthy.
                                  ij^{us} pastor
In feyth sere so we xal · w*ith* all oure sowlys sad.
                                 iij^{us} pastor
I holde it helpfful þ*at* on of vs · w*ith* ȝow · be had.
                                Joachym
Nay abyde w*ith* ȝo*ur* best*ys* sone · in goddys blyssynge.
                               1us pastor
we xal make us so mery · now[3] þis is be-stad   185
þ*at* a myle on ȝo*ur* wey · ȝe xal here us synge.[4]
                                    Anne
¶ Alas ffor my*n* husbond · me is ful wo
   I xal go seke hym · what so evyr be falle

---

[1] *of here* omitted and written above the line.
[2] This is given as the catchword at the bottom of the folio. There is some trace of other rubbed words—which may, however, be scribblings.
[3] *is* written before *þis* and crossed through.
[4] A rough line is drawn from *synge* to the edge of the page—? marking off the shepherds' part.

I wote not in erth which wey is he go
ffadyr of hefne ffor mercy · to ȝour ffete I falle. 190

Angelus

Anne þin husbond ryght now I was withall
þe aungel of god · þat bar hym good tydynge
and as I seyd to hym so to þe sey I xal
god hath herd þi preyour · and þi wepynge

¶ At þe goldyn gate þou xalte mete hym ful mylde 195
and in grett gladnes returne to ȝour hous
So be proces þou xalt conseyve and bere A childe
Whiche xal hyght mary · and mary xal bere jhesus
Which xal be savyour of all þe werd and us
Aftere grett sorwe · evyr grett gladnes is had 200
now myn inbassett I haue seyd to ȝow thus
gooth in oure lordys name and in god beth glad.

Anne

¶ Now blyssyd be oure lorde and all his werkys ay
All heffne and erthe mut blysse ȝow for this
I am so joyful I not what I may say 205
þer can no tounge telle what joye in me is
I to bere a childe · þat xal bere all mannys blys
and haue myn hosbonde a-geyn · ho myth haue joys more
no creature in erth · is grauntyd more mercy i-wys
I xal hyȝe me to þe ȝate · to be þer before. 210

here goth þe Aungel A-ȝen to hefne.

¶ A blyssyd be our lord · myn husbond I se
I xalle on myn knes · and to hym-ward crepe.

Joachym

A gracyous wyff Anne now fruteful xal ȝe be
Ffor joy of þis metyng¹ · in my sowle I wepe
Haue þis kusse · of clennesse · and with ȝow it kepe 215
in goddys name now go we wyff · hom to our hous.

Anne

þer was nevyr² joy sank in me so depe
now may we sey husbond · god is to us gracyous ¶ veryly³.

---

¹ *wepy* first written and crossed through.
² *nevyr* omitted and written above the line in slightly darker ink, but probably by scribe of MS.
³ The paragraph mark before *veryly* has been made over some black sign.

  Joachym

ȝa *and* if we haue levyd wel here be-fore
I pray þe lord þin ore   220
so mote we levyn evyr more
And be þi grace more holyly.

                                                                                 Anne

Fo. 42   Now hom-Ward husbond I rede we gon
Ryth hom al to *our* place
to thank god þat sytt in tron   225
þat þus hath sent us his grace.

                                                                     Contemplacio

¶ Sovereynes ȝe han sen shewyd ȝow be-fore   **9**
Of Joachym *and* Anne · here botherys holy metynge
How *our* lady was conseyvid · *and* how she was bore
We passe ovyr þat · breffnes of tyme consyderynge
And how *our* lady in here tendyr age *and* ȝyng   5
In to þe temple was offryd *and* so forth proced
Þis sentens sayd xal be hire be-gynnyng
now þe modyr of mercy · in þis be *our* sped.

¶ And as a childe of iij ȝere age · here she xal appere
to alle pepyl þat ben here *p*resent   10
and of here grett grace now xal ȝe here[1]
how she levyd evyr to goddys entent
           with grace
That holy matere we wole declare
tyl ffortene ȝere · how sche dyd fare   15
Now of ȝo*ur* speche I pray ȝow spare
        all þat ben in þis place.

here joachym and Anne with oure lady betwen hem beyng al in whyte as a childe of iij ȝere age presente here in to þe temple thus seying joachym *.

---

[1] *here she xal* first written instead of *now xal ȝe here* and crossed through.
* *here jo hym and Anne with*—scribbled underneath this.

¶ Joachim

Blyssyd be oure lord ffayr ffrute haue we now
Anne wyff remembyr wole ȝe
þat we made to god an holy a-vow
þat oure fyrst childe · þe servaunt of god xulde be
The Age of mary oure dowtere is ȝerys thre 5
þer fore to thre personys and on god · lete us here present
þe ȝonger she be drawyn þe bettyr semyth me
and for teryeng of our a-vow · of god we myth be shent.

Anne

¶ it is as ȝe sey · husbond in dede
late us take mary our dowtere us be-twen 10
and to þe temple with here procede
Dowtere þe Aungel tolde us ȝe xulde be a qwen
Wole ȝe go se þat lord ȝour husbond xal ben
and lerne for to love hym · and lede with hym ȝour lyff
telle ȝour ffadyr and me her · ȝour answere let sen 15
Wole ȝe be pure maydyn · and also goddys wyff.

Maria

¶ Ffadyr and modyr if it plesyng to ȝow be
ȝe han mad ȝour a-vow · so sothly wole I
to be goddys chast seruaunt · whil lyff is in me
but to be goddys wyff · I was nevyr wurthy 20
I am þe sympelest þat evyr was born of body
I haue herd ȝow seyd · God xulde haue a modyr swete
þat I may leve · to se hire · god graunt me for his[1] mercy
and Abyl me to ley my handys · vndyr hire fayr fete.
Et genuflectet Ad deum.

Joachym

¶ I-Wys dowtere · it is wel seyd 25
ȝe answere · and ȝe were twenty ȝere olde.

Anne

whith ȝour speche mary I am wel payd
can ȝe gon a-lone · lett se beth bolde.

Maria

To go to goddys hous · wole ȝe now be-holde
I am joyful thedyrward · as I may be. 30

---

[1] ȝour first written, crossed out, and *his* written above the line in fainter ink and possibly by a later corrector. The crossing out by two crossed strokes is not the manner of the MS. scribe, and the *s* of *his* is slightly different.

Mary in the Temple

              Joachym
Wyff I ryght joyful oure dowtere to be-holde.
              Anne
So am I wys husbond · now in goddys name go we.
              Joachym

¶ Sere prince of prestes and it plese ȝow
we þat were barreyn · god hath sent a childe
to offre here to goddys service we mad oure avow   35
here is þe same mayde · mary most mylde.
              Isakar

Joachym I haue good mende · how I ȝow revyled
I am ryght joyful þat god hath ȝove ȝow þis grace
to be amonge fruteful · now be ȝe reconsylid
Comm swete mary comm · ȝe haue a gracyous face.   40

Joachym flectendo Ad deum sic dicens
              Joachym

¶ Now Ffadyr and sone and holy gost
on god and personys thre
we offre to þe lorde of myghtys most
Oure dowtere þi servaunt · evyr more to be.
              Anna

Ther to most bounde evyr more be we   45
Mary in þis holy place leve ȝow we xall
In goddys name · now up go ȝe
oure fadyr · oure prest · lo doth ȝow call.
              Maria

¶ Modyr and it plese ȝow · fyrst wole I take my leve
of my fadyr · and ȝow my modyr i-wys   50
I haue a fadyr in hefne · þis I be-leve
now good ffadyr · with þat fadyr ȝe me blysse.
              Joachym

In nomine patris et filii et spiritus sancti.
              Maria
Amen · Now ȝe good modyr.
              Anne
In nomine patris et filij et spiritus sancti ‖ Maria · Amen.   55
              Maria

¶ Now oure lord thank ȝow for this
here is my fadyr and my modyr bothe
most mekely I beseche I may ȝow kys

---

[1] The writing on this folio is rather careless.

now for-ʒeve me yf evyr I made ʒow wrothe.
Et explexendo osculabit patrem et matrem.

Joachym

¶ Nay dowtere ʒe offendyd nevyr god nor man   60
lovyd be þat lord · ʒow so doth kepe.

Anne

Swete dowtyr thynk on ʒour modyr An
ʒour swemynge smytyht to myn hert depe.

[*Maria*][1]

Ffadyr *and* modyr I xal pray for ʒow *and* wepe
To god wi*th* al myn hert specyaly   65
blysse me day *and* nyght evyr her ʒe slepe
good ffadyr *and* modyr · and beth[2] mery.

Joachym

¶ A ho had evyr suche a chylde
nevyr creature ʒit þat evyr was bore
Sche is so gracyous she is so mylde   70
so xulde childyr to fadyr *and* modyr evyr more.

Anne

Than xulde thei be blyssyd · *and* plese god sore[3]
husbond *and* it plese ʒow not hens go we xal
tyl mary be in þe temple above thore[3]
I wold not for al erthe se here fal.   75

Episcopus

¶ Come gode mary · come babe I þe call[4]
þi pas pratyly to þis plas pretende
þou xalt be þe dowtere · of god Eternall

Fo. 44   If þe fyftene grees · þou may Ascende
It is meracle if þou do · now god þe dyffende   80
Ffrom babylony to hevynly jherusalem þis is þe way
Every man þat thynk[5] his lyff to Amende
þe fiftene psalmys · in memorye of þis mayde say.

Maria

Maria · et sic deinceps usque ad fine[6] *xv^{cim} psalmorum.*
The fyrst degre gostly applyed

---

[1] This name was omitted and is written in different ink and in another hand.
[2] The *th* of *beth* has been crossed through in different ink and with perpendicular strokes, not in the manner of the scribe.
[3] Or *sere, there*; the vowels are not clear.
[4] The writing of the first three lines of this stanza (bottom of folio) is smaller.
[5] Some other word has been altered to *thynk*.   [6] So in MS.

It is holy desyre with god to be 85
In trobyl to god I haue cryed
And in sped · þat lord hath herde me.

Ad dominum cum tribularer clamaui ; et exaudiuit me.
 The secunde is stody · with meke inquysissyon veryly
 How I xal haue knowynge of godys wylle
 To þe mownteynes of hefne I haue lyfte myn ey 90
 Ffrom qwens xal comyn helpe me tylle.

Leuaui oculos meos in montes ; vnde ueniat auxilium mihi.
 The thrydde is gladnes in mende in hope to be
 that we xall be savyd all thus
 I am glad of these tydyngys ben seyd to me
 now xal we go · in to goddys hous. 95

Letatus sum in hijs que dicta sunt mihi: in domum
domini ibimus.
 The fourte is meke obedyence as is dette
 to hym þat is a-bove þe planetys sefne
 to þe I haue myn eyn sette
 þat dwellys above þe skyes in hefne.

Ad te leuaui oculos meos ; qui habitas in celis.
 The ffyfte is propyr confessyon 100
 þat we be nought with-owth god thus
 but god in vs haue habytacion
 Per aventure oure enemyes shulde swelle vs.

Fo. 44ᵛ Nisi quia dominus erat in nobis dicat nunc israel ; nisi
quia dominus erat in nobis.
 The sexte is confidens in goddys strenght A-lon
 Ffor of all grace from hym comyth þe strem 105
 they þat trust in god as þe mownt syon
 he xal not be steryd · endles þat dwellyth in jherusalem.

Qui confidunt in domino sicut mons Syon : non commoue-
bitur in eternum qui habitat in hierusalem.
 The sefte is vndowteful hope of immortalyte
 In oure lorde is as gracy and mercy
 Whan oure lord conuertyth oure captiuite 110
 Than Are we mad as joyful[1] mery.

In conuertendo dominus captiuitatem syon ; facti sumus
sicut consolati.

    [1] A letter has been erased between *joyful* and *mery*.

The eyted is contempt of veynglory in vs
Ffor hym þat Al mankende hath multyplyed
But yf oure lord make here oure hous
they An laboryd in veyn þat it han Edyfied. 115

Nisi dominus edificauerit domum: in uanum laborauerunt qui edificant eam.

The nynte is a childely fer in dede
With A longyng love in oure lord þat ay is
blyssyd Arn All they þat god drede
Whiche þat gon in his holy weys.

Beati omnes qui timent dominum: qui ambulant in vijs eius.

The tende is myghty soferauns of carnal temptacion 120
ffor þe fleschly syghtys ben fers and fel
ofte ȝough is ffowth with · with suech vexacion
þou[1] seynge god say so · clepyd israel.

Sepe expugnauerunt me a iuuentute mea: dicat nunc israel.

The Elefnte is accusatyff confessyon of iniquite
Of which ful noyous is þe noyis 125
Fro depnes lord I haue cryed to the
Lord here in sped my sympyl voys.

Fo. 45[2] De profundis clamaui ad te domine: domine exaudi uocem meam.

The twelfte is mekenes þat is fayre and softe
In mannys sowle with-inne and with-owte
Lord myn herte is not heyued on lofte 130
nyn myn eyn be not lokynge a-bowte.

Domine non est exaltatum cor meum: neque elati sunt oculi mei.

The Threttene is ffeyth þer with
with holy dedys don expresse

---

[1] Halliwell reads þan, but the MS. contraction þᵘ is that always used for þou, the contraction for þan, when it occurs, being þᵃn. The letter u(n) is ambiguous, and a difficulty on Fo. 97 ᵛ would be lessened by expanding þᵘ as þan.

[2] The writing on this folio is noticeably small.

haue mende lorde of davyth
And of All his swettnes. 135

Memento d*omi*ne d*a*u*id ?* et *om*nis mansuetudinis eius.
The ffourtene is brothyrly concorde i-wys
þ*at* norchyth love of creatur*ys* echon
Se how good *and* how glad it is
bretheryn ffor to dwelle in on.

Ecce q*ua*m bonu*m* et q*ua*m jocundum *?* h*a*bitare f*ra*tres
i*n* vnu*m*.
The ffyftene is gracyous · wit*h* on Acorde 140
whiche is syne of godly love semyth me
se now blysse our*e* lord
All þat our*e* lordys servaunt*ys* be.

Ecce nunc benedicite d*omi*n*u*m *?* om*ne*s serui dominj.

             Ep*isco*pus
A gracyous lord þis is A mervelyous thynge
þ*at* we se here all in syght 145
A babe of thre ʒer age so ʒynge
to come vp þese grecys[1] · so vp-ryght
It is An hey meracle *and* by goddys myght
no dowth of she xal be gracyous.

             Maria
Holy ffadyr I be-seche ʒow forth ryght 150
Sey how I xal be rewlyd · in goddys hous.

             Ep*isco*pus
fo. 45ᵛ ¶ Dowter*e* god hath ʒovyn vs comaundement*ys* ten
Which shortely to say be comp*re*hendyd in tweyn
and þo must be kept of all crysten men
or ellys her*e* jugement is p*er*petual peyn 155
ʒe muste love god severeynly[2] · *and* ʒo*ur* evyn crystyn pleyn
god fyrst ffor his hyʒ *and* sovereyn dygnyte
he lovyd ʒow fyrst · love hym a-geyn
ffor of love · to his owyn lyknes · he made the.

¶ Love ffadyr sone *and* holy gost 160
Love god þe fadyr · ffor he gevyth myght
Love god þe sone · ffor he gevyth wysdam þo*u* wost
Love god þe holy gost ffor he gevyth love *and* lyght

---

[1] So in MS. Halliwell gives *greeys*.   [2] So in MS.

thre personys *and* on god · þus love of ryght
wi*th* all þin hert · wi*th* all þi sowle · wi*th* all þi mende[1]   165
and wi*th* all þe strenghthis in þe be-dyght
þan love þi*n* evyn crystyn as þi self wi*th*-owtyn ende.

¶ Thu xalt hate no thynge but þe devyl *and* synne
god byddyth the lovyn þi bodyly enmy
And as for ȝo*ur* self here · þus xal ȝe be-gynne   170
ȝe must serve · *and* wurchep god here dayly
Ffor wi*th* prayȝer · wi*th* grace *and* mercy
Se the[2] haue · A resonable tyme to fede
thanne to haue a labo*ur* bodyly
þat þer in be gostly *and* bodely mede.   175

¶ ȝour A-bydynge · xal be wi*th* ȝour maydenys ffyve
Swyche tyme as ȝe wole haue consolac*i*on.

                                             Maria

This lyff me lyketh as my lyve
of here namys I be-seche ȝow to haue informac*i*on.

                                            Ep*is*copus

Fo. 46   There is þe fyrst meditac*i*on   180
Contryssyon · compassyon · And clennes
And þ*at* holy mayde fruyssyon
Wi*th* these blyssyd maydenes xal be ȝour besynes.

                                             Maria

¶ Here is an holy ffelacheppe I fele
I am not wurthy Amonge he*m* to be   185
Swete syster*ys* to ȝow All I knele
To receyve me I be-seche[3] ȝour charyte.

                                            Ep*is*copus

They xal dowtere · And on þe tothere syde se
ther ben sefne prestys in dede
to schryve · to teche · *and* to mynystryn to the   190
to lerne þe goddys lawys · *and* scrypture to rede.

                                           Maria

¶ Ffadyr knew I here namys wele were I.

                                           Ep*is*copus

Ther is dyscressyon · devoc*i*on · dylexc*i*on · and deliberac*i*on

---

[1] *myght* first written and crossed through; *mende* written above the line.
[2] *Sethe* (in one word) in MS.
[3] *seke* first written and crossed through in red ink; *sethe* then apparently written and corrected by erasure to *seche*. A red dot stands under the *h*.

Mary in the Temple

they xal tende upon ȝow besyly
With declaracion determynacion · dyvynacion 195
Now go ȝe maydenys · to ȝour occupacion[1]
And loke ȝe tende þis childe tendyrly
and ȝe serys knelyth · and I xal gyve ȝow goddys benyson
in nomine patris et filij et spiritus sancti.

Et recedent cum ministris suis omnes virgines dicent· Amen.

¶ To ȝow ffadyr and modyr I me comende   [Maria][2] 200
  blyssyd be þe tyme ȝe me hedyr brought.
                                          Joachym

Dowtere þe ffadere of oure feyth þe mot defende
as he of his myght made all thynge of nowth.
                                          Anne

Mary to þi sowle solas he sende
In whos wysdam all þis werd was wrought. 205
go we now hens husbonde so hende   *Hic joachim et anna recedent*
for owth of care now are we brought.   *domum.*
                                            Maria

Be þe holy gost at hom be ȝe brought
Systerys ȝe may go do what ȝe xall   *Ad virgines*
to serve god fyrst here is al my thought 210
Be-forn þis holy awtere on my knes I fall.

¶ Lord sefne petycions I be-seche ȝow of here
Ffyrst þat I may kepe þi love and þi lawe
Þe secunde to lovyn myn evyn crystyn as my self dere
Þe thrydde from all þat þou hatyst me to with-drawe 215
The fourte All vertuys to þi plesauns knawe
Þe fyfte to obey þe ordenaryes of þe temple echon
Þe sexte and þat all pepyl may serve þe with Awe
þat in þis holy temple fawte be non.

¶ The sefnte lord I haske with grett ffere 220
þat I may se onys in my lyve
þat lady þat xal goddys sone bere
þat I may serve here with my wyttys fyve
If it plese ȝow and ellys · it is not þer with to stryve
with prayers prostrat ffor þese gracys I wepe 225
O my god · devocion depe in me dryve

---

[1] The writing from this line to the bottom of the folio is smaller.
[2] No name of speaker given in MS.

þat myn hert may wake in þe · thow my body slepe.

here þe Aungel bryngyth manna in A cowpe of gold lyke to con-
feccions · þe hefne syngynge · þe Aungel seyth

¶ Merveyle not mekest maydon of my mynystracion
I am a good Aungel sent of god All myght
With Aungelys mete Ffor ȝour Sustentacion 230
ȝe to receyve it · Ffor Natural myght

Fo. 47 We Aungellys xul serve ȝow · day *and* nyght
now fede ȝow þer-with · in goddys name
we xal lerne ȝow þe lyberary[1] of oure lordys lawe lyght
ffor my sawys in ȝow shewyth sygnes of shame. 235

                                                    Maria

¶ To thank oure soveryen lord not sufficyth my mende
I xal fede me of þis fode my lord hath me sent
All maner of savowrys in þis mete I fynde
I felt nevyr non so swete ner so redolent.

                                                    Angelus

Eche day þer with ȝe xal be content 240
Aunge[2] alle howrys xal to ȝow apere.

                                                    Maria

Mercy my makere · how may þis be ment
I am þe sympelest creature · þat is levynge here.

                                                  Angelus

¶ In ȝour name Maria · ffyve letterys we han
M. Mayde most mercyfull *and* mekest *in* mende 245
A. Auerte of þe Anguysch þat Adam began
R. Regina of regyon Reyneng with-owtyn ende
I. Innocent be Influens of Jesses kende
A. Aduocat most Autentyk ȝour Antecer[3] Anna
hefne *and* helle here kneys down bende 250
Whan þis holy name of ȝow is seyd Maria.

                                                    Maria

¶ I qwake grettly for dred · to here þis comendacion
Good swete Aungel why wole ȝe sey thus.

                                                    Aungell

Ffor ȝe xal here aftere[4] haue A salutacion
þat xal þis excede it is seyd Amonge vs 255

---

[1] This word is underlined in pencil and a pencil cross has been made in the margin, probably by some modern reader.
[2] So in MS.; cf. Fo. 70ᵛ. [3] Halliwell reads *autecer*.
[4] Some erasure after *aftere*.

The deyte þat dede xal determyn *and* dyscus
ȝe xal nevyr lady be lefte here A-lone.
                                                    Maria
I Crye þe mercy lorde *and* þin erthe Cus
recomendynge me to þat godhyd · þat is tryne *in* trone.
*hic osculet terra*m · *here xal comyn Allwey An Aungel with dyvers
presentys goynge* and *comyng and in þe tyme þei xal synge in
hefne þis hympne · Jhesu corona virginum · And After þer comyth
A minister fro þe busschop with A present and seyth*
                                                    Minister
¶ Prynce of oure prestes Ysakare be name   260
he hath sent ȝow hym self his servyce *in* dede
And bad ȝe xulde ffede ȝow spare for no shame
In þis tyme of mete · ne lenger ȝe rede.
                                                    Maria¹
Recomende me to my fadyr sere · *and* god do hy*m* mede
These vesselys A-ȝen sone I xal hym sende   265
I xal bere it my systery*s* I trowe þei haue more nede
goddys foyson is evyr to his servaunt*ys* hendyr þan we wende.

¶ Systery*s* oure holy ffadyr isakare
Hath sent us hese servyce here ryght now
Ffede ȝow þer of hertyly · I pray ȝow nat spare   270
and if owght be leve · specyaly I pray ȝow
That þe pore men þe relevys þer of haue now
Ffayn *and* I myth I wolde do þe dedys of mercy
Pore ffolk ffaryn god knowyth how
On hem evyr I haue grett pety.   275
                                                    Contemplacio
¶ lo sofreynes here ȝe haue seyn
in þe temple of oure ladyes presentacion
she was nevyr occapyed in thyng*ys* veyn
but Evyr besy in holy ocupacyoñ
And we be-seche ȝow of ȝour*e* pacyens
þ*at* we pace þese matery*s* so lythly Away
If þei xulde be do w*ith* good prevydens
Eche on wolde suffyce ffor An hool day
Now xal we *p*rocede to here dissponsac*i*on
which Aftere þis was xiiij ȝere   10
tyme sufficyth not to make pawsac*i*on

---
¹ The usual red loop round the name of speaker has been omitted here.

      hath pacyens with vs we be-sech 3ow her
      And in short spas
      The parlement of hefne sone xal 3e se
      and how goddys sone com̅ man xal he   15
      And how þe salutacion Aftere xal be
      be goddys holy gras.

    Remainder of Fo. 48—4¼ inches—left blank. Fo. 48ᵛ has been used for speeches to be inserted between the speeches on Fo. 50 and Fo. 50ᵛ in the next play.
    Their place is made clear by red reference marks and also by rough capitals in black ink of a different shade from that of the MS.

        Passage to be inserted between Fo. 50 and Fo. 50ᵛ.

Fo. 48ᵛ ¶ This Ansuere grettly trobelyth me
      to mak a vow to creaturys it is lefful
      Vovete et reddite · in scripture haue we
      and to observe oure lawe also it is nedful   95
      In þis to dyscerne to me it is dredful
      þerfore to cowcell[1] me in þis cas I calle
      þe holde and þe wyse and swiche as ben spedful
      In þis sey 3our a-vyse · I be-sech 3ow Alle.

                                                  Minister

    ¶ To breke our lawe and custom · it wore hard in dede   100
      And on þat other syde to do a-3en[2] scrypture
      to 3eve sentens in þis degre · 3e must take goo[3] hede
      ffor dowteles þis matere · is dyffuse and obscure
      Myn avyse here in þis I 3ow ensure
      þat we prey all god to haue relacion   105
      ffor be prayour · grett knowlech men recure
      and to þis I counsell 3ow to 3eve assygnacion.

                                                  Episcopus

[E] Trewly 3our counsell is ryght good and eylsum
      And as 3e han seyd so xal it be
      I charge 3ow bretheryn and systerys hedyr 3e com̅   110
      and to-gedyr to god now pray we
      That it may plese his fynyte[4] deyte
      knowleche in þis to sendyn vs   Et hic cantent veni creator
      Mekely eche man ffalle down on kne
      And we xal be-gynne · Veni creator Spiritus   115
    and whan Veni creator is don þe buschop xal seyng
      Now lord god of lordys whysest of alle et cetera.

    ¹ So in MS.   ² The a omitted and written above the line.
    ³ So in MS.   ⁴ So in MS.

The Betrothal of Mary

Tunc venit Abysakar[1] *episcopus*

Listenyth lordyngys bothe hye *and* lowe
And tendyrly takyth heyd on to my sawe
beth buxom *and* benyngne ȝo*ur* busshopp to knowe
Ffor I am þat lord þat made þis lawe
W*ith* hertys so hende herkyn nowe   5
ȝoure damyselys to weddyng ȝa loke þat ȝe drawe
þat passyn xiiij ȝere for what þ*at* ȝe owe
þe lawe of god byddyth þis sawe
þat at xiiij ȝere of age
Euery damesel what so sche be   10
to þe Encrese of mor*e* plente
xulde be browght in good degr*e*
On to her*e* spowsage.

                           Joachym

¶ Herke now Anne my jentyl spowse
how þat þe buschop his lawe hath tolde   15
þat what man hath a dowtyr i*n* his house
þat passyth xiiij ȝer*ys* olde
He muste her*e* brynge I herde hym Rowse
In to þe tempyl a spowse to wedde
wher for oure dowtyr ryth good *and* dowse   20
In to þe tempyl sche must be ledde
nd þat anoon ryght sone.

                           Anne

Ser*e* I grawnt þ*at* it be so
A-ȝen þe lawe may we not do
w*ith* here to-gedyr lete us now go   25
I hold it ryght weyl done.

                           Joachym

¶ Ser*e* busshopp her*e* aftyr þi*n* owyn hest
we haue here brought our*e* dowtyr dere
mary my swete childe she is ful prest
of Age she is ful xiiij ȝer*e*.   30

---

[1] Halliwell reads *ab Ysakar* with the note 'Sic in MS pro Abysakar'; but the MS. has no capital *Y* and the initial *A* is of the form used indifferently for small or capital. The syllable *ab* is not joined to the *y*, but the *v* of the preceding *venit* is similarly detached.

Episcopus

Fo. 49ᵛ
(marked 50)
[B]
Welcome joachym on to my*n* A-reste
bothe anne þ*i* wyff *and* mary cler*e*
now mary chylde to þe lawe þ*ou* leste
and chese þe a spowse to be þ*i* ffer*e*
þ*at* lawe þ*ou* must fful-ffylle.   35

Maria

A-ȝens þe lawe wyl I nevyr be
but mannys ffelachep xal nevyr folwe me
I wyl levyn evyr in chastyte
be þe grace of goddys wylle.

Episcopus

[C] ¶ A Ffayr*e* mayde why seyst þou so   40
what menyth the for to levyn chast
why wylt þ*ou* not to weddyng go
þ*e* cawse þ*ou* telle me *and* þat in hast.

Maria

My ffadyr *and* my modyr sertys also
Er I was born ȝe may me trast   45
thei wer*e* bothe bareyn her*e* frute was do
they come to þe tempyl at þe last
to do her*e* sacryfice
By-cause they hadde nothyr frute ner*e* chylde
reprevyd þei wor*e* of wykkyd *and* wyllde   50
wit*h* grett shame þei were revylyd
Al men dede them dyspyce.

¶ My Ffadyr *and* my modyr thei wepte full sor*e*
fful hevy her*e* hertys wern of þis dede
wit*h* wepynge eyn þei preyd þ*er*for*e*   55
þ*at* god wolde socowr*e* hem *and* sende hem sede
Iff god wold graunt hem a childe be bor*e*
They be-hest þ*e* chylde her*e* lyff xulde lede
In goddys temple to serve evyr-more
and wurchep god in loue *and* drede   60
Than god fful of grace

Fo. 50
he herd her*e* longe p*r*ayo*ur*
and þan sent hem both seed *and* flowr*e* ¹
whan I was born in her*e* bowr*e*
to þe temple offryd I was.   65

¹ The *w* has been adapted from some other letter.

¶ Whan þat I was to þe temple brought
and offerde up to god Above
ther hestyd I as myn hert thought
to serve my god with hertyly love
Clennesse and chastyte myn hert owth 70
Erthely creature nevyr may shoue
Such clene lyff xuld ȝe nouht
In no maner wyse reprove
to þis clennesse I me take
this is þe cawse as I ȝow tell 75
Þat I with man wyll nevyr mell
in þe servyse of god wyl I evyr dwell
I wyl nevyr haue other make.

                                                            Episcopus

[D] ¶ A mercy god þese wordys wyse
of þis fayr mayde clene 80
thei trobyl myn hert in many wyse
her wytt is grett and þat is sene
In clennes to levyn in godys servise
no man here blame non here tene
And ȝit in lawe · þus it lyce 85
þat such weddyd xulde bene
who xal expownd þis oute
þe lawe doth after lyff of clennes
þe lawe doth bydde such maydenes expres
þat to spowsyng they xulde hem dres 90
God help us in þis dowhte.

This Ansuere grettly trobelyth me ? et cetera ut supra [1]

Now lord god of lordys wysest of All

I pray þe lorde knelynge on kne
with carefull herte I crye and calle
þis dowteful dowte enforme þou me.
                                                          Angelus
¶ [2] Thy prayour is herd to hyȝ hevyn halle 120

---

[1] The writing indicates that the scribe went on from *God help us in þis dowhte* to *Now lord god of lordys wysest of All* and inserted the reference to the interpolated speeches later.

[2] This paragraph sign should stand before the line *Now lord god*, &c.

god hath me sent here down to the
to telle þe what þat þou do xalle
and how þou xalt be rewlyd in iche degre
take tent and vndyrstond
This is goddys owyn byddyng 125
þat all kynsmen of dauyd þe kyng
to þe temple xul brynge here du offryng
with whyte ʒardys in þer honde.

¶ loke wele what tyme þei offere there
all here ʒardys in þin hand þou take 130
take heed whose ʒerde doth blome and bere
and he xal be þe maydenys make.

Episcopus

[G] I thank þe lord with mylde chere
thi wurde xal I werkyn with-owtyn wrake
I xal send for hem bothyn fer and nere 135
to werke þi wyl I vndyr-take
A-non it xal be do
herk masangere þou wend þi way
Dauyd · kynsmen as I þe say
byd hem come offyr þis same day 140
and brynge white ʒardys Also.¹

Nuncius ²

¶ Oy · al maner men takyth to me tent³
that be owgth of kynrede to dauid þe kyng
my lord þe busshop hath for ʒow sent
to þe temple þat ʒe come with ʒour offryng.⁴ 145

Joseph

Fo. 51
(E quire)   In gret labore my lyff I lede 155
  †   myne ocupasyoun⁵ lyth in many place
ffor febylnesse of age my jorney I may not spede
I thank the gret god of thi grace.

---

¹ *Also* written twice and the first crossed out.
² Some faint letter written (? scribbled) above the *us* of *Nuncius*.
³ These four lines are written smaller.
⁴ This speech is continued on Ff. 53. Fo. 51 and 52 form a separate interpolated quire E. On Fo. 51 is written the dialogue between Joseph and the generaciones dauid' which follows, in a later cursive hand without rubrication. Ff. 51ᵛ and 52 and 52ᵛ are left blank. Roughly-made reference marks—as above—on Ff. 51 and 53 indicate the place of the interpolated dialogue.
⁵ Or *ocupasyon*.

                                        1ᵘˢ gen*eracionis dauid*
what chere Joseph what ys the case
That ye lye here on this groun*d*¹.   160
                                        Joseph
age *and* febylnesse doth me enbrace
That I may nother well goo ne stoun*d*².
                                        ijᵘˢ generacion[*is*
We be commandy*d* be the beschoppys son*d*
That eue*ry* man of Dauy*d* ³ Kynrede
In the tempyll to offyr a won*d*   165
Therfor in this Jorney let vs *p*rocede.
                                        Joseph⁴
Me to traveyll yt is no nede
I prey you frend*ys* go forth yo*ur* wey.
                                        iijᵘˢ gen*eracion*[*is*
yis com̃ forth Joseph I you rede
an*d* knowyth what the buschop woll sey.   170
                                        iiijᵘˢ gener[*acionis*
Ther ys a may*d* whos name is clepy*d* mary
Doughte*r* to joachym̃ as it is tol*d*⁵
he*r*e to mary thei woll asay
To som̃ many⁶ dowty *and* bol*d*

Fo. 53   He chargight þat ȝe hast ȝow · for he is redy bent
'quire)   ȝow to receyve at ȝo*ur* comyng
          He byddyth ȝow fferthermore in handys þat ȝe hent
          A fay*r*e white ȝerde everych of ȝow ȝe bryng
                    In hyght   150
          Tary not I pray ȝow
          My lord as I say ȝow
 †   now to receyve ȝow
                    is full redy dyght.
                                        Joseph⁷
¶ Benedicite I can not vndyr stande   175
  What our*e p*rince of prest*ys* doth men̄
  þat every man xuld come *and* brynge w*ith* hym a whande

---

¹ Or *groñd*; the stroke may be insignificant.
² Or *stoñd*.   ³ This may be a tailed *d* or *d + ys*.
⁴ iiiᵘˢ generacion first written and crossed through.
⁵ *I haue herd* first written for *it is told*, and crossed through.
⁶ So in MS. for *man*.
⁷ *P*rimus *generacionis* first written and crossed through.

Abyl to be maryed þat is not I so mote I theñ
I haue be maydon evyr *and* evyr more wele beñ
I chaungyd not ȝet of aƚƚ my long lyff   180
and now to be maryed su*m* man wold weñ
it is a straunge thynge An old man to take a ȝonge wyff.

¶ but nevyr-þe-lesse no doute of we must forth to towne
now neybor*ys and* kynnysmen lete us forth go
I xal take a wand in my hand *and* cast of my gowne   185
yf I falle · þan I xalle · gronyn for wo
Ho so take A-way · my staff I say · he were my fo
ȝe be men · þat may wele reñ · go ȝe be-fore
I am old · *and* also colde · walkyng doth me wo
þerfore now wole I · so my staff holde I · þis jurny to wore.   190

             E*pi*sco*pu*s

[H] ¶ Ser*ys* ȝe xal vndyr-stande¹
þat þis is þe cawse of *our* comynge
and why þat ech̄ of ȝow bryngyth a wande
ffor of god we haue knowynge
here is to be maryde a mayde ȝynge   195
aƚƚ ȝo*ur* roddys ȝe xal brynge vp to me
and on hese rodde þat þe holy gost is syttynge   *hic portent virgas*
he xal þe husbond of þis may be.

             Joseph

¶ It xal not be I ley a grote
I xal a-byde be-hynde preuyly   200
now wolde god I were at hom in my cote
I am aschamyd to be seyn veryly.

           P*ri*mus gen*er*acionis dauid'

¶ To² wurchep my lord god hedyr am I come
here ffor to offyr my dewe³ offrynge
A fayr white ȝarde in hand haue I nome   205
* my lord sere busshop at ȝo*ur* byddynge.

          Secu*n*dus gen*er*acionis dauid'

Off dauythis kynred sert*ys* am I com

---

¹ From here to the bottom of the folio the writing is closer.
² *my* first written between *to* and *wurchep* and crossed through.
³ *dw* written before *dewe* and crossed through.
* *To my woursyp* written (scribbled ?) here in different ink and different handwriting—not that of Fo. 51.

The Betrothal of Mary

a ffayr white ȝarde in hand now I bryng
my lord þe busshop after ȝour owyn¹ dom
þis ȝarde do I offre at ȝour chargyng 210
   Ryht here.
       Tercius generacionis dauid
And I a ȝarde haue both fayr and whyght²
here in myn hond it is redy dyght
And here I offre it forth within syght
   Ryght in good manere. 215
       Quartus generacionis dauid
¶ I am þe fourte of Dauid is kyn
and with myn offrynge my god I honoure
þis fayr whyte ȝarde is offryng myn
I trost in god of sum socoure
Com on Joseph with offrynge þin 220
and brynge up þin as we han oure
þou taryst ryth longe be-hynde certeyn
why comyst not fforth to goddys toure
   Com on man for shame.
          Joseph
Com ȝa ȝa · god help full fayn I wolde 225
but I am so Agyd and so olde
þat both myn leggys gyn to folde
   I am ny Almost lame.
         Episcopus
A mercy lord I kan no sygne a-spy
It is best we go a-geyn to prayr. 230
         Vox
he brought not up his rodde ȝet trewly
to whom þe mayd howyth to be maryed her.
         Episcopus
¶ Whath joseph why stande ȝe there by-hynde
I-wys sere ȝe be to blame.
         Joseph
Sere I kan not my rodde ffynde 235
to come þer in trowth me thynkyht shame.
       Episcopus Comyth thens³

---

[1] *owym* first written and corrected.
[2] The *gh* in this word may have been cancelled by the scribe. There appear to be two strokes through them.
[3] So written in the margin in MS.

                              Joseph

Ser*e* he may Euyl go þ*at* is ner lame
in soth I com as fast as I may

                              Ep*isco*pus

offyr¹ up ȝo*ur* rodde ser*e* in goddys name   240
why do ȝe not as men ȝow pray.

                              Joseph

¶ Now in þe wurchep of god of hevyn
I offyr þis ȝerde as lely whyte
prayng þat lord of gracyous stewyñ
With hert · with wytt · with mayn with myght   245
And as he made þe sterr*y*s seveñ
þis sympyl offrynge þat is so lyght
to his wurchep he weldygh evyn
Ffor to his wurchep þis ȝerd is dyght
lord god I þe pray   250
to my herte þ*ou* take good hede
and no thynge to my synful dede
Aftyr my wyl þ*ou* qwyte my mede
          As plesyth to þ*i* pay.

¶ I may not lyfte my*n* handys heye   255
Lo · Lo · Lo · What se ȝe now.

                              Ep*isco*pus

A mercy mercy mercy lord we crye
þe blyssyd of god we se Art thou.   *et clamant omnes* Mercy mercy.

Fo. 54ᵛ ¶ A gracyous god in hevyn trone
ryht wundyrful þ*i* werkys be   260
her*e* may we se A merveyl one
A ded stok beryth flour*y*s ffre
Joseph in hert with-outyn mone
þ*ou* mayst be blyth with game *and* gle
A mayd to wedde þ*ou* must gone   265
be þis meracle I do wel se
Mary is her*e* name.

                              Joseph

What xuld I wedde god for-bede
I am an old man so god me spede

¹ *Helde* first written and crossed through; *offyr* written over the line.

and with a wyff now · to levyn in drede 270
It wore neyther sport nere game.

Episcopus

¶ A-ȝens god joseph þou mayst not stryve
god wyl þat þou a wyff haue
þis fayr mayde xal be þi wyve
she is buxum and whyte as laue. 275

Joseph

A shuld I haue here ȝe lese my lyff
Alas dere god xuld I now rave
An old man may nevyr thryff
With a ȝonge wyff so god me saue
nay nay sere lett bene 280
xuld I now in age begynne to dote
If I here chyde she wolde clowte my cote
blere myn ey and pyke out a mote
and þus oftyn tymes it is sene.

Episcopus

¶ Joseph now as I þe saye 285
God hath assygnyd here to þe
Fo. 55   þat god wol haue do sey þou not nay
oure lord god wyl þat it be so [1].

Joseph

A-ȝens my God not do I may
here wardeyn and kepere wyl I evyr be 290
But fayr maydon I þe pray
Kepe þe clene as I xal me
I am a man of age
therfore sere busshop I wyl þat ȝe wete
þat in bedde we xul nevyr mete 295
Ffor i-wys mayden suete
An Old man may not rage.

Episcopus

¶ This holyest virgyn xalt þou maryn now
ȝour rodde floreschyth fayrest þat man may se 299
þe holy gost we se syttyht on a bow   *Et hic cantent. Bene-*
now ȝelde we all preysyng to þe trenyte.   *dicta sit beata trinitas.*

¶ Joseph wole ȝe haue þis maydon to ȝour wyff
And here honour and kepe as ȝe howe to do.

---

[1] *so* apparently by mistake as it destroys the rhyme.

                                    Joseph

    nay sere so mote I thryff
    I haue ryght no nede þer-to.   305
                                    Episcopus

    Joseph it is goddys wyl it xuld be so
    sey aftyr me as it is skyl.
                                    Joseph

    Sere and to performe his wyl I bow þer-to
    Ffor all thynge owyght to ben at his wyl.
                              Episcopus et idem joseph
¶ Sey þan after me ․ here I take þe Mary to wyff   310
   to hauyn to holdyn ․ as god his wyll with us wyl make [1]
   and as longe as be-thwen us ․ lestyght oure lyff
   to loue ȝow as my selff ․ my trewth I ȝow take.
                         nunc ad mariam sic dicens · Episcopus

Fo. 55ᵛ ¶ Mary wole ȝe haue þis man
    And hym to kepyn as ȝour lyff.   315
                                    Maria

    In þe tenderest wyse fadyr as I kan
    and with all my wyttys ffyff.
                                    Episcopus

¶ Joseph with þis ryng now wedde þi wyff
   and be here hand now þou here take.
                                    Joseph

    Sere with þis Rynge I wedde here Ryff   320
    and take here now here ffor my make.
                                    Episcopus

    Mary mayd with-outyn more stryff
    On to þi spowse þou hast hym take.
                                    Maria

    In chastyte to ledyn my lyff
    I xal hym nevyr for-sake   325
    but evyr with hym a-byde
    And jentyll spowse as ȝe An seyd
    lete me levyn as a clene mayd᾽
    I xal be trewe be not dysmayd᾽
    both terme tyme and tyde.   330
                                    Episcopus
¶ Here is þe holyest matremony þat evyr was in þis werd
   þe hyȝ names of oure lord we wole now syng hy

        [1] *wylmake* (one word) in MS.

we all wole þis solempn dede record
devowtly · Alma chorus *domi*ni · *nunc* pangat *nomina* summi.

¶ Now goth hom all in godys name 335
Where as ȝo*ur* wonyng was be-fore
Maydenys to lete her*e* go alone it wor*e* shame
It wold hevy ȝo*ur* herty*s* sore
ȝe xal blysse þe tyme þat sche was bor*e*
now loke ȝe at hom here brynge. 340

Maria

To haue ȝo*ur* blyssyng ffadyr I falle ȝow be-for*e*.

*Ep*i*scopus*

he blysse ȝow þat ha*t*h non hendyng
In no*m*i*n*e patris *et* filij *et* sp*iritu*s s*an*cti.

*Ep*i*scopus*

fo. 56 ¶ Joseph þi selph art old of Age
And þi wyff of Age is ȝonge 345
and as we redyn in old sage
many man is sclepyr of tonge
Þ*er* fo*r*e Euyl langage for to swage
þat ȝo*ur* good fame may leste longe
iij damysellys xul dwelle w*ith* ȝow i*n* stage 350
W*ith* þi wyff to be evyr mor*e* a-monge
I xal these iij her*e* take
Susanne þ*e* fyrst xal be
Rebecca þe secunde xal go w*ith* the
Sephor*e* þe thrydde · loke þat ȝe thre 355
þis maydon nevyr ȝe for-sake.

Susanne

¶ Ser*e* I am redy Att ȝo*ur* wyll
w*ith* þis maydon for to wende.

Rebecca

ȝo*ur* byddyng ser*e* xall fful-ffyl
and ffolwe þis maydon ffayr *and* hende. 360

Sephor

To ffolwe hyre it is good skyl
And to ȝo*ur* byddynge wole I bende.

Joseph

Now ser*e* buschop hens go I wyl
for now comyth on to my mende
A mater*e* þat nedful is. 365

Episcopus

Ffare wel joseph and mary clere
I pray god kepe 3ow all in fere
and sende 3ow grace in good manere
to serve þe kynge of blysse.

Maria

¶ Ffadyr and modyr 3e knowe þis cas 370
how¹ þat it now doth stonde with me
with myn spowse I must forth passe
and wott nevyr whan I xal 3ow se
Therfore I pray 3ow here in þis plas
of 3our blyssynge for charyte 375
and I xal spede þe betyr and haue more gras
In what place þat evyr I be
On knes to 3ow I falle
I pray 3ow fadyr and modyr dere
to blysse 3our owyn dere dowtere 380
and pray ffor me in all manere
and I ffor 3ow all.

Joachym

¶ Almyghty god he mote þe blysse
and my blyssynge þou haue Also
In all godnesse god þe wysse 385
on londe or on watyr wher evyr þou go.

Anna

Now god þe kepe from every mysse
and saue þe sownd in welth from wo
I pray þe dowtyr þou onys me kys
or þat þi modyr parte þe fro 390
I pray to god þe saue
I pray þe mary my swete chylde
be lowe and buxhum meke and mylde
Sad and sobyr and no thyng wylde
and goddys blyssyng þou haue. 395

Joachym

¶ Ffor² wel joseph and god 3ow spede
wher so 3e be in halle or boure.

Joseph

Almyghty god 3our weys lede
and saue 3ow sownd from all douloure.

---
¹ *Whow* first written, the *W* crossed through. ² So in MS.

	Anna
Goddys grace on ȝow sprede [1]	400

Ffare wel mary my swete fflowre
Fare weyl joseph and god ȝow rede
Ffare weyl my chylde and my tresowre
Ffare wel my dowtere ȝyng.

           Maria

Ffare wel fadyr and modyr dere   405
at ȝow I take my leve ryght here
god þat sytt in hevyn so clere
haue ȝow in his kepyng.

           Joseph

¶ Wyff it is ful necessary þis ȝe knowe
þat I And my kyn-rede go hom be-fore   410
for in soth we haue non hous of oure owe
þer fore I xal gon ordeyn · and thanne come ȝow fore
We Ar not ryche of werdly thynge
and ȝet of oure sustenauns we xal not mys
therfore A-bydyth here stylle to ȝour plesynge   415
to worchep ȝour god is all ȝour blysse.

¶ he þat is and evyr xal be
Of hefne and helle Ryche kynge
in erth hath chosyn poverte
and all Ryches and welthis refusynge.   420

           Maria

Goth husbond in oure lordys blyssynge
he mote ȝow spede in all ȝour nede
and I xal here a-byde ȝour A-ȝen comynge
and on my sawtere book I xal rede
Now blyssyd be oure lord ffor this   425
of hefne and erthe and all þat beryth lyff
I am most bound to ȝow lord i-wys
ffor now I am bothe mayde and wyff.

¶ Now lord god dyspose me to prayour
þat I may sey þe holy psalmes of dauyth   430
wheche book is clepyd þe sawtere
þat I may preyse the · my god þer with

---

[1] This and following three lines written smaller.

Fo. 57ᵛ
 Of þe vertuys þer of þis is þe pygth
 It makyht sowles fayr þat doth it say
 Angelys be steryd to help us þer with   435
 it lytenyth therkeness and puttyth develys Away.

 ¶ þe song of psalmus is goddys dete
 synne is put A-wey þer by
 It lernyth A man vertuysful to be
 It feryth mannys herte gostly   440
 who þat it vsyth customably
 it claryfieth þe herte and charyte makyth cowthe
 he may not faylen of goddys mercy
 þat hath þe preysenge of god evyr in his mowthe.

 ¶ O holy psalmys · O holy book   445
 Swetter to say than Any ony
 þou lernyst hem love lord þat on þe look
 and makyst hem desyre thyngys celestly
 With these halwyd psalmys lord I pray the specyaly
* ffor all þe creatures qwyke and dede   450
 þat þou wylt shewe to hem þi mercy
 and to me specyaly þat do it rede.

 ¶ I haue seyd sum of my sawtere and here I am
 at þis holy psalme in dede
 Benedixisti domine terram tuam   455
 In this holy labore · lord me spede.
              Joseph

 ¶ Mary wyff and mayd most gracyous
 displese ȝow not I pray ȝow so long I haue be
 I haue hyryd for us a lytyl praty hous
 and þer-in ryght hesely levyn wole we   460
 Come forth mary and folwe me
 to Nazareth now wele we go

Fo. 58
 And all þe maydonys bothe ffayr and fre
 with my wyff comyth forth also
 Now lystenyth well wyff what I tell þe   465
 I must gon owth hens fer þe fro
 I wyll go laboryn in fer countre

* Here some words are scribbled in faint ink in the margin : of þe Vyn. R. Wych.

with trewth to maynteyn oure housholde so
þis ix monthis þou seyst me nowth
Kepe þe clene my jentyl spowse 470
and all þin maydenys in þin howse
þat evyl langage I here not rowse
Ffor hese love þat all hath wrought.

                                             Maria

¶ I pray to god he spede ȝour way
and in sowle helth he mote ȝow kepe 475
and sende ȝow helth bothe nyth and day
he shylde and saue ȝow from al shenschepe
Now lord of grace to þe I pray
with morny mood on kne I krepe
me saue from synne from tene and tray 480
with hert I mourne with eye I wepe
lord god of pete
Whan I sytt in my conclaue
all myn hert on þe I haue
Gracyous god my mayden-hed saue 485
Euyr clene in chastyte.

                                             contemplacio

1us [1]

11

¶ Ffowre thowsand · sex vndryd · foure ȝere I telle
Man ffor his offens and ffowle foly
Hath loyn ȝerys · in þe peynes of helle
And were wurthy to ly þer-in endlesly
But thanne xulde perysche ȝour grete mercye 5
good lord haue on man pyte
haue mende of þe prayour seyd by Ysaie
lete mercy meke þin hyest mageste.

¶ wolde god þou woldyst breke þin hefne myghtye
and com down here in to[2] erth 10
And levyn ȝerys thre and threttye
thyn famyt ffolke with þi fode to fede

---

[1] This figure is in red.
[2] *to* twice written, the first crossed through.

To staunche þi thryste lete þi syde blede
ffor erste wole not be mad redempcion
Cum vesyte vs in þis tyme of nede 15
of þi careful creaturys haue compassyon.

2¹ ¶ A woo to vs wrecchis of² wrecchis be
ffor god hath haddyd³ ssorwe to sorwe
I prey þe lord þi sowlys com se
How þei ly and sobbe ffor syknes and sorwe⁴ 20
With þi blyssyd blood ffrom balys hem borwe
thy careful creaturys cryenge in captyvyte
A tary not gracyous lord tyl it be to-morwe
The devyl hath dysceyved hem be hys iniquite.

¶ A quod Jeremye · who xal gyff wellys to myn eynes 25
Þat I may wepe bothe day and nyght
to se oure bretheryn in so longe peynes
here myschevys Amende · may þi mech myght
Fo. 59 As gret as þe se lord · was Adamys contryssyon ryght
Ffrom oure hed is falle þe crowne 30
Man is comeryd in synne · I crye to þi syght
Gracyous lord · Gracyous lord · Gracyous lord come downe.
                                                    Virtutes

¶ Lord plesyth it þin hyȝ domynacion
On man þat þou made to haue pyte
Patryarchys⁵ and prophetys han made⁶ supplycacion 35
oure offyse is to presente · here⁷ prayerys to the
Aungelys · Archaungelys we thre
þat ben in þe fyrst ierarchie
Ffor man to þin hy mageste
Mercy · mercy · mercy we crye. 40

---

¹ This figure is in red.

² The original word *of* has been crossed out and *that* written in a different hand above.

³ The initial *h* has been crossed through with a thick perpendicular stroke of different ink.

⁴ The original phrase *ffor sykness* and *sorwe* has been altered in this different ink and presumably the same hand that wrote *that* for *of* above, to *bothe eve and morewe*. [Not impossibly the hand of the scribe of Ff. 95, 96.]

⁵ Or *kys*.   ⁶ Or *mad*.

⁷ Some word was written between *here* and *prayerys* and crossed through.

¶ The Aungel lord þou made so gloryous
  whos synne[1] hath mad hym a devyl in helle
  he mevyd man to be so contraryous
  man repentyd · and he in his obstynacye doth dwelle
  Hese grete males good lord repelle 45
  And take man on to þi grace
  lete þi mercy make hym with Aungelys dwelle
  of locyfere to restore þe place.

              Pater

 PRopter miseriam inopum ⎱ nunc exurgam
  et gemitum pauperum ⎰

¶ Ffor þe wretchydnes of þe nedy
  And þe porys lamentacion 50
  now xal I ryse þat am Almyghty
  tyme is come of reconsyliacion
  My prophetys with prayers haue made supplicacion
  my contryte creaturys crye all for comforte
  All myn Aungellys in hefne · with-owte cessacion 55
  they crye þat grace to man myght exorte.

              Veritas

¶ Lord I am þi dowtere trewth
  þou wylt se I be not lore
  thyn vnkynde creaturys to saue were rewthe
  the offens of man hath grevyd þe sore 60
  Whan Adam had synnyd þou seydest þore
  þat he xulde deye and go to helle
  And now to blysse hym to resstore
  twey contraryes mow not to-gedyr dwelle.

¶ Thy trewthe lord xal leste with-owtyn ende 65
  I may in no wyse ffro þe go
  þat wretche þat was to þe so vnkende
  he may not haue to meche wo
  He dyspysyd þe and plesyd þi ffo
  þou art his creatour · and he is þi creature 70
  þou hast lovyd trewthe · it is seyd evyr mo
  þerfore in peynes · lete hym evyr more endure.

---

[1] A blotted *hath* between *synne* and *hath* crossed through.

Misericordia

¶ O Ffadyr of mercy · *and* god of comforte
þat counsell us in eche trybulacion
lete ȝo*ur* dowtere mercy to ȝow resorte 75
And on man þat is myschevyd haue compassyon
hym grevyth fful gretly his transgressyon
All hefne *and* erthe crye ffor mercy
Fo. 60 Me semyth þer xuld be non excepcion
ther prayers ben offeryd so specyally. 80

¶ Threwth Sseyth she hath evyr be than
I graunt it wel she hath be so
and þou seyst endlesly · þat mercy þou hast kept ffor man
than mercyabyl lorde kepe us bothe to
Thu seyst · Veritas mea *et* m*isericord*ia mea cu*m* ipso 85
Suffyr not þi sowlys than in sorwe to slepe
þat helle hownde þat hatyth þe byddyth hy*m* ho
þi love man no lenger*e* lete hy*m* kepe.

Justicia

¶ Mercy me merveylyth what ȝow movyth
ȝe know wel I am ȝo*ur* systere ryghtwysnes 90
God is ryghtful *and* ryghtffulnes lovyth
man offendyd hym þat is endles
Ther fore his endles punchement may nevyr sees
Also he forsoke his makere þat made hym of clay
And þe devyl to his mayst*er* he ches 95
xulde he be savyd · nay nay nay.

¶ As wyse as is god he wolde A be
this was þe Abhomynabyl *p*resu*m*pcion
it is seyd ȝe know wel þis of me
þat þe ryghtwysnes of god hath no diffynici*o*n 100
Therffore late þis be oure conclusyon
he þat sore synnyd ly stylle in[1] sorwe
he may nevyr make A seyth be reson
whoo myght thanne thens hym borwe.

*Misericord*ia

Fo. 60ᵛ ¶ Systyr Ryghtwysnes ȝe Are to vengeabyl 105
Endles synne god endles may restere

[1] A thin stroke in darker ink above this *n* noted by Manly appears to be an accidental mark.

Above aĦ hese werkys god is mercyabyl
þow he for-sook god be synne · be feyth he for-sook hym never
 þe more
And þow he presumyd nevyr so sore
ȝe must consyder þe frelnes of mankende   110
lerne[1] and ȝe lyst þis is goddys lore
þe mercy of god is with-owtyn ende.

               Pax

¶ To spare ȝour speches systerys it syt
 It is not onest in vertuys to ben dyscencion
 the pes of god ovyr comyth aĦ wytt   115
 þow trewth and ryght sey grett reson
 ȝett mercy seyth best to my pleson
 ffor yf mannys sowle xulde abyde in helle
 be-twen God and man evyr xulde be dyvysyon
 And than myght not I pes dwelle.   120

¶ Therefore me semyth best ȝe thus Acorde
 than hefne and erthe ȝe xul qweme
 putt bothe ȝour sentens in oure lorde
 And in his hyȝ wysdam lete hym deme
 This is most syttynge me xulde seme   125
 And lete se how we ffowre may aĦ A-byde
 þat mannys sowle it xulde perysche it wore sweme
 or þat ony of vs ffro othere xulde dyvyde.

               Veritas

¶ In trowthe here-to I consente
 I wole prey oure lord it may so be.   130

               Justicia

 I Ryghtwysnes am wele contente
 Ffor in hym is very equyte.

               Misericordia

Fo. 61 ¶ And I mercy ffro þis counsel wole not fle
 tyl wysdam hath seyd I xal ses.

               Pax

 Here is god now · here is vnyte   135
 hefne and erth is plesyd with pes.

               ffilius

¶ I thynke þe thoughtys of pes · and nowth of wykkydnes
 this I deme to ses ȝour contraversy

[1] The original first word of this line—probably a miswritten *lerne*—has been obliterated.

If Adam had not deyd · peryschyd had ryghtwysnes
And Also trewth had be lost þer-by 140
Terrewth¹ and ryght wolde chastyse ffoly
ȝiff a-nother deth come not · mercy xulde perysch
þan pes were exyled ffynyaly
So tweyn dethis must be ȝow fowre to cherysch.

¶ But he þat xal deye ȝe must knawe 145
þat in hym may ben non iniquyte
þat helle may holde hym be no lawe
But þat he may pas at hese lyberte
Qwere swyche on his prevyde and se
And hese deth · for mannys deth xal be redempcion 150
All hefne and erth seke now ȝe
Plesyth it ȝow þis conclusyon.

Veritas

¶ I trowthe haue sowte þe erthe · with-owt and with-inne²
and In sothe þer kan non be fownde
þat is of o day byrth · with-owte synne 155
nor to þat deth wole be bownde.

Misericordia

I mercy haue ronne · þe hevynly Regyon rownde
and þer is non of þat charyte
þat ffor man wole suffre A deddly wounde
I I³ can nott wete how þis xal be. 160

Justicia

Fo. 61ᵛ⁴ ¶ Sure I can fynde non sufficyent
Ffor servauntys vn-profytable we be ech on
He⁵ love nedyth to be ful Ardent
that for man to helle wolde gon.

Pax

That god may do is non but on 165
þer fore þis is Pesys⁶ A-vyse

---

¹ Or *Trewth*; the stroke marking the contraction may be a flourish.
² *with inne and with owt* first written.   ³ So repeated in MS.
⁴ The writing becomes more spaced and remains so till Fo. 66. It is the hand of the same scribe.
⁵ The original *he* has been corrected by a different hand to *hes* with a very faint *s*.
⁶ The original *Pesys* has been corrected by a different hand to *be hys*. [Not the hand of the scribe of Ff. 95, 96.]

he þat ȝaff þis counsell · lete hym ȝeve þe comforte A-lon
Ffor þe conclusyon · in hym · of all þese lyse.

                 Ffilius

¶ It peyneth me · þat man I mad
þat is to seyn peyne I must suffre fore   170
A counsel of þe trinite must be had
Whiche of vs xal man restore.

                 Pater

In ȝour wysdam son · man was mad thore
And in wysdam¹ was his temptacion
þerfor sone sapyens ȝe must ordeyn here-fore   175
and se how of man may be salvacion.

                 Filius

¶ Ffadyr he þat xal do þis must be both god and man
lete me se how I may were þat wede
And syth in my wysdam he be-gan
I am redy to do þis dede.   180

             Spiritus Sanctus

I the holy gost · of ȝow tweyn do procede
this charge I wole take on me
I love to ȝour lover xal ȝow lede ²
þis is þe Assent of oure vnyte.

               Misericordia

Fo. 62 Now is þe loveday mad of us fowre fynialy   185
now may we leve in pes · as we were wonte
Misericordia et veritas obviauerunt sibi
Justicia et pax · osculate sunt
 et hic osculabunt pariter omnes.

                 Pater

¶ Ffrom vs god Auɴgel Gabryel þou xalt be sende
In to þe countre of Galyle   190
the name of þe cyte Nazareth is kende
to A mayd · w[e]ddyd ³ to A man is she
Of whom þe name is joseph se
of þe hous of davyd bore
The name of þe mayd ffre   195
Is Mary þat xal Al Restore.

   ¹ A miswritten letter before *was*.
   ² *procede* first written for *ȝow lede*.
   ³ The *e* has disappeared through a flaw in the MS.

                                                                Ffilius
¶ Say þat she is with-owte wo and ful of grace
 And þat I þe son of þe godhed of here xal be bore
 Hyȝe þe þou were there A pace
 ellys we xal be there the be-ffore   200
 I haue so grett hast to be man thore
 In þat mekest and purest virgyne
 Sey here she xal restore
 Of ȝow Aungellys þe grett Ruyne.
             Spiritus Sanctus
Fo. 62ᵛ ¶ And if she Aske þe how it myth be   205
 telle here I þe holy gost xal werke al this
 Sche xal be savyd thorwe oure vnyte
 In tokyn here bareyn cosyn Elyzabeth is
 Qwyk with childe · in here grett Age i-wys
 Sey here to vs is no thynge impossyble ¹   210
 Here body xal be so ful-fylt with blys
 Þat she xal sone thynke · þis sownde credyble.
             Gabriel
¶ In thyn hey inbassett lord I xal go
 It xal be do with a thought
 be-holde now lord I go here to   215
 I take my fflyth and byde nowth

Ave maria ² gratia plena Dominus tecum.
 Heyl fful of ³ grace god is with the
 Amonge All women blyssyd art thu ·
 here þis name Eva · is turnyd Aue
 þat is to say with-owte sorwe ar ȝe now.   220

¶ Thow sorwe in ȝow hath no place
 ȝett of joy lady ȝe nede more
Fo. 63 Therfore I Adde And sey Fful of grace
 Ffor so Ful of grace was nevyr non bore
 ȝett who hath grace he nedyth kepyng sore 225
 therfore I sey god is with the

---

¹ The words *no thynge impossyble* are written in larger form and, to a less degree, ll. 208–16.
² *Maria* is crossed through in a different ink, possibly to make the line correspond exactly with the translation in the line following.
³ The word *of* written twice, the second crossed through.

Whiche xal kepe ȝow endlesly thore
So amonge All women blyssyd¹ are ȝe.

Maria

¶ A mercy god þis is a mervelyous herynge²
In þe Aungelys wordys I am trobelyd her   230
I thynk how may be þis gretynge
Aungelys dayly to me doth Aper
But not in þe lyknes of man þat is my fer³
And Also thus hyȝly to comendyd be
and am most vn-wurthy I can-not Answere   235
grett shamfastnes and grett dred is in me.³

Gabryel

¶ Mary in þis take ȝe no drede
Ffor At God · grace ffownde haue ȝe
ȝe xal conceyve in ȝour wombe in dede
A childe þe sone of þe trynyte   240
His name of ȝow · jhesu · clepyd xal be
He xal be grett · þe son of þe hyest · clepyd of kende
and of his ffadyr davyd · þe lord xal ȝeve hym þe se
Reynyng in þe hous of jacob · of which regne xal be no⁴ ende.

Maria

¶ Aungel I sey to ȝow   245
In what manere of wyse xal þis be
Ffor knowyng of man I haue non now
I haue evyr more kept and xal my virginyte
I dowte not þe wordys ȝe han seyd to me
But I Aske how it xal be do.   250

Gabryel

The holy gost xal come fro A-bove to the⁵
and þe vertu of hym hyest xal schadu þe so.

¶ Ther fore þat holy gost of þe xal be bore
he xal be clepyd þe son of god sage
And se Elyzabeth ȝour cosyn thore   255

---

¹ The double *s* is a correction in different ink—*blylled* first written.
² *thynge* first written and corrected by scribe.
³ From *But not in þe* to *is in me* the writing is still larger.
⁴ MS. ñ.
⁵ This speech *The holy gost . . . wyl seyn* and especially the words *impossyble to goddys vsage* are written in larger form.

She hath conseyvid A son in hyre Age
This is þe sexte monyth of here passage ·'
Fo. 64 Of here þat clepyd was bareyn
no thynge is impossyble to goddys vsage
they thynkyth longe to here what ȝe wyl seyn 260
here þe Aungel makyth a lytyl restynge and *mary be-holdyth hy*m
and *þe Aungel seyth*

¶ Mary come of *and* haste the
And take hede in thyn entent
Whow þe holy gost · blyssyd he be
A-bydyth þin answere *and* þin assent
Thorwe wyse werke of dyvinyte 265
the secunde *per*sone verament
is mad man by fraternyte
With-inne þi self in place present.

¶ Fferther more take hede þis space
Whow all þe blyssyd spyrytys of vertu 270
þat are in hefne by-ffore goddys face
And all þe gode levers *and* trew
That Are here · in þis erthely place
thyn owyn kynrede · þe sothe ho knew
And þe chosyn sowlys · þis tyme of grace 275
þat Are in helle *and* byde rescu¹.

Fo. 64ᵛ ¶ As Adam · Abraham · *and* davyd in fere
And many othere of good reputacion
þat þin Answere desyre to here
and þin Assent to þe incarnacion 280
In which þou standyst · As *per*severe²
of All man-kende savacion
Gyff me myn Answere · now lady dere
to All these creaturys comfortacion.
Maria

¶ With All mekenes I clyne to þis A-corde 285
Bowynge down my face with All benyngnyte

¹ *rescu* roughly scratched through in different ink and three dots above and three below in this ink; *þer* (?) *rescu* roughly written in yet another ink and not the scribe's hand—possibly that of writer of Ff. 95–6.
² p*er*syvere first written, the *e* thickly written over the *y*.

The Salutation and Conception

Se here þe hand-mayden of oure lorde
Aftyr þi worde · be it don to me.

Gabryel

Gramercy my¹ lady ffre
Gramercy · of ȝour Answere on hyght 290
Gramercy · of ȝour grett humylyte
Gramercy · ȝe lanterne off lyght.

here þe holy gost discendit with iij bemys to our lady · the sone of þe godhed nest with iij bemys · to þe holy gost · the fadyr godly with iij bemys to þe sone · And so entre Att thre to here bosom · and Mary seyth

Maria

Fo. 65   A now I ffele in my body be
parfyte god *and* parfyte man
havyng Al schappe · of chyldly carnalyte 295
Evyn Al at onys · þus god be-gan.

¶ Nott takynge ffyrst o membyr *and* sythe A-nother
but parfyte childhod ȝe haue A-non
of ȝour hand-mayden · now ȝe haue mad ȝour modyr
With-owte peyne in Fflesche *and* bon 300
Thus conceyved nevyr woman non
þat evyr was beynge in þis lyff
O myn hyest ffadyr in ȝour tron
It is worthy ȝour son · now my son · haue A prerogatyff.

¶ I can not telle what joy what blysse 305
now I fele in my body
Aungel Gabryel I thank ȝow for thys
most mekely recomende me · to my faderys mercy
To haue be þe modyr of god fful lytyl wend I
Now myn cosyn Elyzabeth ffayn wold I se 310
how sche hath conseyvid as ȝe dede specyfy
Now blyssyd be þe hyȝ trynyte.

Gabryel

Fo. 65ᵛ ¶ Ffare weyl turtyl · goddys dowtere dere
Ffare wel goddys modyr · I þe honowre
Ffare wel goddys sustyr · *and* his pleynge fere 315
Ffare wel goddys chawmere *and* his bowre.

---

¹ *my* omitted and written above the line.

                                                            Maria
¶ Ffare wel Gabryel specyalye
   Ffare wel goddys masangere expresse
   I thank ȝow for ȝour traveyl hye
   Gramercy of ȝour grett goodnes.   320
¶ And namely of ȝour comfortabyl massage
   Ffor I vndyrstande by inspyracion
   þat ȝe knowe by syngulere preuylage
   most of my sonys incarnacion
   I pray ȝow take it in to vsage   325
   be A custom ocupacion
   to vesyte me ofte be mene passage
   ȝour presence is my comfortacion.

                                                           Gabriel
¶ At ȝour wyl lady so xal it be
   ȝe gentyllest of blood · and hyest of kynrede   330
   þat reynyth in erth in ony degre
   be pryncypal incheson of þe god-hede.

¶ I comende me on to ȝow · þou trone of þe trinyte[1]
   O mekest mayde now þe modyr of jhesu
Fo. 66 qwen of hefne · lady of erth · and empres of helle be ȝe   335
   socour to All synful · þat wole to ȝow sew
   Thour[2] ȝour body beryth þe babe · oure blysse xal renew
   to ȝow modyr of mercy · most mekely I recomende
   and as I began I ende · with An Ave new
   Enjoynyd hefne and erth · with þat I Ascende.   340

   Ave maria gratia plena   ⎫ Angeli cantando istam
   Dominus tecum · uirgo sesena[3] ⎭ sequenciam.[4]

   Remainder of Fo. 66—5½ inches—and Fo. 66ᵛ left blank.

[1] The more cramped writing is resumed with this line.   [2] MS. Thð.
[3] sesena first written; trace of correction to serena.
[4] The words And þan mary seyth follow on as part of this direction, but have been crossed through in red ink.

12

                                                                      ¶ Joseph

How dame how · vn-do ȝoure dore vn-do
Are ȝe at hom why speke ȝe notht [1]

Susanna

Who is ther why cry ȝe so
telle us ȝour herand wyl ȝe ought.

Joseph

Vn-do ȝour dore I sey ȝow to
Ffor to com in is all my thought.   5

Maria

it is my spowse þat spekyth us to
On-do þe dore his wyl were wrought.

¶ well-come hom myn husbond dere
how haue ȝe ferd in fer countre.   10

Joseph

To gete oure levynge with-owtyn dwere
I haue sore laboryd ffor þe and me.

Maria

husbond ryght gracyously now come be ȝe
it solacyth me sore · sothly to se ȝow [2] in syth.

Joseph

Me merveylyth wyff surely · ȝour face I can not se   15
but as þe sonne with his bemys · quan he is most bryth.

Maria

¶ Husbond it is as it plesyth oure lord þat grace of hym grew
Who þat evyr be-holdyth me veryly
they xal be grettly steryed to vertu
ffor þis ȝyfte and many moo good lord gramercy.   20

Joseph

¶ How hast þou ferde jentyl mayde
whyl I haue be out of londe.

Maria

Sekyr sere beth nowth dysmayde
Ryth aftyr þe wyl of goddys sonde.

joseph

That semyth evyl I am afrayd   25
þi wombe to hyȝe doth stonde
I drede me sore I am be-trayd

---

[1] The words *how hast* follow as the beginning of a third line, but have been crossed out.
[2] ȝw written before ȝow and crossed through.

　　　　　Sum other man þe had in honde
Fo. 67ᵛ　Hens sythe þat I went
　　　　　Thy Wombe is gret it gynnyth to ryse　　　　　30
　　　　　than hast þou be-gownne a synfull gyse
　　　　　telle me now in what wyse
　　　　　thy self þou Ast þus schent.

　　　　¶ Ow dame what þinge menyth this
　　　　　with childe þou gynnyst ryth gret to gon　　　35
　　　　　Sey me mary þis childys fadyr ho is
　　　　　I pray þe telle me and þat anon

　　　　　　　　　　　　　　　　　　　　　　　Maria

　　　　　The fadyr of hevyn and ȝe it is
　　　　　other fadyr hath he non
　　　　　I dede nevyr forfete with man i-wys　　　　　40
　　　　　Wher-fore I pray ȝow amende ȝour mon
　　　　　this childe is goddys and ȝour.

　　　　　　　　　　　　　　　　　　　　　　　Joseph

　　　　　Goddys childe þou lyist in fay
　　　　　God dede nevyr jape so with may
　　　　　And I cam nevyr ther I dare wel say　　　　　45
　　　　　ȝitt so nyh þi boure
　　　　　But ȝit I sey mary whoos childe is this.

　　　　　　　　　　　　　　　　　　　　　　　Maria

　　　　　Goddys and ȝoure I sey i-wys.

　　　　　　　　　　　　　　　　　　　　　　　Joseph

　　　　¶ ȝa ȝa all Olde men to me take tent
　　　　　and weddyth no wyff in no kynnys wyse　　　50
　　　　　þat is a ȝonge wench be myn a-sent
　　　　　ffor doute and drede and swych servyse
　　　　　Alas Alas my name is shent
　　　　　all men may me now dyspyse
　　　　　and seyn olde cokwold þi bow is bent　　　　55
　　　　　newly now after þe frensche gyse
　　　　　Alas and welaway
Fo. 68　Alas dame why dedyst þou so
　　　　　Ffor þis synne þat þou hast do
　　　　　I the for-sake and from þe go　　　　　　　60
　　　　　Ffor onys evyr and Ay.

Maria

¶ Alas gode spowse why sey ȝe thus¹
Alas dere hosbund a-mende ȝour mod
It is no man but swete jhesus
he wylt be clad in flesch and blood   65
and of ȝour wyff be born.

Sephor

Ffor sothe þe Aungel þus seyd he
þat goddys sone in trynite
Ffor mannys sake a man wolde be
to save þat is for-lorn.   70

Joseph

¶ An Aungel allas alas fy for schame
ȝe syn now in þat ȝe to² say
to puttyn an Aungel in so gret blame
Alas alas let be do way
It was sum boy be-gan þis game   75
þat clothyd was clene and gay
and ȝe ȝeve hym now an Aungel name
Alas alas and wel away
þat evyr this game be-tydde
A dame what thought haddyst þou   80
Here may all men þis proverbe trow
þat many a man doth bete þe bow
Another man hath þe brydde.

Maria

A gracyous god in hefne trone
comforte my spowse in þis hard cas   85
mercyful god A-mend his mone
as I dede nevyr so gret trespas.

Joseph

¶ Lo · Lo Serys · What told I ȝow
þat it was not for my prow
     A wyff to take me to   90
An þat is wel sene now
Ffor mary I make god A-vow
     is grett with childe lo
Alas why is it so
to þe busshop I wole it telle   95

¹ *So* first written instead of *thus*.   ² *to* miswritten before *to* and crossed through.

þat he þe law may here do
With stonys here to qwelle.

¶ Nay nay ȝet god ffor-bede
þat I xuld do þat vegeabyl[1] dede
   but if I wyst wel qwy   100
I knew never with here so god me spede
tokyn[2] of thynge in word nor dede
   þat towchyd velany
nevyr þe les what for thy
þow she be meke and mylde   105
With-owth mannys company
she myght not be with childe.

¶ But I ensure myn was it nevyr
thow þat she hath not don here devyr
   rather than I xuld[3] pleynyn opynly   110
Serteynly ȝitt had I levyr
Ffor sake þe countre ffor evyr
   and nevyr come in here company
Ffor and men knew þis velany
In repreff þei wolde me holde   115
and ȝett many bettyr than I
ȝa · hath ben made cokolde

Fo. 69 * Now alas whedyr xal I gone
I wot nevyr whedyr nor to what place
ffor oftyn tyme sorwe comyth sone   120
and longe it is or it pace
no comforte may I haue here
I-wys wyff þou dedyst me wronge
Alas I taryed from þe to longe
All men haue pety on me[4] amonge   125
Ffor to my sorwe is no chere.
            Maria

¶ God þat in my body Art sesyd
þou knowist myn husbond is dysplesyd
   to se me in þis plight

---

[1] So in MS.
[2] *nevyr þe les* first written as beginning of this line, and crossed through.
[3] A contraction mark stands above this word by a scribal slip.
* No paragraph sign in MS.   [4] MS. *onime* corrected by dot under *i*.

Ffor vnknowlage he is desesyd 130
and þerfore help þat he were esyd
  þat he myght knowe þe ful perfyght
Ffor I haue levyr abyde respyt
to kepe þi sone in priuite
grauntyd · by þe holy spyryt 135
þan þat it xulde be opynd by me.

          Deus

¶ Descende I sey myn Aungelle
on to joseph for to telle
  such as my wyl is
byd hym with mary A-byde and dwelle 140
Ffor it is my sone fful snelle
  þat she is with i-wys.

          Angelus

Almyghty god of blys
I am redy ffor to wende
wedyr as þi wyl is 145
to go · bothe fer and hynde

* Joseph Joseph þou wepyst shyrle
ffro þi wyff why comyst þou owte.

          Joseph

Good sere lete me wepe my ffylle
Go forthe þi wey and lett me nowght. 150

          Angelus

In þi wepynge þou dost ryght ylle
A-ȝens god þou hast mys-wrought
Go chere þi wyff with herty wylle
and chawnge þi chere Amende þi thought
Sche is a ful clene may 155
I telle þe god wyl of here be born
And sche clene mayd as she was be-forn
to saue mankynd þat is for-lorn
Go chere hyre þerfore I say.

          Joseph

¶ A lord god benedicite 160
of þi gret comforte I thank the
  þat þou sent me þis space
I myght wel A wyst parde

   * No paragraph sign in MS.

So good a creature as she
  wold nevyr A done trespace   165
For sche is ful of grace
I know wel I haue myswrought
I walk to my pore place
and Aske ffor-gyfnes I haue mys-thought.

¶ Now is þe tyme sen At eye   170
þat þe childe is now to veryfye
  which xal saue mankende
As it was spoke be prophesye
I thank þe god þat syttys on hye
  with hert wyl and mende   175
þat evyr þou woldyst me bynde
to wedde · mary to my wyff
þi blysful sone · so nere to fynde
In his presens · to lede my lyff.

¶ Alas ffor joy I qwedyr and qwake   180
Alas what hap now was this
Fo. 70   A mercy mercy my jentyl make
mercy I haue seyd al Amys
All þat I haue seyd here I for-sake
зour swete fete now lete me kys.   185
          Mary

Nay lett be my fete not þo зe take
my mowthe зe may kys i-wys
and welcom on to me.
          Joseph

Gramercy myn owyn swete wyff
gramercy myn hert my love my lyff   190
xal I nevyr more make suche stryff
be-twyx me and þe.

¶ A mary mary wel þou be
and blyssyd be þe frewte in the
  goddys sone of myght   195
now good wyff fful of pyte
as be not evyl payd with me
  þow þat þou haue good ryght
As for my wronge in syght

Joseph's Return 115

```
    ,o wyte þe with ony synne                              200
       had þou not be A vertuous wythe
       god wold not A be þe with-inne.

¶ I knowlage I haue don A-mys
     I was never wurthy i-wys
           ffor to be þin husbonde                         205
     I xal amende Aftere thys
     ryght as þin owyn wyl is
           to serve þe at foot and honde
     and þi chylde bothe to vndyr-stonde
     to wurchep hym with good Affeccion                    210
     and þerfore telle me and nothynge whonde
     the holy matere of ȝour concepcion.
                                                  Maria
```

fo. 70ᵛ
```
¶ At ȝowre owyn wyll as ȝe bydde me
     ther cam An Aunge¹ hyght Gabryell
     and gret me ffayr and seyd Aue                        215
     And ferther more to me gan tell
     God xulde² be borne of my bode
     þe ffendys pouste ffor to ffelle
     þorwe þe holy gost as I wel se
     þus god in me · wyl byde and dwelle.                  220
                                                  Joseph

¶ Now I thank god with spech and spelle
     þat Euyr mary I was weddyd to the.
                                                  Mary
     it was þe werk of³ god as I ȝow telle
     now blyssyd be þat lord · so purveyd for me.
```
           Remainder of Fo. 70ᵛ—4¾ inches—left blank.

                                                  Maria
fo. 71⁴
BVtt husbond of oo thyng I pray ȝow most mekely   13
     I haue knowyng þat oure Cosyn Elizabeth with childe is
     þat it plese ȝow to go to here hastyly
     If owught we myth comforte here it wore to me blys.
                                                  Joseph
     A godys sake · is she with childe sche   5
     than wole here husbond zakarye be mery

---
¹ So in MS.; cf. Fo. 47. ² *xulde* is written twice in the MS.
³ *of* omitted and written above the line.
⁴ The writing on Ff. 71, 71ᵛ, and 72 is slightly less cramped.

The Visit to Elizabeth

In Montana they dwelle · fer hens so moty the
In þe cety of juda [1] I knowe it veryly
It is hens I trowe myles two and ffyfty
We are lyke to be wery or we come at þat same   10
I wole [2] with a good wyl · blyssyd wyff mary
now go we forthe than in goddys name.

                                           Maria

¶ Goth husbond þow it be to ȝow peyne
this jurny I pray ȝow · lete us go fast
ffor I am schamfast of þe pepyl to be seyne   15
and namely of men þer of I am A-gast
Pylgrymagys and helpyngys wolde be go in hast
þe more þe body is peynyd þe more is þe mede
Say ȝe ȝour devocionys and I xal myn I cast [3] (?)
now in þis jurny god mote us spede.   20

                                           Joseph

Amen Amen · and evyr more
lo wyff lo · how starkly I go be-fore.   *et sic transient circa placeam*.

                                           Comtemplacio [4]

Fo. 71ᵛ ¶ Sovereynes vndyrstondyth þat kynge davyd here
Ordeyned ffoure and twenty prestys of grett devocion
In þe temple of god · Aftere here let [5] apere   25
þei weryd clepyd summi sacerdotes · ffor here mynistracion
And on was prynce of prestys · havynge dominacyon
A-monge whiche was An old prest clepyd zakarye
and he had An old woman [6] to his wyff of holy conversacion
whiche hyth Elizabeth þat nevyr had childe verylye.   30

¶ In hese mynistracion the howre of incense
the Aungel Gabryel · Apperyd hym to
þat hese wyff xulde conseyve he ȝaff hym intelligence
hese juge · hese vnwurthynes and Age not be-levyd so
The plage of dompnesse · hise lippis [7] lappyd lo [8]   35

---

[1] A miswritten letter after *juda* crossed through.
[2] *wyl* first written and crossed through.
[3] Halliwell reads *reast* (?). There is a small stroke between the *j* and *cast* as if the scribe had begun to write *jn*. Cf. l. 148.
[4] So in MS.   [5] So in MS. for *lot*?
[6] *wyff* first written and crossed through.
[7] *lippis* omitted and written above the line.
[8] *to* first written for *lo* and crossed through.

thei wenten hom *and* his wyff was conseyvenge
this concep*ci*on gabryel¹ tolde our*e* lady to
and in soth sone After*e* · þat sage sche was sekynge
And of her*e* tweyners metyng
her*e* gynnyth þe proces   40
now god be our*e* be-gynnynge
and of my tonge I wole ses.
                                                    Joseph

¶ A · A · Wyff Infeyth I am wery
therfor*e* I wole sytt downe *and* rest me ryght her*e*
lo Wyff · here is þe hous of zakary   45
Wole ʒe I clepe Elyzabeth to ʒow to A-pere.
                                                    Maria

Fo. 72 Nay husbond And it plese ʒow I xal go ner
now þe blyssyd trynite · be in þis hous
A cosyn Elizabeth · swete modyr what cher
ʒe grow grett · A my god how ʒe be gracyous.   50
                                                    Elizabeth

¶ A-non as I herd of ʒow þis holy gretynge
mekest mayden *and* þe modyr of god mary
be ʒo*ur* breth þe holy gost vs was inspyrynge
þat þe childe in my body enjoyd gretly
And turnyd down on his knes · to our*e* god reverently   55
whom ʒe bere in ʒo*ur* body þis veryly I ken
ffulfyllyd wi*th* þe holy gost þus lowde I cry
blyssyd be þo*u* A-monge All women.

¶ And blyssyd be þe frute of þ*i* wombe also
þo*u* wurthyest virgyne *and* wyff þat evyr was wrought   60
how is it þat þe modyr of god me xulde come to
þat wrecche of all wrecchis · A whyght wers þan nought
And þo*u* art blyssyd þat be-levyd veryly in þ*i* thought
þat þe wurde of god xulde profyte in the
but how þis blyssydnes A-bought was brought   65
I can not thynk nyn say how it myght be.
                                                    Maria

¶ To þe preysynge of god cosyn this seyd mut be
whan I sat in my lytyl hous · on to god praynge
Gabryel come *and* seyde to me Ave

¹ The first syllable of *Gabryel* repeated before *tolde* and crossed through.

    ther I conceyvyd god · At my consentynge   70
Fo. 72ᵛ Parfyte god · and parfyte man · At onys beynge
    than þe Aungel seyd on to me
    þat it was sex monethys syn ȝour conseyvynge
    Þis cawsyth my comynge cosyn[1] ȝow to comforte and se.
                   Elizabeth

¶ Blyssyd be ȝe cosyn · ffor ȝour hedyr comynge   75
 How I conseyvyd I xal to ȝow say
 þe Aungel Apperyd · þe howre of[2] incensynge
 Seynge I xulde conseyve · and hym thought nay
 Sethe ffor his mystrost · he hath be dowm̅ Al-way
 and þus of my concepcion I haue tolde ȝow sum.   80
                   Maria

ffor þis holy psalme I be-gynne here þis day.
  M Agnificat,' anima mea dominum
    Et exultauit spiritus meus ? in deo salutari meo.
                   Elizabeth

Be þe holy gost with joye goddys son is in þe cum
þat þi spyryte so injouyid þe helth of þi god so.
                   Maria

Quia respexit humilitatem ancille sue
ecce enim ex hoc beatam me dicent omnes generaciones.
                   Elizabeth

Ffor he be-held þe lownes of hese hand-maydeȝe[3]
so ferforthe ffor þat · Aɫɫ generacionys blysse ȝow in pes.   85
                   Maria

Quia fecit mihi magna qui potens est
et sanctum nomen eius.
                   Elizabeth

Ffor grett thyngys he made and also myghtyest
And ryght holy is þe name of hym in vs.*
                   Maria

Fo. 73 Et misericordia eius a progenie in progenies
   timentibus eum.
                   Elizabeth

ȝa þe mercy of hym · fro þat kynde in to þe kynde of pes
Ffor aɫɫ þat hym drede · now is he cum.

---

[1] Some miswritten letter obliterated before *cosyn*.
[2] *of* omitted and written above the line.
[3] The *de* is in darker ink. It looks as if *may ȝe* was originally written.
* *Et* written at the bottom of the folio.

The Visit to Elizabeth

Maria

Fecit potenciam in brachio suo ⁖
disspersit superbos mente cordis sui.

Elizabeth

The pore in his ryght Arme · he hath mad so   90
þe prowde to dyspeyre · And þe thought of here hertys only.

Maria

Deposuit potentes de sede
et exaltauit humiles.

Elizabeth

The prowde men · fro hey setys put he
And þe lowly vpon heyth · in þe sete of pes.

Maria

Esurientes impleuit bonis ⁖
et diuites dimisit inanes.

Elizabeth

Alle þe pore and þe nedy · he fulfyllyth with his goodys
And þe Ryche · he fellyth to voydnes.   95

Maria

Suscepit israel puerum suum ⁖
recordatus est misericordie sue.

Elizabeth

Israel ffor his childe · vp-toke he to cum
On his mercy to thynk · ffor hese þat be.

Maria

Sicut locutus est ad patres nostros
abraham et semini eius in secula.

Elizabeth

As he spak here to oure forfaderys in clos
Abraham and to All hese sed of hym in þis werd sa.

Maria

Gloria patri et filio ⁖
et spiritui sancto.

Elizabeth

Preysyng be to þe fadyr in hevyn lo   100
þe same to þe son · here be so
       þe holy gost Also to ken ⁖

Maria

Sicut erat in principio et nunc et semper ⁖
et in secula seculorum amen.

Elizabeth
As it was in þe begynnynge · *and* now is · *and* xal be forevyr
and in this werd · in aȸ good werkys to abydyn then.

Maria
¶ This psalme of *pro*phesye seyd be-twen vs tweyn   105
In hefne it is wretyn *with* Aungellys hond
evyr to be songe · *and* Also to be seyn
Euery day Amonge us at our*e* eve song.

¶ but cosyn Elyzabeth I xal ȝow her*e* kepe
and þis thre monethis Abyde here now   110
tyl ȝe han childe to wasche skore *and* swepe
and in aȸ þat I may to comforte ȝow.

Elizabeth
A ȝe modyr of god · ȝe shewe us here how
we xulde be meke þat wrecchis her*e* be
Aȸ hefne *and* herthe wurcheppe ȝow mow   115
þat are trone *and* tabernakyl of þe hyȝ trinite.

Joseph
¶ A how do ȝe · how do ȝe · ffadyr zacharye
we ffalle ffast in Age *with*-owte oth
why shake ȝe so ȝo*ur* hed · haue ȝe þe palsye
Why[1] speke ȝe not sere I trowe ȝe ar*e* not wroth.   120

Elizabeth
Nay wys ffadyr joseph · þ*er* to he wer*e* ful loth
it is þe vesytac*i*on of god · he may not speke veryly
lete us thank god · þerffor both
he xal remedy it · whan it plesyth his mercy.[2]   *

Joseph
¶ Of ȝo*ur* dissese thynkys no greff[3]   125
thank god of al adu*er*syte
Ffor he wyl chastyse *and* repreff

---

[1] Some miswritten letter at the beginning of the line obliterated.
[2] The asterisk here indicates the following alternative continuation with *Contemplacio* as name of next speaker written in the bottom margin :

                 * his mercy
                   ⎧ come I pray ȝow specialy
    Si placet. ⎨ I-wys ȝe Are welcome mary   *Elizabeth*
                   ⎩ ffor þis comfortabelest comynge good god gramercy.
                                                     Contemplacio.

[3] From here to the end of the play, with the exception of the first eight lines of Contemplacio's speech, two lines are written as one divided by two parallel black strokes or the stop : and a red stroke.

The Visit to Elizabeth

  þo þat he lovyth most hertyle
  Mary I hold best þat we go hens
  we haue fer hom with-owt fayl.   130
               Maria

  Al redy husbond without defens
  I wyl werke be ȝour counsayl
  Cosyn be ȝour leve · and ȝour lycens
  for homward now us must travayl
  of þis refreschynge in ȝour presens   135
  god ȝeld ȝow þat most may Avayl.
             Elizabeth

Fo. 74[1] ¶ Now cosynes bothe god ȝow spede
  and wete ȝow wele with-owtyn mo
  ȝour presens comfortyth me in dede
  and þer fore now am I ryght wo   140
  That ȝe my ffrendys and my kynrede
  þus sone now xul parte me fro
  but I pray god he mote ȝow lede [2]
  in every place wher so ȝe go.

here mary and elizabet partyn and elizabeth goth to zakarie and seyth

  ¶ Good husbond ryse up I be-seke ȝow   145
  and go we to þe temple now fast
  to wurchep god with þat we mow
  and thank hym bothe this is my cast
  Of þe tyme þat is comynge now
  ffor now is cum mercy and venjauns is past   150
  God wyl be born for mannys prow
  to brynge us to blysse þat euer xal last.

             Contemplacio
  ¶ lystenyth sovereynys here is conclusyon
  how þe Aue was mad · here is lernyd vs
  þe Aungel seyd · Ave gratia plena dominus tecum ·
  benedicta tu in mulieribus ·
  Elizabeth seyd · et benedictus ·   5
  fructus uentris tui· thus þe chirch addyd Maria And Jhesus· her

---

[1] The writing becomes more cramped and the text extends into the right-hand margin on this folio.
[2] *spede* first written.

who seyth oure ladyes sawtere dayly ? ffor A ȝer þus
he hath pardon · ten thousand And eyte hundryd ȝer.

¶ Than ferther to oure matere for to procede
   Mary with elizabeth abod þer stylle   10
   'iij monthys fully as we rede
   thankynge god with hertly wylle
ɑ ¹ A lord god what hous was þis on
   þat þese childeryn and here moderys to
   as mary and elizabeth jhesus and john   15
   and joseph and zakarye Also.

¶ And evyr oure lady a-bod stylle þus
   tyl johan was of his modyr born
   and þan zakarye spak i-wus
   þat had be dowm and his spech lorn   20
   he and Elizabeth prophesyed as þus
   they mad Benedictus · them be-forn
   and so Magnificat · And · Benedictus
   ffyrst in þat place þer made worn.

¶ Whan all was don oure lady fre   25
   toke here leve than aftere this
   At Elizabeth and at zakarie
   And kyssyd johan and gan hym blys
   Now most mekely we thank ȝou of ȝour pacyens
   and beseke ȝou of ȝour good supportacion   30
   If here hath be seyd ore don Any inconuenyens
   we Asygne it to ȝour good deliberacion
   Be-sekynge to crystys precious passyon
   conserve and rewarde ȝour hedyr comynge
   with Aue we be-gunne · and Aue is oure conclusyon   35
     Ave regina celorum · to oure lady we synge.

---

¹ This red ink sign is otherwise only found in this MS. (in larger form) in the second Passion Play and in the Assumption Play, where it is used to mark couplet rhymes. In black it is used in both Passion Plays and occasionally elsewhere to mark stage directions.

¶ A-voyd Ser*ys* · And lete my lorde þe buschop come
And syt in þe courte · þe lawes ffor to doo
And I xal gon in þis place · them for to somowñe
tho þat ben in my book · þe court ȝe must com too
I Warne ȝow here all abowte 5
þat I somown ȝow all þe rowte
loke ȝe fayl for no dowte
   at þe court to per*e*
bot**h** Johan Jurdoñ *and* Geffrey Gyle
Malkyn mylkedoke *and* fayr mabyle 10
Stevyn sturdy *and* Jak at þe style
   and sawdyr sadeler*e*.

¶ Thom tynker*e and* betrys belle
peyrs potter*e and* whatt at þe welle
Symme Smalfeyth *and* kate kelle 15
   and bertylmew þe bocher*e*
kytt cakelere *and* colett crane
gylle fetyse *and* fayr jane
powle pewter*ere and* pernel prane
   and phelypp þe good flecchere. 20

¶ Cok crane *and* davy drydust
Luce lyer*e and* letyce lytyl trust
Miles þe myller*e and* colle Crake crust
   bothe bette þe baker*e and* Robyn rede [1]
And loke ȝe rynge wele in ȝo*ur* purs 25
Ffor ellys ȝo*ur* cawse may spede þe wurs
þow þ*at* ȝe slynge goddys curs
   Evy*n* at my*n* hede Ffast com A-way [2]
Bothe boutyng þe browster*e and* sybyly slynge
Megge mery wedyr *and* sabyn [3] sprynge 30
Tyffany Twynkeler*e* ffayle ffor no thynge
The courte xal be þis day. [4]

---

[1] The words 'and Robyn rede' are enclosed in a red loop as if the name of a speaker.
[2] The words *Ffast com A-way* are similarly enclosed in a red loop.
[3] *spy* miswritten before *sprynge* and crossed through.
[4] The following note is written at the foot of the page in ink of the same

Fo. 75 *hic intrabit pagetum de purgacione Marie et joseph · hic dicit primus detractor*

A A · serys god saue ʒow alt
here is a fayr pepyl in good ffay
Good ser*ys* telle me what me*n* me calle
I trowe ʒe kan not be þis day
ʒitt I walke wyde *and* many way 5
but ʒet þ*er* I come I do no good
to reyse slawdyr is al my lay
bakbytere is my brother of blood.

¶ Dede he ought come hedyr in al þis day
now wolde god þ*at* he wore here 10
and be my trewtħ I dare wel say
þat Yf we tweyn to-gedyr a-pe*r*e
More slawndyr we to xal a-rere
wit*h*-in an howre thorwe-outh this town
than evyr þ*er* was þis thowsand ʒere 15
and ellys I¹ shrewe ʒow bothe vp *and* dowɲ.

¶ Now be my trewtħ I haue a syght
Evyn of my brother lo where he is
Welcom dere brother my trowtħ I plyght
ʒowre jentyl mowth let me now kys. 20

Secundus detractor

Gramercy brother so haue I blys
I am ful glad we met þis day.

1ᵘˢ *detractor*

Ryght so am I brothyr i-wys
mecħ gladdere than I kan say.

¶ but ʒitt good brother I ʒow pray 25
telle alt þese pepyl what is ʒo*ur* name.
Ffor yf þei Knew it my lyf I lay
they wole ʒow wurchep *and* speke gret fame.

colour and by the same or a contemporary hand, in the same script as the note on the five Annes, Fo. 37ᵛ, but neater and not rubricated:
   a. 14. kal. ap*r*ilis · Translat*io* Sa*n*c*t*e Marie Magdalene · Et s*a*n*ct*i Joseph' sponsi d*ei* genit*r*icis Marie.
   e. x. kal. ap*rilis* Adam creatus est.

¹ *s* converted to *I*.

ij̄ᵘˢ detractor

I am bakbytere þat spyllyth all game
bothe kyd *and* knowyn in many a place.   30

1ᵘˢ detractor

be my trowth I seyd þe same
and ȝet sum seyden þou xulde haue evyl grace.

ij̄ᵘˢ detract[or

¶ Herk reyse sclaundyr canst þou owth telle
of Any newe thynge þat wrought was late.

1ᵘˢ detract[or

With-in a short whyle a thynge be-felle   35
I trowe þou wylt lawhȝ ryght wel þer Ate
Ffor be trowth ryght mekyl hate
If it be wyst þer of wyl growe.

ij̄ᵘˢ detractor

If I may reyse þer with de-bate
I xal not spare þe seyd to sowe.   40

1ᵘˢ detractor

¶ Syr in þe tempyl a mayd þer was
calde mayd mary þe¹ trewth to tell
Sche semyd so holy withinne þat plas
men seyd sche was ffedde with holy Aungell
Sche made A vow with man nevyr to melle   45
but to leve chast *and* clene virgine
How evyr it be · here wombe doth swelle
and is as gret as þinne or myne.

ij̄ᵘˢ detractor

¶ ȝa þat old shrewe joseph my trowth I plyght
was so Anameryd upon þat mayd   50
þat of hyre bewte whan he had syght
He sesyd nat tyll had here a-sayd.

1ᵘˢ detractor

A nay nay wel wers she hath hym payd
Sum fresch ȝonge galaunt she loveth wel more
þat his leggys to here hath leyd   55
and þat doth greve þe old man sore.

ij̄ᵘˢ detractor

¶ be my trewth al may wel be
ffor fresch *and* fayr she is to syght

¹ Some miswritten letter before *þe* crossed through.

And such a mursel as semyth me
Wolde cause A ȝonge man to haue delyght.   60

            1us detractor

Such a ȝonge damesel of bewte bryght
And of schap so comely Also
Of hire tayle ofte tyme be lyght
and rygh¹ tekyl vndyr þe too.

            ij^{us} detractor

¶ that olde cokolde was evyl be-gylyd   65
to þat fresche wench whan he was wedde
now muste he faderyn A-nothyr mannys chylde
and with his swynke he xal be fedde.

            1us detractor

* A ȝonge man may do more chere in bedde
to A ȝonge wench þan may An olde   70
þat is þe cawse such lawe is ledde
þat many a man is a kokewolde.

*hic sedet episcopus Abiȝachar inter duos legis doctores et audientes
hanc de-famocionem vocat ad se detractores dicens*

            Episcopus

¶ Herke ȝe felawys why speke ȝe such schame
of þat good virgyn ffayr mayd mary
ȝe be a-cursyd · so hire for to defame   75
She þat is of lyff so good and holy
Of hire to speke suche velany
ȝe make myn hert ful hevy of mood
Fo. 76ᵛ I charge ȝow sese of ȝoure fals cry
Ffor sche is sybbe of myn owyn blood.   80

            ij^{us} detractor

¶ Syb of þi kyn þow þat she be
all gret with chylde hire wombe doth swelle
Do calle here hedyr þi-self xal se
þat it is trewthe þat I þe telle.

            1us detractor

Sere ffor ȝour sake I xal kepe cowncelle   85
ȝow for to greve I am ryght loth
But lest syrys lyst what seyth þe belle
Oure fayr mayd now gret with childe goth.

---

¹ So in MS. * A paragraph mark has been erased here.

The Trial of Joseph and Mary

         primus doctor legis

¶ Take good heed serys what ȝe doth say
A-vyse ȝow wele what ȝe present   90
ȝyf þis be fownd fals a-nothyr day
Ful sore ȝe xal ȝour tale repent.

         ij^{us} detractor

 Sere þe mayd for sothe is good and gent
bothe comely and gay and a fayr wench
And feetly with help sche can consent   95
to set A cokewolde on þe hye benche.

         ij^{us} doctor legis

¶ ȝe be to besy of ȝour langage
I hope to god ȝow fals to preve
It were gret rewthe she xulde so outrage
or with such synne to myscheve.   100

         Episcopus

 This evy talys my hert doth greve
of hire to here such fowle dalyawnce
If she be fowndyn in suche repreve
she xal sore rew here governawns.

¶ Sym̄ somnore in hast wend þou þi way   105
byd Joseph and his wyff be name
At þe coorte to Appere þis day
here hem to pourge of here defame
Sey þat I here of hem grett schame
and þat doth me gret hevynes   110
if þei be clene with-owtyn blame
byd hem come hedyr and shew wyttnes.

         Den̄

¶ Aƚƚ redy sere I xal hem calle
here at ȝour courte for to appere
And yf I may hem mete with aƚƚ   115
I hope ryght sone þei xal ben here
A-wey serys lete me com nere
A man of wurchep here comyth to place
of curtesy me semyth ȝe be to lere
Do of ȝour hodys with an evyl grace.   120

¶ Do me sum wurchep be-for my face
or be my trowth I xal ȝow make
If þat I rolle ȝow up in my race
Ffor fere I xal do ȝour ars qwake

But ȝit sum mede *and* ȝe me take 125
I wyl wi*th*-drawe my gret rough toth
gold or sylvyr I wol not for-sake
but evyn as all somnor*ys* doth.

¶ A Joseph good day wi*th* þi ffayr spowse
my lorde þe buschop hath for ȝow sent 130
it is hym tolde þat in þi*n* house
A cuckolde is bowe is ech nyght bent
he þat shett þe bolt · is lyke to be schent
Ffayr*e* mayde þat tale ȝe kan best telle
now be ȝour*e* trowth telle ȝo*ur* entent 135
dede not þe Archere plese ȝow ryght well.

Maria

¶ Of god in hevyn I take wyttnes
þat synful werk was nevyr my thought¹
I am a mayd ȝit of pure clennes
lyke as I was in to þis werd brought. 140

Den

Othyr wyttnes xal non be sought
þou art wi*th* childe eche man may se
I charge ȝow bothe ȝe tary² nought
but to þe buschop com forth wi*th* me.

Joseph

¶ To þe buschop wi*th* ȝow we wende 145
of oure purgac*i*on hawe we no dowth.

Maria

Almyghty god xal be oure frende
Whan þe trewthe is tryed owth.

Den

ȝa on þis wyse excusyth here every scowte
Whan here owyn synne hem doth defame 150
but lowly þan þei gyn to lowth
Whan þei be gylty *and* fowndyn in blame.

¶ Ther-fo*re* com forth cokewolde be name
þe busschop xal ȝo*ur* lyff appose
Com forth Also ȝe goodly dame 155
A clene huswyff as I suppose
I xal ȝow tellyn wi*th*-owtyn glose.

---

¹ *werk* first written and crossed through.
² *t* written over some other letter.

and ȝe were myn with owtyn lak
I wolde ech day be-schrewe ȝour nose
and ȝe dede brynge me such a pak. 160

¶ My lord þe buschop here haue I brought
þis goodly copyl at ȝour byddyng
and as me semyth as be here fraught
Ffayr chylde lullay sone must she syng.

         1us detractor

Fo. 78 To here a credyl and ȝe wolde brynge 165
ȝe myght saue mony in here purse
be-cawse she is ȝour cosyn ȝynge
I pray ȝow sere lete here nevyr fare þe wers.

         Episcopus

¶ Alas mary what hast þou wrought
I am a schamyd evyn for þi sake 170
how hast þou chaungyd þin holy thought
dude old joseph with strenght þe take
Or hast þou chosyn a-nother make
by whom þou art þus brought in schame
telle me who hath wrought þis wrake 175
how hast þou lost þin holy name.

         Maria

¶ My name I hope is saff and sownde
god to wyttnes I am a mayd
of ffleschly lust and gostly wownde
In dede · nere thought · I nevyr a-sayd. 180

         1us doctor legis

how xulde þi wombe þus be arayd
So grettly swollyn as þat it is
but if sum man þe had ovyr-layd
þi wombe xulde never be so gret i-wys.

         ij^{us} doctor legis

¶ Herke þou joseph I am afrayd 185
þat þou hast wrought þis opyn synne
þis woman þou hast þus be-trayd
with gret flaterynge or sum fals gynne.

         ij^{us} detractor

Now be myn trowth ȝe hytte þe pynne
with þat purpose in feyth I holde 190

   telle now how þou þus hire dudyst wynne
   Ore knowlych þi self ffor a cockewold.

              Joseph

Fo. 78v [1] ¶ Sche is for me a trewe clene mayde
   And I for hire am clene Also
   of ffleschly synne I nevyr a-sayde   195
   Sythyn þat sch was weddyd me to.

              Episcopus

   Thu xalt not schape from vs ȝitt so
   Ffyrst þou xalte tellyn us a-nother lay
   Streyt to þe Awter þou xalt go
   þe drynge of vengeawns þer to a-say.   200

  ¶ here is þe botel of goddys vengeauns
   this drynk xal be now þi purgacion
   þis [hath] [2] suche vertu by goddys ordenauns
   þat what man drynk of þis potacion
   And goth [3] serteyn in processyon   205
   here in þis place þis Awtere abowth
   If he be gylty sum maculacion
   Pleyn in his face xal shewe it owth.

  ¶ Iff þou be gylty telle us lete se
   Ouer godys myght be not to bolde   210
   If þou presume [4] *and* gylty be
   god þou dost greve many afolde.

              Joseph

   I am not gylty as I fyrst tolde
   All myghty god I take wytnes.

              Episcopus

   than þis drynke in hast þou holde   215
   and on processyon anon þe dresse.

  hic joseph bibit et sepcies circuiuit altare dicens   *Joseph*

  ¶ This drynk I take *with* meke entent
   as I am gyltles to god I pray

Fo. 79 Lord as þou art omnypotente
   on me þou shewe þe trowth þis day. *modo bibit.* 220
   About þis Awtere I take þe way

---

[1] The writing on this page is freer and less neat.
[2] The word *hath* is written above the line in another ink—but probably by the scribe.
[3] MS. *Angoth* with *d* written above the line.
[4] *presue* first written and crossed through.

The Trial of Joseph and Mary

  O gracyous god help þi servaunt
  As I am gyltles A-ȝen ȝon may
  þin hand of mercy þis tyme me graunt.
            Den

¶ This olde shrewe may not wele gon   225
  longe he taryeth to go A-bowth
  lyfte up þi feet sett forth þi ton
  or be my trewth þou getyst a clowte.
           ij*us* detractor
  now sere evyl Thedom com to þi snowte
  What heylyght þi leggys now to be lame   230
  þou dedyst hem put ryght freschly owte
  Whan þou dedyst pley with ȝon ȝonge dame.
           1 detractor
¶ I pray to god gyf hym myschawns
  hese leggys¹ here do folde for Age
  but with þis damysel whan he dede dawns   235
  þe olde charle had ryght gret corage.
            Den

  The shrewe was þan sett in a dotage
  and had good lust þat tyme to pleyn
  ȝaff sche not ȝow cawdel to potage
  whan ȝe had don to comforte ȝour brayn.   240
           Joseph
¶ A gracyous god help me þis tyde
  ageyn þis pepyl þat me doth fame
  as I nevyr more dede towch here syde
  þis day help me fro werdly schame
  A-bowte þis awtere to kepe my fame   245
  vij tymes I haue gon rownd abowte
  If I be wurthy to suffyr blame
  O ryghtful god my synne shewe² owughte.
           Episcopus
¶ Joseph with hert thank god þi lorde
  Whos heyȝ mercy doth the excuse   250
  Ffor þi purgacion we xal recorde
  With hyre of synne þou dedyst never muse
  But Mary þi-self mayst not refuse
  All grett with chylde we se þe stonde

---

¹ The *y* is written over an original *e* (in the same ink).
² *sw* miswritten before *shewe*.

What mystyr man dede þe mys-vse 255
Why hast þou synned Ageyn þin husbonde.

Maria

¶ I trespacyd nevyr with erthely wyght
Þer of I hope þurowe goddys sonde
Here to be purgyd be-fore ȝour syght
Ffrom all synne clene · lyke as myn husbonde 260
Take me þe botel out of ȝour honde
Here xal I drynke be-forn ȝour face
A-bowth þis Awtere than xal I fonde
Vij tymes to go by godys grace.

1us doctor legis

¶ Se þis bolde bysmare wolde presume 265
Ageyn god to preve his myght
þow goddys vengeauns hyre xuld consume [1]
Sche wyl not telle hyre fals delyght
þou art with chylde we se in syght
to us þi wombe þe doth accuse 270
Þer was nevyr woman ȝitt in such plyght
þat ffrom mankynde hyre kowde excuse.

1us detractor

¶ in Ffeyth I suppose þat þis woman slepte
Fo. 80   With-owtyn all coverte whyll þat it dede snowe
And a flake þer of in to hyre mowthe crepte 275
and þer of þe chylde in hyre wombe doth growe.

ij^{us} detractor

Than be-ware dame for this is wel i-knowe
whan it is born yf þat þe sunne shyne
it wyl turne to watyr ageyn as I trowe
ffor snow on to watyr doth evyr more reclyne [2]. 280

ij^{us} doctor legis

¶ with goddys hyȝ myght loke þou not jape
of þi purgacion wel þe Avyse
yf þou be gylty þou mayst not schape
be-ware evyr of god þat ryghtful justyce
if god with vengeauns set on þe his syse 285
not only þou but all þi kyn is schamyd
bettyr it is to telle þe trewth devyse

---

[1] Some word—possibly *persume*—first written and crossed through.
[2] *ren* first written and crossed through.

than god for to greve and of hym be gramyd.

Maria

¶ I trostyn in his grace I xal hym nevyr greve
his servaunt I am in worde dede and thought 290
A mayd vndefyled I hope he xal me preve
I pray ʒow lett me nought.

Episcopus

Now be þat good lord þat all þis werd hath wrought
If god on þe shewe ony manyr tokyn
purgacion I trowe was nevyr so dere bowth 295
If I may on the in Any wyse be wrokyn.

¶ holde here þe botel and take a large draught
and abowth the Awtere go þi processyon.

Maria

to god in þis case my cawse I haue be-taught
lorde thorwe þin helpe I drynke of þis potacyon. 300

hic beata uirgo bibit de potacione et postea circuiuit altare dicens

Maria

God as I nevyr knew of mannys maculacion
but evyr haue lyued in trewe virginite
send me þis day þin holy consolacion
þat all þis fayr peple my clennes may se.

¶ O gracyous god as þou hast chose me 305
ffor to be þi modyr of me to be born
saue þi tabernacle þat clene is kepte for þe
which now am put at repref and skorn
Gabryel me tolde with wordys he be-forn
þat ʒe of ʒour goodnes wold be-come my chylde 310
help now of ʒour hyʒness my wurchep be not lorn
A dere sone I pray ʒow help ʒour modyr mylde.

Episcopus

¶ Almyghty god what may þis mene
Ffor all þe drynke of goddys potacyon
þis woman with chylde is fayr and clene 315
with-owtyn fowle spotte or maculacion
I can nat be non ymagynacion
Preve hyre gylty and synful of lyff
it shewith opynly by here purgacion
Sche is clene mayde bothe modyr and wyff. 320

                                                                1ᵘˢ detracto[r
¶ Be my fadyr sowle here is gret gyle
  be-cawse sche is syb of ȝour kynreed
  þe drynk is chaungyd by sum fals wyle
  þat sche no shame xuld haue þis steed.
                                                                Episcopus
  Be-cawse þou demyst þat we do falshede   325
  and for þou dedyst hem fyrst defame
  þou xalt ryght here magre þin heed
  beforn all þis pepyl drynk of þe same.
                                                                1ᵘˢ detractor
Fo. 81 ¶ Syr in good ffeyth oo draught I pulle
  If these to drynkerys haue not All spent.   330

hic bibit et scenciens dolorem in capite cadit et dicit

  Out out Alas what heylith my sculle
  A myn heed with ffyre me thynkyht is brent
  Mercy good mary I do me repent
  of my cursyd and ffals langage [1].
                                                                Maria
  Now god lord in hevyn omnypotent   335
  of his gret mercy ȝour seknes aswage.
                                                                Episcopus
¶ We all on knes fall here on grownd
  þou goddys hande-mayd prayng for grace
  all cursyd langage and schame on sownd [2]
  good mary ffor-ȝeve us here in þis place.   340
                                                                Maria
  Now god for-ȝeve ȝow all ȝowre trespace
  and also for-ȝeve ȝow all defamacion
  þat ȝe haue sayd both more and lesse
  to myn hynderawnce and maculacion.
                                                                Episcopus
¶ Now blyssyd virgyne we thank ȝow alle   345
  of ȝoure good hert and gret pacyens
  we wyl go with ȝow hom to ȝour halle
  to do ȝow servys with hyȝ reverens
  I thank ȝow hertyly of ȝoure benevolens   Maria
  On to ȝour owyn hous I pray ȝow ȝe goo   350

[1] A thick line has been drawn over the original line dividing this speech and the next.
[2] The *w* has been written by the scribe over another letter—possibly a *u*.

The Trial of Joseph and Mary

and take þis pepyl hom with ȝow hens
I am not dysposyd to passyn hens froo.

Episcopus

* Than ffare wel mayden and pure virgyne
fare wel trewe hand-mayd of god in blys
we aH to ȝow lowly inclyne   355
and take oure leve of ȝow as wurthy is.

Maria

o. 81ᵛ   AH myghty god ȝour weys wys se¹
Ffor þat hyȝ lord is most of myght
he mote ȝow spede þat ȝe not mys
In hevyn of hym to haue A syght.   360

Joseph

¶ Honouryd in hevyn be þat hyȝ lord
whos endles grace is so habundaunt
þat he doth shewe þe trewe recorde
of iche wyhgte þat is his trewe servaunt
That lord to wurchepe with hert plesaunt   365
we bothe be bownd ryght on þis place
which oure purgacyon us dyde graunt
and prevyd us² pure by hieȝ grace.

Maria

¶ Ffor sothe good spowse I thank hym hyȝly
Of his good grace for oure purgacion   370
oure clennesse is knowyn ful opynly
be vertu of his grett consolacion

      Remainder of Fo. 81ᵛ—4⅞ inches—left blank.

exxplicit cum gaudio amen.³

Fo. 82

¶ Joseph

L Ord What travayl to man is Wrought
  Rest in þis werd be-hovyth hym non
Octauyan oure Emperour sadly hath be-sought
Oure trybute hym to bere · ffolk must forth ichon
It is cryed in every bourgh and cety be name   5

---

* No paragraph sign here in MS.
¹ This *se* stands after *wys* (apparently in the same ink and the same hand) as a separate word.
² Some miswritten word or a second *pure* between *us* and *pure* crossed through.
³ Scribbled in a later hand enclosed in rudely scribbled flourishes.

I þat am a pore tymbre wryth · born of þe blood of dauyd
þe Emperorys comawndement I must holde with
And ellys I were to blame.

¶ Now my wyff mary · what sey ȝe to this
For sekyr nedys I must fforth wende 10
On to þe cyte of bedleem ffer hens i-wys
þus to labore I must my body bende.

Maria

Myn husbond and my spowse with ȝow wyl I wende
A syght of þat cyte ffayn wolde I se
If I myght of myn Alye ony þer ffynde 15
It wolde be grett joye on to me.

Joseph

¶ My Spowse ȝe be with childe . I fere ȝow to kary
Ffor me semyth it were werkys wylde
but ȝow to plese ryght ffayn wold I
ȝitt women ben ethe to greve · whan þei be with childe 20
now latt us Fforth wende as ffast as we may
and al-myghty god spede us in oure jurnay.

Maria

¶ A my swete husbond · wolde ȝe telle to me
What tre is ȝon standynge vpon ȝon hylle.

Joseph

Ffor sothe mary it is clepyd A chery tre 25
In tyme of ȝere · ȝe myght ffede ȝow þeron ȝour Ffylle.

Maria

¶ Turne A-geyn husbond and beholde ȝon tre
how þat it blomyght now so swetly.

Joseph

Fo. 82ᵛ Cum on Mary þat we worn At ȝon Cyte
or ellys we may be blamyd I telle ȝow lythly. 30

Maria

¶ Now my spowse I pray ȝow to be-hold
How þe cheryes growyn vpon ȝon tre
Ffor to haue þer-of ryght ffayn I wold
and it plesyd ȝow to labore so mech for me.

Joseph

¶ ȝour desyre to ffulfylle I xal Assay sekyrly 35
Ow · to plucke ȝow of these cheries · it is a werk wylde
Ffor þe tre is sò hyȝ · it wol not be lyghtly
þerfore lete hym pluk ȝow cheryes · be-gatt ȝow with childe.

The Birth of Christ

               Maria
¶ Now good lord I pray þe · graunt me þis boun
 to haue of þese cheries *and* it be ȝo*ur* wylle   40
 now I thank it god · þis tre bowyth to me dowñ
 I may now gaderyn A-nowe · *and* etyn my ffylle.
               joseph
¶ Ow · I know weyl I haue offendyd my god i*n* trinyte
 Spekyng to my spowse these vnkynde wurdys
 Ffor now I beleue wel it may non other be   45
 but þat my spowse beryght þe kyngys son of blys
 He help us now at oure nede
 of þe kynrede of jesse · worthely were ȝe bore
 Kyng*ys and* patryarkys ȝow be-ffore
 aH þese wurthy · of ȝo*ur* kynred · wore   50
 as clerkys in story rede.
               Maria
¶ Now gramercy husbond for ȝo*ur* report
 in oure weys wysely late us forth wende
 þe fadyr aH myghty he be oure comfort
 þe holy.gost gloryous · he be oure frende.   55
               Joseph
Fo. 83 ¶ Heyl wurchepful sere *and* good day
 A ceteceyn of þis cyte · ȝe seme to be
 of herborwe ffor spowse *and* me I ȝow p*r*ay
 Ffor trewly þis woman is fful were
 And fayn At reste sere wold she be   60
 We wolde ffulffylle þe byddynge of oure emp*er*o*ur*
 Ffor to pay trybute As ryght is oure
 and to kepe oure self ffrom dolowre
 We are come to þis cyte.
               Ciues
¶ Sere ostage in þis town know I non   65
 þin Wyff *and* þo*u* in for to slepe
 this cete · is be-sett · wi*th* pepyl every won
 And ȝett þei ly wi*th*-owte fful every strete.

¶ Wi*th*-inne no waH man comyst þou nowth
 Be þou onys wi*th*-inne þe cyte gate   70
 on-ethys in þe strete A place may be sowth
 þ*er* on to rest wi*th*-owte debate.

Joseph

¶ Nay sere debate · þat wyl I nowth
all such thyngys passyn my powere
but ȝitt my care and all my thought   75
is for Mary my derlynge dere.

¶ A Swete Wyff · what xal we do
wher xal we logge þis nyght
on to þe ffadyr of heffne pray we so
vs to kepe ffrom every wykkyd whyt.   80

Ciues

¶ Good man O word I wyl þe sey
If þou wylt do by þe counsel of me
ȝondyr is An hous of haras þat stant be þe wey
Amonge þe bestys herboryd may ȝe be.

Maria

Fo. 83ᵛ ¶ Now þe fadyr of hefne he mut ȝow ȝelde   85
His sone in my wombe forsothe he is
He kepe þe and þi good be fryth and ffelde
go we hens husbond for now tyme it is.

* But herk now good husbond a new relacyon[1]
which in my self I knew ryght well   90
Cryst in me hath take in-carnacion
Sone wele be borne þe trowth I fele.

¶ In þis pore logge my chawmere I take
here for to A-byde þe blyssyd byrth
of hym þat all þis werd dude make   95
be-twyn myn sydys I fele he styrth.

Joseph

¶ god be þin help spowse it swemyth me sore
þus febyly loggyd and in so pore degre
goddys sone amonge bestys to be bore
his woundyr werkys ffulfyllyd must be.   100

¶ In An hous þat is desolat with-owty Any wall
Ffyer nor wood non here is.

Maria

Joseph myn husbond a-bydyn here I xal
ffor here wyl be born þe Kyngys sone of blys.

---

* No paragraph sign in the MS. here.
[1] *ral* first written (for *rel*) and crossed through.

The Birth of Christ

Joseph

¶ Now jentyll wyff be of good myrth
and if ȝe wyl owght haue telle me what ȝe thynk
I xal not spare for schep nor derth¹
now telle me ȝour lust of mete and drynk.

Maria

¶ Ffor mete and drynk lust I ryght nowth
all-myghty god my fode xal be
now þat I am in chawmere brought
I hope ryght well my chylde to se
Therfore husbond of ȝour honeste
A-voyd ȝow hens out of þis place
And I a-lone with humylite
here xal abyde goddys hyȝ grace.

Joseph

¶ All redy wyff ȝow for to plese
I wyl go hens out of ȝour way
and seke sum mydwyuys ȝow for to ese
Whan þat ȝe trauayle of childe þis day
Ffare well trewe wyff and also clene may
God be ȝour comforte in trinyte.

Maria

to god in hevyn for ȝow I pray
He ȝow preserve wher so ȝe be.
hic dum joseph est Absens parit Maria filium vnigenitum.

Joseph

¶ Now god of whom comyth all releffe
And as all grace in þe is grownde
So saue my wyff from hurt and greffe
tyl I sum mydwyuys for here haue fownde
Travelynge women in care be bownde
with grete throwys whan þei do grone
god helpe my wyff þat sche not swownde
I am ful sory sche is a-lone.

¶ It is not conuenyent a man to be
þer women gon in travalynge
wher fore sum mydwyff fayn wold I se
my wyff to helpe þat is so ȝenge.

---

¹ *derke* first written and crossed through.

zelomy

¶ Why makyst þou man suche mornyng
 tell me sum dele of ȝour gret mone.

Joseph

 My wyf is now in gret longynge
 trauelyng of chylde and is a-lone   140
 Ffor godys loue þat sytt in trone
 as ȝe myd-wyuys þat kan ȝour good
 help my ȝonge spowse in hast a-none
 I drede me sore of þat fayr food.

Salome

 be of good chere and of glad mood   145
 we ij mydwyuys with þe wyll go
 þer was nevyr woman in such plyght stood
 but we were redy here help to do.

¶ My name is Salomee all men me knowe
 Ffor a mydwyff of wurthy fame   150
 Whan women travayl grace doth growe
 þer as I come I had nevyr shame.

zelomye

 And I am zelomye men knowe my name
 We tweyn with the wyl go togedyr
 and help þi wyff fro hurt and grame   155
 Com forth joseph go we streyth thedyr.

Joseph

¶ I thank ȝow damys · ȝe comforte my lyff
 streyte to my spowse walke we þe way
 In þis pore logge lyght mary my wyff
 hyre for to comforte gode frendys a-say.   160

Salome

 We dare not entre þis logge in fay
 þer is þer-in so gret bryghtnes
 mone be nyght nor sunne be day
 Shone nevyr so clere in þer lyghtnesse.

zelomye

¶ In to þis hous dare I not gon   165
 þe woundyrffull lyght doth me affray.

Joseph

 than wyl my-self gon in Alon
 and chere my wyff if þat I may

The Birth of Christ 141

    Aħ heyl maydon *and* wyff I say
    how dost þou fare telle me þi chere   170
    the for to comforte in gesyne þis day
    tweyn gode mydwyuis I haue brought here.

¶ The for to helpe þat art in harde bonde
    zelomye · And salomee be com̃ wit*h* me
    Ffor dowte of drede wit*h*-owte þei do stond   175
    and dare not come in for lyght þat they se.
                                            Maria

hic maria subridendo dicat · Maria.

    The myght of þe godhede in his mageste
    Wyl not be hyd now at þis whyle
    the chylde þat is born wyl preve his modyr fre
    A very clene mayde *and* þerfore I smyle.   180
                                          Joseph

¶ Why do ȝe lawghe[1] wyff ȝe be to blame
    I pray ȝow spowse do no more so
    In happ þe mydwyuys wyl take it to grame
    and at ȝo*ur* nede helpe wele non do *
    Iff ȝe haue nede of mydwyuys lo   185
    per auenture thei wyl gon hens
    þ*er*for be sad *and* ȝe may so
    And wynnyth aħ þe mydwyuis good diligens.
                                           Maria

Fo. 85ᵛ ¶ husbond I p*r*ay ȝow dysplese ȝow nowth
    þow þat I lawghe · *and* gret joye haue   190
    here is þe chylde þis werd hath wrought
    born now of me þat aħ thynge xal saue.
                                          Joseph

    I Aske ȝow grace for I dyde raue
    O gracyous childe I aske mercy
    As þou art lord *and* I but knaue   195
    Ffor-ȝeue me now my gret foly.

¶ Alas mydwyuis what haue I seyd
    I pray ȝow com to us more nere
    Ffor here I fynde my wyff a mayd
    and in here Arme a chyld hath here   200

      [1] A miswritten final letter (? *te* or *e*) deleted and *e* written above.
       * ȝelloñy roughly scribbled here in faint ink.

bothe mayd *and* modyr scħ is in ffer*e*
þat god wole haue · may nevyr mor*e* fayle
modyr on ertħ was nevyr non cler*e*
Wi*th*-owth sche had in byrtħ travayle.

zelomy

¶ In byrth trauayle muste sche nedys haue   205
or ellys no chylde of her*e* is born.

Joseph

I pray ȝow dame *and* ȝe vowcħ-saue
com se þ*e* chylde my wyff be-forn.

Salome

Grete god be in þis place
Swete systyr how far*e* ȝe.   210

Maria

I thank þ*e* fadyr of his hyȝ grace
his owyn son *and* my chylde her*e* ȝe may se.

zelomye

¶ Aħ heyl mary *and* ryght good morn̄
Who was mydwyfe of þis ffayr chyld.

Maria

he þat no thynge wyl haue for-lorn̄   215
Sent me þis babe *and* I mayd mylde.

zelomye

Fo. 86 ¶ Wi*th* honde lete me now towcħ *and* fele
(H quire)   yf ȝe haue nede of medycyn̄
I xal ȝow comforte *and* helpe ryght wele
As other women yf ȝe haue pyn̄.   220

Maria

Of þis fayr byrtħ þat her*e* is myn̄
Peyn*e* ner*e* grevynge fele I ryght non̄
I am clene mayde *and* pure virgyn̄
tast wi*th* ȝou*r* hand ȝo*ur*-self a-lon

hic palpat zelomye beatam mariam virginem dicens

zelomy

¶ O myghtfuħ god haue mercy on me   225
A merveyle þat nevyr was herd be-forn
Her*e* opynly I fele *and* se
A fayr chylde of a maydon is born̄
And nedyth no waschynge as other don̄
Fful clene *and* pur*e* for sotħ is he   230
wi*th*-outyn spot or ony polucyon̄
his modyr nott hurte of virgynite.

The Birth of Christ

¶ Coom nere gode systyr Salome
be-holde þe brestys of þis clene mayd
Fful of fayr mylke how þat þei be   235
And hyre chylde clene as I fyrst sayd
As other ben · nowth fowle arayd
but clene and pure bothe modyr and chylde
Of þis matyr I am dysmayd[1]
to se them both thus vndefyled.   240
          Salome

¶ It is not trewe it may nevyr be
þat bothe be clene I can not be-leve
a mayde mylke haue[2] · never man dyde se
ne woman bere chylde with-owte grett greve.

fo. 86v ¶*I xal nevyr trowe it but I it preve   245
With hand towchynge but I Assay
in my conscience it may nevyr cleue
þat sche hath chylde and is a may.
          Maria

¶ ȝow for to putt clene out of dowth
towch with ȝour hand and wele a-say   250
Wysely ransake and trye þe trewthe owth
Whethyr I be fowlyd or a clene may.[3]

hic tangit salomee marie et cum arescerit manus eius vlulando[4] et quasi flendo dicit
          Salomee

¶†Alas Alas and Wele a-Waye
Ffor my grett dowth and fals beleve
myne hand is ded and drye as claye   255
my fals vntrost hath wrought myscheve.

¶ Alas þe tyme þat I was born
thus to offende A-ȝens goddys myght
myn handys power is now all lorn

---

[1] *dyf* first written for *dys* and crossed through.
[2] The word *haue* is crossed through in darker ink and the contraction for *ys* has been added, also in darker ink and another hand, to *mayd*.
* This paragraph sign seems inserted by mistake.
[3] *ransak* roughly scribbled in the margin here.
[4] There is an erasure in this word; the second *l* is adapted from some other letter.
† This paragraph sign again seems inserted by mistake.

Styff as a stykke *and* may nowth plyght 260
Ffor I dede tempte þis mayde so bryght
And helde A-ȝens here pure clennes
In grett myscheff now am I pyght
Alas alas ffor my lewdnes.

¶ O lord of myght þou knowyst þe trowth 265
þat I haue evyr had dred of þe
on every power whyght evyr I haue rowthe
and ȝove hem almes for loue of þe
Bothe wyff *and* wedowe þat Askyght for the
And frendles chylderyn þat haddyn grett nede 270
Fo. 87   I dude them cure *and* all for the
and toke no rewarde of them nor mede.

¶ Now as a wrecch ffor fals be-leve
þ*at* I shewyd in temptynge þis mayde
my hand is ded *and* doth me greve 275
Alas þat evyr I here assayde.

          Angelus[1]

Woman þi sorwe to haue de-layde[2]
wurchep þat childe þat þer is born
towch þe clothis þer he is layde
ffor he xal saue all þat is lorn. 280
          Salomee

¶ O gloryous chylde *and* kynge of blysse
I aske ȝow mercy for my trespace
I knowlege my synne · I demyd a-mys
O blyssyd babe grawnt me su*m* grace
Of ȝow mayde Also here in þis place 285
I aske mercy knelynge on kne
moste holy mayde grawnt me solace
Su*m* wurde of comforte sey now to me.
          Maria

¶ As goddys Aungel to ȝow dede telle
my chyld is medycyn ffor every sor 290
towch his clothis be my cowncelle
ȝowre hand ful sone he wyl restor.

---

[1] From here to the end of this play the initial of the name of speaker is marked with a red stroke.

[2] *leyde* originally written, the *e* altered in slightly darker ink to *a*.

hic salomee tangit fimbriam Christi dicens   Salomee

    A now blyssyd be þis chylde euer more
    þe sone of god for sothe he is
    Hath helyd myn hand þat was for-lore   295
    thorwe ffals be-leve and demynge a-mys.

fo. 87ᵛ ¶ In every place I xal telle þis
    Of a clene mayd þat god is born
    And in oure lyknes god now clad is
    Mankend to saue þat was for-lorn   300
    His modyr a mayde as sche was be-forn
    natt fowle polutyd as other women be
    but fayr and fresch as rose on thorn
    Lely wyte · clene with pure virginyte.

  ¶ Of þis blyssyd babe my leve now do I take   305
    and also of ȝow · hyȝ modyr of blysse
    Of þis grett meracle more knowlege to make
    I xal go telle it in iche place i-wys.

                                             Maria

    Ffare wel good dame and god ȝour wey wysse
    In all ȝour jurnay god be ȝour spede   310
    and of his hyȝ mercy þat lord so ȝow blysse
    þat ȝe nevyr offende more in word thought nore dede.

                                             zelomy

  ¶ And I Also do take my leve here
    Of all þis blyssyd good company
    praynge ȝour grace bothe fere and nere   315
    On us to spede ȝour endles mercy.

                                             Joseph

  ¶ The blyssyng of þat lord þat is most myghty
    mote sprede on ȝow in every place
    of all ȝour enmyes to haue þe victory
    God þat best may grawnt ȝow his grace.   Amen.   320

                Fo. 88 left blank.

146 *The Adoration of the Shepherds*

Fo. 88ᵛ   *Angelus ad pastores dicit*   [Gloria in excelsis deo.¹]

                  Angelus

JOye to god þat sytt in hevyn̄
 And pes to man on erthe grownde
A chylde is born be-nethe þe levyn
thurwe hym many ffolke xul be vn-bownde   4
Sacramentys þer xul be vij
Wonnyn þurowe² þat childys wounde   16
Therfore I synge A joyful stevene
þe flowre of frenchep now is founde
God þat wonyght on hyʒ³
he is gloryed mannys gost to wynne   10
he hath sent salue to mannys synne
Pes is comyn to mannys kynne
thorwe goddys sleytys slyʒ⁴.

              1ᵘˢ pastor

¶ Maunfras maunfras felawe myne
 I saw a grett lyght with shene⁵ shyne   15
 ʒit saw I nevyr so selkowth⁶ syne
 Shapyn vpon þe skyes
 It is bryghtere þan þe sunne bem
 It comyth ryght ouer all þis rem
 Evyn above bedleem̄   20
 I saw it brenne thryes.

              ijᵘˢ pastor

¶ Thu art my brother boosras
 I haue beholdyn þe same pas
 I trowe it is⁷ tokenynge of gras

---

¹ The words *Gloria in excelsis deo* have been crossed through in a different ink.

² This word has been crossed out in darker ink, and *thorough* (?) or *therogh* (?)—there has been some double correction—written above in another hand.

³ An *e* has been added to *hyʒ* in different ink.

⁴ *sleytys* is partly erased, and over *slyʒ* is written *hye*, and the words *wysdam̄ I saye* are added in the writing of the scribe of the interpolated Ff. 95, 96, and 112.

⁵ *shene* crossed out and *bryght* written above in the hand of the same later scribe.

⁶ *mervelus a* written by the same hand over *selkowth*, which is barely decipherable.

⁷ Some miswritten letter after *is* crossed through by scribe.

Fo. 89   þat shynynge shewyght be-forn   25
          Balaam spak in prophesye
          A lyght xuld shyne vpon þe skye
          Whan A sone[1] of a mayd marye
          In bedleem were i-born.

                                                    iij⁽ᵘˢ⁾ pastor
     ¶ thow I make lyty noyse   30
       I am an herde þat hattyht moyse
       I herde carpynge of a croyse
       of Moyses in his lawe
       Of a mayd a barne born *
       On a tre he xulde be torn   35
       delyver folkes þat arn forlorn
       The chylde xulde be slawe.

                                                    1⁽ᵘˢ⁾ pastor
     ¶ Balaam spak in prophecie
       out of jacob xuld shyne a skye
       many ffolke he xulde bye   40
       with his bryght blood
       Be þat bryght blod þat he xulde blede
       he xal us brynge fro þe develys drede
       as a duke most dowty in dede
       thorwe his deth on rode.   45

                                                    ij⁽ᵘˢ⁾ pastor
     ¶ Amos spak with mylde meth
       A frute swettere than bawmys breth
       His deth · xulde slen oure sowlys deth
       And drawe us all from helle
       Ther fore such lyght goth be-forn   50
Fo. 89ᵛ In tokyn þat þe childe is born
          Whiche xal saue þat is for-lorn
          As prophetys gonne spelle.

---

[1] *sone* crossed out and *chyld* written above in the later hand.
* These lines have been altered by the second hand thus :
      thow I make lytyl noyse of this
      I am an herde mañ þat hattyht sayyng Amys
      I herde spekyng of a chyld of blys
      of Moyses in his lawe
      Of a mayd a child xuld be borne.

                                                    iij^us pastor
¶ Danyel þe prophete þus gan speke
   wyse god from woo us wreke   55
   þi bryght hevyn þou to-breke
   and medele þe with a mayde
   This prophecye is now spad
   Cryst in oure kend is clad
   þerfore mankend may be glad   60
   As prophetys be-forn han seyd.
   Gloria in excelsis deo *cantent.*

                                                    1us pastor
B¹ ¶ The prophecye of boosdras is spedly sped
   now leyke we hens as þat lyght us lede
   myght we se onys þat bryght on bed
   oure bale² it wolde vnbynde   65
   We xulde shadyr³ for no shoure
   buske⁴ we us hens to bedleem boure
   to se þat fayr fresch flowre
   the mayde mylde in mynde.

                                                    ij^us pastor
¶ Lete us ffolwe with all oure myght   70
   With songe and myrth we xul us dyght
   and wurchep with joye þat wurthy wyght
   þat lord is of mankynne
   Lete us go fforthe⁵ fast on hye   74
Fo. 90   And honowre þat babe wurthylye   *tunc pastores cantabunt · stella*
   with merthe songe and melodye   *celi extirpauit · quo facto ibunt*
   haue do þis songe be-gynne.   *ad querendum christum.*

                                                    1us pastor
A   Ey · Ey · þis was a wondyr note

---

¹ The marginal letters B, A, C indicate the order of the passages. They are slightly rubricated. It will be noted that a third shepherd's speech is missing here. The disturbance of the text is connected with the cancelling of the angels' song at the beginning of the play. It is not irrelevant that the humorous passage marked A closely resembles a passage in the Chester Shepherds' play. It would appear to be a borrowed interpolation replacing a speech from the third shepherd.

² *bale* crossed out and *sorow* written above by the later hand.

³ *shadyr* crossed out and *not let* written above by the later hand.

⁴ *buske* crossed out and *go* written against it in the margin by the later hand.

⁵ This *ff* is miswritten with three strokes.

The Adoration of the Shepherds 149

  þat was now songyn above þe sky
  I haue þat voys fful wele I wote   80
  þei songe gle glo glory.
            ij^{us} pastor

¶ Nay so moty the so was it nowth
  I haue þat songe fful wele i-nvm̃ [1]
  In my wytt weyl it is wrought
  It was gle glo glas glum.   85
            iij^{us} pastor

¶ The songe me thought it was glory
  and aftyr-warde he seyd us to
  þer is a chylde born xal [2] be a prynce myghty
  Ffor to seke þat chylde I rede we go.
            1us pastor

C ¶ Heyle floure of flourys fayrest i-fownde   90
  Heyle perle peerles prime rose of prise
  heyl blome on bedde we xul be vn-bownde
  with þi blody woundys and werkys full wyse
  heyl god grettest I grete þe on grownde
  þe gredy devyl xal grone grysly as a gryse   95
  whan þou wynnyst þis worlde with þi wyde wounde
  and puttyst man to paradys with plenty of prys
  to loue þe is my delyte
[fo. 90^v marked 90] Heyl floure fayr and fre
  Lyght from þe trynyte   100
  Heyl blyssyd mote þou be
    heyl mayden fayrest in syght.
            ij^{us} pastor

¶ Heyl floure ovyr fflowrys fowndyn in fryght [3]
  Heyl Cryst kynde in oure kyth
  Heyl werker of wele to wonyn us wyth   105
  Heyl wynnere i-wys
  Heyl fformere and ffrende
  Heyl ffellere of þe fende
  Heyl clad in oure kende
    heyl prince of paradys.   110

---

[1] The v has been written by the scribe (?) over some other letter—? o or a half-formed u.

[2] xal written twice in MS. crossed through in different ink.

[3] So in MS. for fryth.

                                                    iij^us pastor

¶ Heyl lord ou*er* lordys þat lyggyst ful lowe
  Heyl kynge ovyr kyng*ys* þ*i* kynrede to knowe
  Heyl comely knyth þe deuyl to ou*er* throwe
  Heyl flowr*e* of alle
  Heyl werker*e* to wynne   115
  bodyes bowndyn in synne
  Heyl in a bestys bynne
  Be-stad¹ in a stalle.

                                                    joseph
¶ Herdys on hylle ²
  beth not stylle   120
  but seyth ȝo*ur* wylle
  to many A man
  How god is born
  þis mery morn
  þat is for-lorn   125
  fyndyn he can.

                                                    1us pastor
¶ We xull telle
  be dale *and* hylle
  How harwer*e* of helle
  was born þis nyght   130
  myrthis to melle
  and fendys to quelle
  þat wer*e* so felle
  Aȝens his ryght.

                                                    ij^us pastor
Fo. 91 ¶ Ffare wel babe *and* barne of blys   135
  Ffare wel lord þat lovely is
  þe to wurchep þ*i* feet I kys
  on knes to þe I falle
  The to wurchepe I falle on kne
  all þis werd may joye of þe   140
  now fare wel lorde of grett pouste
  ȝa fare wel kynge of alle.

¹ This *d* is adapted from some miswritten letter.
² From line 119 to line 134 two lines are written as one divided by two black strokes and one red.

The Adoration of the Shepherds

iij^us pastor

¶ Thow I be þe last þat take my leve
ȝit fayre mullynge take it nat at no greve
now fayre babe wele mut þou cheve 145
Ffayr chylde now haue good day
Ffare weyl myn owyn dere derlyng
I-wys þou art a ryght fayr thyng
Ffare wel my lorde and my swetyng
Ffare wel born in pore Aray. 150

Maria

¶ Now ȝe herd-men wel mote ȝe be
Ffor ȝoure omage and ȝour syngynge
my sone xal a-qwyte ȝow in hefne se
and ȝeue ȝow all ryght good hendynge.

Amen.

Remainder of Fo. 91—3⅔ inches—left blank.

Fo. 91ᵛ blank except for the following scribblings in different hands of a later date than the MS. :
1. W. William Dere
   Polerd.
2. Wylliam Dere.
3. John Hasycham.
4. John Taylphott of Parish Bedonson
   wee that will not when we paie
   when we would we shall saie* nay.

¶ Herode[

AS a lord in ryalte in non Regyon[2] so ryche
And rulere of all remys I ryde in ryal a-ray
Ther is no lord of lond in lordchep to me lyche
non lofflyere[3] non lofsummere evyr lastyng is my lay.

*¹ Of bewte and of boldnes I bere ever-more þe belle
Of mayn and of myght I master every man
I dynge with my dowtynes þe devyl down to helle
Ffor bothe of hevyn and of herth I am kyng sertayn.

18

* Or *find.*
¹ Ff. 92 and 93 are narrower than the rest, more margin having been cut away. Half the number 18 has been thus clipped.
² *Regyon* omitted and written above the line.
³ The *ff* is written over some other letter.
*¹ No paragraph sign in MS. here.

¶ I am þe comelyeste kynge clad in gleterynge golde
ȝa′and þe semelyeste syre þat may be-stryde a stede 10
I welde att my wyłł ałł wyghtys upon molde.

* ¶ ȝa and wurthely I am wrappyd in a wurthy wede
ȝe knyghtys so comely bothe curteys and kene
to my paleys wyl I passe · fułł prest I ȝow plyth
ȝe dukys so dowty ffolwe me be-dene 15
On to my ryal paleys þe wey lyth ful ryght.

¶ Wyghtly fro my stede I skyppe dowñ in hast
to myn heyȝ hallys I haste me in my way
ȝe mynstrełł of myrth blowe up a good blast
Whyłł I go to chawmere and chaunge myn array. 20

1us. rex

¶ Heyl be ȝe kyngys tweyne
Fferre rydyng out of ȝour regne
me thynkyth be ȝour presentys seyne
    ȝe sekyn oure sauyour
Ffro Saba haue I folwyd fferre 25
the glemynge of ȝon gay sterre
a chyldys blood xal bye vs dere
    þat þer is born in bestys boure.

¶ My name is kynge Baltazare
Fo. 92^v Of prophetys speche I am ware 30
Therfore a ferre wey I fare
    A maydenys childe to seche
Ffor he made man of þe moolde
and is kynge of hevyn holde
I wyl hym offere þe rede golde 35
    as reson wyl me teche.

Secundus rex

¶ Melchizar þat my name is kydde
in hote loue myn hert is hydde
to þe blosme upon his bedde
born by bestys bynne 40
in tarys I am kynge with crowne
by bankys and brymmys browne
I haue trauaylid by many a towne
my lordys love to wynne.

* Paragraph sign inserted a line too high.

The Adoration of the Magi

¶ I seke hym with ensens sote 45
of all prestys he xal be rote
his bryght blood xal be oure bote
to brynge vs out of bende
The childe xal be chosyn a preste
in all vertuys ffowndyn meste 50
be-forn his fader*ys* fayr breste
Ensens he xal up sende.

Tercius Rex

¶ˑIn ypotan And Archage
I am kynge knowyn in kage
to seke a childe of Semlant sage 55
I haue faryn ryght fferre
Jasper is my name knowyn
in many countres þat are myn owyn
thorwe byttyr blastys þat gyn blowyn
I stryke aftere þe sterre. 60

Fo. 93 ¶ I brynge myrre to my present
A byttyr lycour verament
Ffor he xal th[o]lyn[1] byttyr dent
In a maydonys flesch is clad
on byttyr tre he xal be bent 65
Man and god omnypotent
with byttyr betynge his fflesch be rent
tyl all his blood be bledde.

Her[ode

¶ Now I regne ... þe[2] kynge A-rayd ful Rych
Rollyd in rynggys and robys of array 70
Dukys with dentys I dryve in to þe dych
my dedys be ful dowty demyd be day
I xall marryn þo men þat r ... n on a myche[3]
and þer-inne sette here sacrement*ys* sottys ... say[4]

---

[1] *tholyn* has been in great part erased and the word *suffyr* written over it by the second (Ff. 95, 96) hand.

[2] Over an illegible word and þe, *lyk a* has been written by the later hand.

[3] The latter part of this line has been altered in the second hand to (þo) heretykys (þat) beleuen a-mysse.

[4] þer-*inne* has been changed to þerin ; *here* to *there* ; and from *sottys* to the end the line has been altered by erasure and adaptation to : *falsse* þey are

þer is no lorde in þis werde¹ þat lokygh me lyche² 75
Ffor to lame l . . . rys³ of þe lesse lay
I am jolyere than þe jay
Stronge thevys to steke
þat wele oure lawys breke
on þo wrecchis I wyll be wreke 80
      and hont hem vndyr ha[y.

¶ In kyrtyl of cammaka kynge am I cladde
Cruel and curryd in myn crowne knowe
I sytt in vndyr⁴ sesar in my se . . e⁵ sadde
Sorwyn to sottys such seed wyll I sowe 85
Boys now blaberyn bostynge of a baron bad
In bedde (?)⁶ is born be bestys suche bost is blowe
I xal prune þat paphawk and prevyn hym as a pad⁷
Scheldys and shaftys sh . . . lh . . . y sowe⁸
my knyghtys xaln rydyn on rowe⁹ 90
knave¹⁰ chylderyn ffor to qwelle
be mahound dyngne duke of helle
Sowre deth his lyff xall selle
Such thrett wolde me throw¹¹.

¶ Styward¹² bolde 95*
walke þou on wolde¹³

I say. The words cannot be deciphered with certainty as the adaptation is confused.
¹ An *l* has been inserted in this word.
² *lyche* has been altered to *lyke i-wysse.*
³ This word, altered by the reviser to *heretykkys,* may be lo[ve]rys or le[ve]rys.
⁴ *in* has been altered to *here,* and the *v* of *vndyr* to *o.*
⁵ The later writer has *sette,* the *tt* being written over an erasure.
⁶ This word is uncertain; the reviser has altered it to *bedlem.*
⁷ *prune* altered to *prycke,* and *paphawk* to *paddoke.*
⁸ Two or three words of some thirteen letters indecipherable in this line. The reviser alters *shaftys* to *sperys* and continues: *schall I ther (sowe).*
⁹ *xaln* (or *xuln*) altered to *xalle,* and an *a* inserted between *on* and *rowe.*
¹⁰ A *k* between *knave* and *chylderyn* crossed through by original scribe.
¹¹ *thrett* altered to *threttys* and *throw* to *ouerthrow.*
¹² *Styward* crossed through. Only the last letters *or* of the substituted word remain, the rest having been cut away with the margin.
* From line 95 to line 150 two short lines are printed in one, divided as before.
¹³ *ouer mowlde* substituted by second hand for *on wolde.*

The Adoration of the Magi

 and wisely beholde
Aƚƚ Aboute
Iff Any thynge
shuld greve þe kynge   100
brynge me tydynge
if þer be ony dowte.

         Senescall*us*

[¶]¹ Lord kynge in crowne
I go fro towne
by bankys browne   105
I wyƚƚ a-byde
And w*ith* erys lyste
est *and* west
iff any geste
on grownde gynnyth glyde.   110

       [*H*]*erodes nup*²...

Tunc ibit senescallus et obuiabit tribus regibus et dicit eis

[¶]¹ Kyn*gys* iij
vndyr þis tre
In þis countre
why wyƚƚ ȝe abyde
Herowde is kynge   115
of þis wonynge
On to his dwellynge
now xul ȝe glyde.

         1*ᵘˢ* Rex

[¶]¹ Now lede us alle
to þe kyn*gys* halle   120
how it befalle
we pray to the
wyttys to wete
he may us pete
In flesshe be glete   125
godys frute fre.

---

¹ These paragraph marks have been cut away with the margin.
² This, which appears to be a stage direction, is faintly written in the later hand in the margin, and has been partly cut away. The second word (or words) is not certainly decipherable—? nup*er*. . . .

  Senescallus

[¶]¹ Ffolw*ith* in stownde
vpon þis grownde
to þe castel rownde
I xal ȝow tech   130
Where² kynge gynny³ wyde
vp in þis tyde
in pompe *and* pryde
his myght gynnyth reche.

[¶]¹ Ser*e* kyng in trone   135
her*e* comyth a-none
Fo. 94 by strete *and* stone
kyng*ys* thre
They ber*e* present
what thei haue ment   140
ne whedyr thay arn bent
I can not se.
                                            Herodes Rex

¶ I xal hem craue
what they haue
Iff they raue   145
or waxyn wood
I xal hem reve
her*e* wyttys deve
her*e* hedys cleve
And schedyn her*e* blood.   150
                                            1ᵘˢ Rex

¶ Heyl be þ*ou* kynge in kage ful hye
Heyl we nyghe þin halle ryght nye
Knowyst þ*ou* ought þat chylde slye
He is born her*e* a-bowth
He is born of a mayd ȝynge   155
he xal be kynge ou*er* every kynge
We go to seke þat louely thynge
to hym ffayn wolde I lowth.

---

¹ Paragraph marks cut away with the margin.
² The *h* omitted and written small above the line by the scribe.
³ *gynny* altered by reviser to *wonyt*.

The Adoration of the Magi

ij^us Rex

¶ Balaam spak in prophecy
A sterre xulde ful louelye   160
lythtyn vpon mayd marye
  comyn of jacobys kynne
þe childe is born and lyth here by
Blomyd in a madenys body
A sterre hath strekyn upon þe sky   165
  and ledde us fayr be fenne.

iij^us Rex

¶ The sterre hath ledde us out of þe est
to seke a baron born best
he xal be kynge of myghtys mest
  as prophecy gynneth spelle   170
we be kyngys in wey wery
Syr kynge ffor þi curtesy
telle us to þat childe so louely
  in what town gynnyth he dwelle.

herodes Rex

¶ ȝe thre kyngys rekenyd be rowe   175
ley now downe ȝour wurdys lowe
Such a carpynge is vnknowe
  on-rekenyd in my regne
I am a kynge of hyȝ degre
þer xal non ben above me   180
I haue florens and fryhthis fre
  parkys and powndys pleyne.

¶ But goth to fynde þat ȝe sech
and yf ȝe knowe such a lech
and ȝe hym fynde I ȝow be-sech   185
  comyth a-ȝen be me
And I xal be both blyth and bowne[1]
þat all worchep to hym be done
with reuerens I xal seke hym sone
  and honour hym on kne.   190

¶ and þerfore kyngys I ȝow pray
whan ȝe haue don ȝour jurnay

---

[1] *do* first written for *bo* and crossed through.

come aȝen þis same way
þe trewth to me to telle
come and telle me as ȝe spede195
and I xal qwyte ryght wel ȝour mede
with gold and tresour and rych wede
with furrys rych and wurth pelle.

1ᵘˢ Rex

¶ Kynge haue good day
I go my way200[1]
to seche
lord of myght
he xal be ryght
oure leche.

ijᵘˢ Rex

¶ Kynge fful sterne205
be felde and ferne
I goo
to sekyn a kynge
he takyth wonynge
in woo.210

iijᵘˢ Rex

¶ If we hym fynde
oure kynge ful kynde
be a may
Ffrom kynge and qwen
we comyn aȝen215
þis day.

transient.[2] Herodes Rex

¶ A Fy Fy on talys þat I haue ben tolde
here be-forn my cruel kne
how xulde a barn wax so bolde
be bestys yf he born be220
Fo. 95[3] he is yong and I am old
An hardy kyng of hye degre

---

[1] From line 199 to line 216 three lines are written in one, divided as before.
[2] This word is in a later hand, but not apparently that of the Ff. 95-6 scribe.
[3] Ff. 95 and 96 are interpolated and the writing is different, being the same as that of Fo. 112 and as that in which alterations are made in several plays, notably in the Shepherds and Magi plays and the Resurrection play. Ff. 95, 96, and 112 are not rubricated.

The Adoration of the Magi 159

This daye the kynggys xal be¹ kold
If þey cum ageyne be me
My goddys I xall vp-reyse 225
A derke devyll with falsnese I saye
Shall cast a myst in þe kynggys eye
Be bankys and be² dalys drey·
þat be derk þei xall cum this weyys.
          primus [Rex
Go we to sek owr lord and our lech 230
yon stere will us tech þe weyis full sone
To saue vs from myschyff god I here be-sech
on to his joyis þat we may rech
I pray hem of this bone.

☧ *Tunc ibunt reges cum muneribus ad ihesum et primus rex dicit*
Heyle be þou kyng Cold clade 235
heyll with maydynnys mylk fade
heyll I cum to þe with gold glade
As wese wrytyng bere it record
gold is þe³ rycheste metall
And to weryng most ryall 240
gold I gyff þe in this hall
And know þe for my lorde.
          ij*us* r[ex
Lorde I knele vpon my kne⁴
Sote encence I offere to the
Thow xalte be þe fyrst of hyȝ degre⁵ 245
non so mekell of myght
In goddys howse as men xall se
Thow xalt honour þe trynite
iij personys in oon gode free
And all oo lord of myght. 250

Lord I knele downe be thy bede   iij*us* r[ex
In maydyns fleshe þou arte hede
Thy name xal be⁶ wyde rede
And kyng ouer all kynggys

---

¹ *xalbe* (one word) in M.S.
² A miswritten letter before *be* crossed through.
³ Omitted and written above the line.
⁴ *Knele* first written and crossed through.
⁵ *dr* first written and crossed through.   ⁶ *xalbe* as above in MS.

Byttyr myre to þe I brynge	255
ffor bytter dentys on þe þei xall dyng	
and byttyr deth xall be þy endyng	
And þerfor I make mornyng [1].	

Maria

Kynggys kynde	
ffrome þe fende	260
god yow defende	
homwarde ȝe wende	
and to your placys ȝe lende	
þat ȝe xulde tende.	

primus rex

Now haue we þe place fownde	265
To Herode go we this stownde	
with owr wordys we were bownde	
That we xulde cum a-geyne	
go we a pace and sey owr spech	
ffor we haue fownde our lord and lech	270
All þe truth we wyll hem tech	
how þe kyng is borne of a quene.	

ij^us rex

Myn hede is [2] hevy as lympe of leede	
But yf I slepe I am a-drede	
My witt xall fare þe wurse	275
I wax hevy in lyme and flamke [3]	
Downe I ley me vpon this banke	
Vnder this bryght sterre i-wys.	

iij^us rex

Broþer I must lye þe bye	
I will go neuer ouer þis stye	280
Tyll I haue a slepe	
The yong kyng and his moþer mary	
Saue vs all [4] frome euery velany	
Now Cryst vs save and kepe.	

Primus rex

Such hevynese haue vs cawght [5]	285
I must drynk with yow a drawght	

---

[1] First written *mornyngys* as the rhyme demands. The *ys* contraction has, however, been erased.

[2] *is* omitted and written above the line.   [3] So in MS.

[4] *all* omitted and written above the line.

[5] This line, written too far to the left, is crossed out and rewritten.

The Adoration of the Magi

To slepe a lytyll whyle [1]
I am hevy heed and footte
I xulde stumbyll at resch and root
And I xuld goo a myle.   290

α [2] *Hic dormiunt reges et venit Angelus et dicit eis*

                                         Ange[lus

ȝe kynggys on this hill
Werk ȝe not aftyr herodys wyll
for [3] yf ȝe do he wyll yow kyll
þis day or nyght
My lorde yow sent this tydyng   295
to [4] Rest yow kynggys in rych clothyng
And when ȝe rysyn and goo to your dwellyng
Tak home þe wey full ryght

whether þat ȝe be [5] wakyn or slepe [6]
My lorde god xall yow keppe   300
In goode tyme ȝe dede down drepe
To take yowr rest
herode to þe devyll he tryst.

                                         loke ferþer

            Fo. 96ᵛ—left blank.

97 [7] herowdys to þe devyl he tryste
to marre ȝow in a myrke [8] myste
my lord god is ful of lyste   305
    to glathe ȝow for his geste.

¶ And þerfore kyngys whan ȝe ryse
Wendyth ferth be weys wyse
þer ȝour hall be [9] sett in syse
    in dyverse londe   310

---

[1] *wyll* first written and crossed through.
[2] This and the similar sign on Fo. 95 are very roughly made.
[3] *for* omitted and written in the margin against *yf*.
[4] *to* omitted and written in the margin against *Rest*.
[5] *be* omitted and written above the line.
[6] *slepyne* first written and crossed through.
[7] With Fo. 97 the original hand is resumed.
[8] *myrke* altered by reviser to *thyrke*.
[9] Some erasure between *hall* and *be*.

ffadyr¹ of god in all thynge
hath ȝow grawntyd his swete blyssynge
he xal ȝow saue ffrom all shendynge
   with his ryght honde.
    Tunc surgant reges et dicat

        Primus Rex

¶ A bryght sterre ledde us in to bedleem   315
A bryghtere thynge I saw in drem
bryghtere than þe sunne beem
   An Aungell I saw ryght here
þe fayre flowre þat here gan falle
ffrom herowdys kynge he gan vs kalle   320
he taught vs hom tyll oure halle
   A wey by a-nother mere.

        2ᵘˢ Rex

¶ I saughe a syght²
myn hert is lyght
to wendyn home   325
god fful of myght
hath us dyght
ffro develys dome.

        3ᵘˢ Rex

¶ Oure god I blysse
he sent us i-wys   330
his Aungel bryght
now we wake³
þe wey to⁴ take
home full ryght.

Fo. 97ᵛ   ¶ Symeon Justus

19

I Haue be prest in jherusalem here
 And tawth goddys lawe many A ȝere
desyrynge in all my mende
þat þe tyme we⁵ neyhand nere
In which goddys son xul Apere   5

---

¹ þe has been added before *ffadyr* by the reviser.
² From line 323 to line 334 three lines are written in one, divided as before.
³ *be* inserted before *wake* by the Ff. 95-6 hand.
⁴ *to* crossed out and *whe* written above the line by the Ff. 95-6 hand.
⁵ So in MS. for *were*.

The Purification

 in erthe to take mankende
 Or I deyd þat I myght fynde
 my savyour with myn ey to se
 but þat it is so longe be-hynde
 it is grett dyscomforte · on to me.   10

¶ Ffor I wax old and wante my myght
 and be-gynne to fayle my syght
 þe more I sorwe þis tyde
 save only as I telle ȝow ryght
 god of his grace · hath me hyght   15
 þat blysful byrth to byde
 Wherfore now here be-syde
 to sancta sanctorum · wyl I go
 to pray god to be my gyde
 to comfort me aftyr my wo   20
 here Symeon knelyth and seyth

¶ A gode god · in trinite
 Whow longe xal I abyde the
 tyl þat þou þi son þou[1] doth sende
 þat I in erth myght hym se
 good lord consydyr to me   25
 I drawe fast to An ende
 þat or my strenthis fro me wende
 gode lorde send dow[2] þi son
 þat I with my ful mende
 myght wurcheppe hym if I con̄[3].   30

¶ Bothe with my fete · and hondys to
 to go to hym and handele also
   my eyn to se hym in certayn
 my tonge for to speke hym to
 and all my lemys to werke and do   35
   In his servyse to be bayn.
 Send forth þi son · my lord sovereyn
 hastely A-non with-owte teryenge.
 Ffor fro þis world · I wolde be ffayn
 It is contrary to my levynge.   40

---

[1] So in MS. Should one þⁿ (þᵘ) be expanded to þan?
[2] So in MS.   [3] Angelus scribbled here in the margin.

Angelus[1]

¶ Symeon leff þi careful stevene
 Ffor þi prayer is herd in hevene
   to jher*us*alem ffast now wynne
 And þer xalt se ful evene
 he þat is goddys son ffor to nemene[2]   45
   in þe templ[3] þer þou dwellyst inne
 The dyrknes of orygynal synne
 he xal make lyght *and* clarefye
 and now þe dede xal be-gynne
 whiche hath be spokyn be prophecye.   50

Symeon

¶ A· I thanke þe lord of grace
 þat hath grauntyd me tyme *and* space
   to lyve *and* byde thys
 and I wyl walk now to þe place
 Where I may se þi sonys face   55
   which is my joye *and* blys
 I was nevyr lyghtere i-wys
 to walke · nevyr here be-forn
 Ffor a mery tyme now is
 Whan god my lord is born.   60

Anna prophetessa

¶ Al heyl symeon what tydyng*ys* wit*h* ȝow
 Why make ȝe al þis myrth now
   telle me whedyr ȝe far*e*.

Symeon

 Anne prophetes · *and* ȝe wyst whov
 So xulde ȝe · I make A-vow   65
   and all maner men þat Are
 Ffor goddys son as I declare
 Is born to bye mankende
 Oure savyour is come to sesyn oure care
 þerfore haue I grett merth to wende.   70

¶ And þat is þe cawse I hast me
 On to þe temple hym to se
   and[4] þerfore lett me not good frende.

---

[1] Some words in another ink are written against *Angelus*, and have been clipped with the margin: *Cum do* . . .
[2] So in MS.   [3] So in MS.   [4] A miswritten *and* deleted.

Anna

Now blyssyd be god in trinyte
syn þat tyme is come to be 75
  and with ʒow wyl I wende
To se my savyour ende
and wurcheppe hym Also
with all my wyll and my ful mende
As I am bound · now wyl I do. 80
Et tunc ibunt Ambo ad templum. et prophetissa.

Symeon

¶ In þe temple of god · who vndyrstod
þis day · xal be offeryd with mylde mood
which þat is kynge of Alle.
þat xal be skorgyd and shedde his blood
And Aftyr dyen on þe rood 85
With-owtyn cawse to calle
Ffor whos passyon þer xal be-ffalle
Swych a sorwe bothe sharpe and smerte
þat as a swerd perce it xalle
ʒevene thorwe his moderys herte. 90

Anna prop[hetissa

¶ ʒa þat xal be as I wel fende
Ffor redempcion of All mankende
þat blysse ffor to restore
Whiche hath be lost fro oute of mende
As be oure fadyr of oure owyn kende 95
Adam and Eue be-ffore.

Maria

¶ Joseph my husbond · with-owtyn mys
ʒe wote þat ffourty days nere is
Sythe my sonys byrth fful ryght
Wherfore we must to þe temple i-wys 100
Þer fore to offre oure sone of blys
  up to his fadyr in hyght
And I in goddys syght
puryfyed ffor to be
in clene sowle with al my myght 105
in presence of þe trinyte [2]

---

[1] *fende* has been altered to *fynde* by another hand.
[2] Lines 103-6 are written in two long lines divided by black strokes without rubrication.

Joseph

¶ To be purefyed · haue ȝe no nede
ne þi son · to be offeryd · so god me spede
ffor fyrst þou art ful clene
Vndefowlyd in thought *and* dede 110
and a-nothyr þi son wit*h*-owtyn drede
is god *and* man to mene
Wherefore it nedyd not to bene
but to kepe þe lawe · on moyses wyse
Whereffore we xal take us be-twene *Et ibunt ad templum.* 115
Dowys *and* turtelys ffor sacrefyce.

                                            Symeon

¶ All heyl my kyndely comfortour

                                         Anna *prophetissa*

All heyl mankyndys creatoure

                                             Symeon

All heyl þou god of myght

                                         Anna *prophetissa*

All heyl mankyndys savyo*ur* 120

                                             Symeon

All heyl bothe kynge *and* emp*erour*

                                         Anna *prophetissa*

All heyl As it is ryght

                                             Symeon

All heyl · Also mary bryght

                                         Anna *prophetissa*

All heyl · Salver of Seknes

                                         Symeon

All heyl lanterne of lyght 125

                                       Anne *prophetissa*

All heyl þou modyr of mekenes.

                                          Maria

¶ Symeon · I vndyrstand *and* se
þat bothyn of my sone *and* me
       ȝe haue knowynge cler*e*
and also in ȝo*ur* compane 130
my sone desyryth for to be
       and þer-ffore haue hym here.

                                         Symeon
Welcome prynce wit*h*-owte pere   *et Accipiet Jhesum.*
Welcome goddys owyn sone
Welcome my lord so der*e* 135
Welcome wit*h* me to wone.

The Purification

Suscepimus deus m*isericord*iam tuam.

¶ Lord god in mageste
we haue receyvyd þis day of þe
in myddys of þi temple here
thy grett mercy As we may se   140
therfore þi name of grett degre
be wurchepyd in aīl manere
Over aīl þis werde bothe fer *and* nere
ȝevyn on to þe vnterest ende
Ffor now is man out of daungere   145
and rest *and* pes to Aīl man-kende.

Nunc dimittis seru*um* tuum domine ,' etcet*era*. *The psalme songyn every vers* and *þer qwyl' Symeon pleyth with* þe *child'* and *qwhan* þe *psalme is endyd · he seyth*

¶ Now lete me dye lorde *and* hens pace
Ffor I þi servaunt in þis place
haue sen my savyo*ur* dere
Whiche þo*u* hast ordeyned be-forn þe face   150
Of Al mankynde þis tyme of grace
opynly to Appere
þi lyth is shynand' clere
to Aīl mankyndys savacion
mary take ȝo*ur* childe now here   155
and kepe wel · þis man is savacion.

         Anna p**r**oph*etissa*

l. 100 ¶ Ne I rowth nere to dye Also [1]
Ffor more than ffowre skore ȝere *and* to
   þis tyme hath bede to se
And sythe þat it is come þer to   160
What goddys wyl is w*ith* me to do
   ryght evyn [2] so mot it be.

         Joseph

Take here these candelys thre
Mary · Symeon · And Anne
And I xal take þe fowrte to me   165
to offre oure child up thanne.

---

[1] The words *to Aīl mankyndys* are written as beginning of the next line and crossed through.

[2] First written ȝeryn, the initial ȝ being crossed through.

Maria

¶ Hyest Ffadyr god of powere
 ȝour owyn dere son I offre ȝow here
  as I to ȝour lawe Am sworñ
Receyve þi childe in glad manere   170
Ffor he is þe fyrst þis childe so dere
 þat of his modyr is born
but þow I offre hym ȝow be-forn
good lord ȝit ȝyf me hym A-ȝen   174
Ffor my comforte were fully lorn   *Mari leyth þe childe*
If we xuld longe A-sondyr ben.   *on þe Autere*

Joseph

¶ Sere prest of þe temple now
haue he¹ ffyff pens · vnto ȝow
  oure childe Aȝen to take
It is þe lawe As ȝe woot how.   180

Capellanus

Joseph · ȝe An do ryght A-now
  As for ȝour childys sake
But othere offerynge ȝett must ȝe make
and þerfore take ȝour sone mary
In meche joye ȝe may A-wake   185
Whylys he is in ȝour company.

Maria

Fo. 100ᵛ ¶ Ther to I am ful glad and fayn
Ffor to receyve my childe Agayn
ellys were I to blame.
And Aftere-warde · ffor to be bayñ   190
to offre to god in ful certayñ
As in my sonys name
With ffowlys bothe wylde and tame
Ffor in goddys servyse² I xal nevyr irke.

Joseph

Lo mary · haue here tho same   195
to do þi dewtys · of holy kyrke.
And þer mary offeryth ffowlys on to þe Autere and seyth

Maria

All-myghty-fful fadyr · mercyful kynge
Receyvyth now þis lytyl offerynge

---

¹ So in MS. for *here*.
² The word *servyse* omitted and written above the line.

Ffor it is þe fyrst in degre
þat ȝour lytyl childe so ȝynge 200
presentyth to-day be my shewyng
to ȝour hyȝ mageste
Of his sympyl poverte
be his devocion and my good wylle
Vpon ȝour Awtere receyve of me 205
ȝour sonys offrynge As it is skylle. 1468.[1]

Remainder of 100ᵛ—2⅝ inches—left blank.

Tunc respiciens senescallus vadyt ad herodem dicens

¶ Senescallus

Lord I haue walkyd be dale and hylle
And wayted as it is ȝour wyll.
The kyngys iii. stelyn awey full styll
 thorwe bedleem londe
They wyl nevyr so moty the 5
Com in þe lond of Galyle
Ffor to se ȝour fay[2] cete
 ne dedys of ȝour honde.

herodes Rex[3]

¶ I ryde on my rowel ryche in my regne 20[4]
Rybbys fful reed with rape xal I rende 10
popetys and paphawkys I xal puttyn in peyne
with my spere prevyn pychyn and to pende
The gomys with gold crownys ne gete nevyr ageyn[5]
to seke þo sottys sondys xal I sende
Do howlott howtyn hoberd and heyn 15
Whan here barnys blede vndyr credyl bende
Sharply I xal hem shende
The knaue childeryn þat be
in all israel countre
thei xul haue blody ble 20
 ffor on I calde vnkende.

¶ It is tolde in grw
his name xulde be jhesu

---

[1] This date appears to be in the same hand as the text, and it is looped in red as are the names of the speakers.

[2] So in MS. for *fayr*. [3] Red name loop omitted.

[4] A *1* has been erased here before the *2*. Cf. *Introduction, Numbering of Plays*.

[5] *gomys* has been altered to *gowys* (or *gollys*), *ne* has been crossed through, and *þei* inserted after *gete* in this line in darker ink in another hand.

The Massacre of the Innocents

  i-fownde
  to haue hym ʒe gon   25
  hewe þe flesch with þe bon
  and gyf hym wownde*
  Now kene knyghtys kythe[1] ʒoure craftys[2]
  and kyllyth knaue chylderyn and castyth hem in clay
  Shewyth on ʒour shulderys scheldys and schaftys   30
  Shapyht amonge schel chownys[3] ashyrlyng shray[4]
  doth rowncys[5] rennyn with rakynge[6] raftys
  tyl rybbys be to-rent with a reed ray
  lete no barne[7] beleve on bete baftys
  tyl a beggere blede be bestys baye   35

Fo. 101ᵛ Mahound þat best may
  I warne ʒow my knyghtys
  A barn is born I plyghtys
  Wolde clymbyn[8] kynge and knytys[9]
  and lett my lordly lay.   40

 ¶ knyghtys wyse
  chosyn[10] ful chyse
  A-ryse a-ryse
  and take ʒoure tolle
  And every page   45
  of ii · ʒere Age
  or evyr ʒe swage
  sleyth ilke a fool[11].

 ¶ On of hem alle
  was born in stalle   50
  ffolys hym calle

---

* From line 22 to line 27 three lines are written in one; from line 37 to line 72 two lines are written in one.

[1] Altered by the later hand of Ff. 95-6 to *scharpe*.
[2] Altered by the later hand to *knyvys*.
[3] Altered by the later hand to *schel chowthys*.
[4] *shar* first written and crossed through by scribe.
[5] Crossed out and *your speris* written above by the later hand.
[6] Crossed out and *longe* (?) written above by the later hand.
[7] Crossed out and *chyld* written above by the later hand.
[8] *clyuer* ... first written and crossed through in red ink.
[9] *ky* before *knytys* crossed through by scribe.
[10] Some miswritten letters (*cof* ?) before *chosyn* crossed through in red ink.
[11] Or *foal*.

The Massacre of the Innocents 171

 kynge in crowne [1]
 With byttyr galle
 he xall down falle
 my myght in halle   55
 xal nevyr go down.

          i^{us} miles
¶ I xall sle scharlys
 And qwenys with therlys
 here knaue gerlys
 I xal steke   60
 Fforth wyl I spede
 to don hem blede
 thow gerlys grede
 we xul be wreke.

         ij^{us} miles
¶ Ffor swerdys sharpe   65
 as An harpe
 quenys xul karpe
 and of sorwe synge
 barnys ȝonge
 they xul be stunge   70
 thurwe levyr and lunge
 we xal hem stynge.

         Angelus
¶ A-wake joseph and take þi wyff
 thy chylde also ryd be-lyff
 ffor kynge herowde with sharpe knyff 75
   his knyghtys he doth sende
 The fadyr of hevyn hath to þe sent
 In to Egypte þat þou be bent
 Ffor cruel knyghtys þi childe haue ment
   with swerde to sle and shende. 80

         Joseph
¶ Awake good wyff out of ȝour slepe
 and of ȝour childe takyght good kepe
 Whyl I ȝour clothis ley on hepe
   and trus hem on þe asse
 kynge herowde þe chylde wyl scloo 85
 þerfore to Egypte muste we goo

---

[1] *Angelus* scribbled here in the margin in a later hand.

An Aungel of god seyd me soo
  and þer-fore lete us passe.

Fo. 102 *Tunc ibunt milites ad pueros occidendos et dicat prima femina*

            i^a femina

¶ Louge lullynge haue I lorn
 Alas qwhy was my baron born   90
 With swappynge swerde now is he shorn
  þe heed ryght fro þe nekke
 Shanke and shulderyn is al to torn
 Sorwyn I se be-hyndyn and be-forn
 both mydnyth mydday and at morn 95
  of my lyff I ne recke.

           ij^a femina

¶ Serteynly I sey þe same
 gon is all my good game
 my lytyll childe lyth all lame
  þat lullyd on my pappys   100
 my Ffourty wekys gronynge
 hath sent me sefne ȝere sorwynge
 mykyl is my mornynge
  and ryght hard arne myn happys.

           i^{us} miles

¶ Lorde in trone   105 [1]
 makyght no mone
 qwenys gyn grone
 in werdl aboute [2]
 upon my spere
 a gerle I bere   110
 I dare well swere
 lett moderys howte.

           ij^{us} miles

¶ Lord we han spad
 as ȝe bad
 barnis ben blad   115
 and lyne in dych
 Fflesch and veyn
 han tholyd peyn

---

[1] From line 105 to line 128 two lines are written in one.
[2] The words *go hom ward*ys are written here in the margin in the later hand.

and ȝe xul reyne
euer more rych.   120
                                    herodes rex

¶ ȝe xul haue stedys
to ȝour medys
londys and ledys
ffryth and ffe
wele haue ȝe wrought   125
my ffo is sought
to deth is he brought
now come up to me.

¶ In sete now am I sett as kynge of myghtys most
All þis werd ffor þer loue to me xul þei lowt   130
both of hevyn and of erth and of helle¹ cost
Ffor dygne of my dygnyte þei haue of me dowt
102ᵛ þer is no lord lyke on lyve to me wurth a toost
nother kyng nor kayser in all þis worlde abought
If any brybour² do bragge or blowe a-ȝens my bost 135
I xal rappe þo rebawdys and rake þem on rought
With my bryght bronde
þer xal be neythey³ kayser nere knyge
But þat I xal hem down dynge
lesse þan he at my byddynge   140
       be buxum to myn honde.

¶ Now my jentyll and curteys knyghtys herke to me þis stownde
Good tyme sone me thynkygh at dyner þat we were
Smertly þerfore sett a tabyll a-non here fful sownde
Couerid with a coryous cloth and with rych wurthy fare   145
Servyse ffor þe lovelyest lorde þat levynge is on grownde
Beste metys and wurthyest wynes loke þat ȝe non spare
þow þat a lytyl pynt xulde coste a Ml. pownde
brynge alweye of þe beste for coste take ȝe no care
Anon þat it be done.   150
                                    Senescallus
My lorde þe tabyl is redy dyght
here is watyr now wasch forthryght

¹ hevyn first written and crossed through.
² or bragges first written after brybour and crossed through.
³ So in MS.

now blowe up mynstrall with all ȝour myght
þe servyse comyth in sone.

herodes rex

¶ Now am I sett at mete 155
and wurthely servyd at my degre
Com forth knyghtys sytt down and ete
and be as mery as ȝe kan be.

i^{us} miles

Lord at ȝowre byddynge we take oure sete
with herty wyl obey we the 160
þer is no lorde of myght so grett
thorwe all þis werde in no countre
Fo. 103 In Wurchepp to a-byde.

Herodes

(J quire) I was nevyr meryer here be-forn
Sythe þat I was fyrst born 165
than I am now ryght in þis morn
in joy I gynne to glyde.

Mors

¶ Ow I herde a page make preysyng of pride
all prynces he passyth he wenyth of powste
he wenyth to be þe wurthyest of all þis werde wyde 170
kynge ovyr All kyngys þat page wenyth to be
He sent into bedlem to seke on every syde
Cryst for to qwelle yf þei myght hym se
but of his wykkyd wyl lurdeyn ȝitt he lyede
goddys sone doth lyve þer is no lorde but he 175
Ouer all lordys he is kynge
I am deth goddys masangere
All myghty god hath sent me here
ȝon lordeyn to Sle with-owtyn dwere
ffor his wykkyd werkynge. 180

¶ I am sent fro god deth is my name
All thynge þat is on grownd I welde at my wylle
both man and beste and byrdys wylde and tame
Whan þat I come them to · with deth I do them kylle
Erbe gres and tres stronge · take hem all in same 185
ȝa þe grete myghty Okys · with my dent I spylle
what man þat I wrastele with he xal ryght sone haue schame
I ȝeve hym such a trepett · he xal evyr more ly stylle

*Ffor deth kan no sporte
wher I smyte þer is no grace 190
Ffor aftere my strook man hath no space
to make amendys ffor his trespace
but god hym graunt comforte

103v ¶ Ow se how prowdely ȝon kaytyff sytt at mete
of deth hath he no dowte he wenyth to leve evyr-more 195
to hym wyl I go and ȝeve hym such An hete
þat all þe lechis of þe londe his lyf xul nevyr restore
A-ȝens my dredful dentys it vaylyth nevyr to plete
or I hym part fro I xal hym make ful pore
Aḻḻ þe blood of his body I xal hym owt swete 200
Ffor now I go to sle hym with strokys sad and sore
Bothe hym and his knyghtys aḻḻ   ¶ þis tyde[1]
I xal hem make to me but thraḻḻ
with my spere sle hem I xaḻḻ. 205
  and so cast down his pride
                                 herodes Rex

¶ Now kende knyghtys be mery and glad
With aḻḻ good diligens shewe now sum myrth
Ffor be gracyous mahound more myrth never I had
ne nevyr more joye was inne from tyme of my byrth 210
Ffor now my fo is ded and prendyd as a padde
aboue me is no kynge · on grownd nere on gerth[2]
merthis þerfore make ȝe and be ryght no thynge sadde
spare nother mete nor drynke and spare for no dyrthe
of wyne nor of brede 215
Ffor now am I a kynge alone
So wurthy as I · may þer be none
þerfore knyghtys be mery echone
  ffor now my ffo is dede.
                               i^us miles

¶ Whan þe boys sprawlyd at my sperys hende 220
by sathanas oure syre it was a goodly syght
A good game it was þat boy for to shende
þat wolde abene oure kynge and put ȝow from ȝour ryght

---

* Paragraph mark erased here.
[1] Halliwell prints this line before the line *Bothe hym etc.* which is its right place in the stanza.
[2] The *e* written by the scribe over some other letter *r* or *?y*.

Fo. 104

ij^us miles

Now trewly my lorde þe kynge we had ben vn-hende [1]
and nevyr non of us Able for to be a knyght   225
If þat Any of us to hem had ben a frende
and aȝavyd Any lyff a-ȝen þi mekyl myght
Ffrom deth hem to flytt.

herodes Rex

Amonges all þat grett rowthte
he is ded I haue no dowte   230
þerfore menstrell rownd a-bowte
blowe up a mery fytt.

Hic dum buccinant mors interficiat herodem et duos milites subito et diabolus recipiat eos.

Diabolus

¶ All oure all oure þis catel is myn
   I xall hem brynge on to my celle
   I xal hem teche pleys fyn   235
   and showe such myrthe as is in helle
It were more bettyr Amonges swyn
þat evyr more stynkyn þer be to dwelle
ffor in oure logge is so gret peyn
þat non erthely tonge can telle   240
   with ȝow I go my way
I xal ȝow bere forth with me
and shewe ȝow sportys of oure gle
of oure myrthis now xal ȝe se   245
   and evyr synge welawey.

Mors

¶ Off kynge herowde all men beware
þat hath rejoycyd in pompe and pryde
Ffor all his boste of blysse ful bare
he lyth now ded here on his syde
Ffor whan I come I can not spare   250
Fro me no whyht may hym hyde
now is he ded and cast in care
In helle pytt evyr to A-byde
   his lordchep is al lorn

---

[1] Altered by the later hand to *vnkende*.

The Death of Herod

104ᵛ
Now is he as pore as I 255¹
wormys mete is his body
his sowle in helle ful peynfully
of develis is al to-torn.

¶ All men dwellyng upon þe grownde
Be-ware of me be myn councel 260
Ffor feynt felachep in me is fownde
I kan no curtesy as I ȝow tel
Ffor be a man nevyr so sownde
of helth in herte nevyr so wel
I come sodeynly with-in a stownde 265
me with-stande may no castel
my jurnay wyl I spede.
of my comyng no man is ware
Ffor when men make most mery fare
þan sodeynly I cast hem in care 270
and sle þem evyn in dede.

¶ Thow I be nakyd and pore of array
and wurmys knawe me al a-bowte
ȝit loke ȝe drede me nyth and day
Ffor whan deth comyth ȝe stande in dowte 275
Evyn lyke to me as I ȝow say
shull all ȝe be here in þis rowte
Whan I ȝow chalange at my day
I xal ȝow make ryght lowe to lowth
and nakyd for to be 280
Amonges wormys as I ȝow telle
Vndyr þe erth xul ȝe dwelle
and thei xul Etyn both flesch and felle
As þei haue don me.

Remainder of Fo. 104ᵛ—2¼ inches—and Ff. 105–105ᵛ left blank.

[1] From line 255 to line 258 two lines are written in one, divided by black strokes without rubrication.

Christ and the Doctors

Fo. 106

Modo de doctoribus disputantibus cum jhesu in templo.[1]

¶ Primus doctor

¶ Scripture sacre esse dinoscimur doctos
We to · bere þe belle of all manere clergyse.

ij[us] doctor

Velud rosa omnium florum flos
lyke on to us was nevyr clerke so wyse.

i[us] doctor

Loke what scyens ȝe kan devyse
of redynge wrytynge and trewe ortografye
Amonges all[2] clerkys we bere þe prysse
of gramer cadens and of prosodye.

ij[us] doctor

¶ No clerke Abyl to bere oure book
of versyfyeng nor of other scyens
of Swete musyke who so wyll look
seke no ferther but to oure presens
Of dyaletyk we haue þe hyȝ excellence
of sophestrye · logyk and phylosophye
Ageyn oure argemente[3] is no recystence
In metaphesyk[4] ne astronomye.

i[us] doctor

¶ Of calculacion and negremauncye
Also of Augrym̄ and of asmatryk
O[5] lynyacion þat longyth to jematrye
of dyetis and domys þat longyth to phesyk
In all þis scyens is non us lyke
In caton̄ gryscysme nor doctrynal
and ffor endytynge with retoryke[6]
þe hyest degre is oure be call[7].

5

10

15

20

---

[1] This title is not written in larger form in the MS.

[2] The *ll* is written over some other letter, and some letter (? i) is written small above the *a*.

[3] *argeru* first written and crossed through.

[4] An erasure in the word indicates some difficulty with 'trewe ortografye'.

[5] So in MS. for *Of*.

[6] *retr* first written; some letter written above between *t* and *r* and then the whole crossed through.

[7] *be call* crossed out and *over all* written above by another hand, possibly the Ff. 95–6 scribe.

ij^us doctor

¶ In grett canon *and* in Cevyle lawe
 Also in scyens of polycye
 Is non to us wurthe An hawe
 of all cunnynge we bere þe maystrye
 Therfore in þis temple we sytt on hye
 and of most wurchep kepe þe souereynte
 þer is on erthe no man so wurthye
 þe hy3 stat to holdyn as we tweyn be.

Jhesus

¶ Omnis sciencia a *domino* deo est
 Al wytt *and* wysdam of god it is lent[1]
 Of all 3o*ur* lernynge wit*h*-inne 3o*ur* brest
 thank hyghly þat lord þat hath 3ow sent
 thorwe bost *and* pryde 3o*ur* soulys may be shent
 Of wytt *and* wysdome 3e haue not so mech
 but god may make at hese entente
 of all 3o*ur* connynge many man 3ow lech.

i^ms doctor

¶ Goo hom lytyl babe *and* sytt on þi moder*ys* lappe
 and put a mokador[2] a-forn þi brest
 and pray þi modyr to fede þe wit*h* þe pappe
 of þe for to lerne we desyre not to lest[3].

ij^ms doctor

Go to þi dyner for þat be-hovyth the best
whan þ*ou* art a-threste þan take þe A sowke
Aftyr go to cradyl þer-in to take þi rest
Ffor þat canst þ*ou* do bettyr þan for to loke on book.

Jhesus

¶ Stondynge þat 3e be so wytty *and* wyse
 Can 3e owth tellyn how þis werde was wrought
 how longe xal it laste can 3e devyse
 wit*h* all þe cunnyge þat 3e han sought.

i^ms doctor

Nay all erthely clerkys þat telle can nought
it passyth oure wytt þat for to contryve

---

[1] *sent* first written and crossed through.
[2] The *o* in *or* is written over some other letter.
[3] *lyest* first written and crossed through.

Fo. 107 It is not possyble A-bought to be brought   55
         þe worldys endyng no man kan dyscryve [1].

                                                                    Jhesus

    ¶ How it was wrought and how longe it xal endure
      þat I telle [2] be good delyberacion
      not only þer-of but of every creature
      How it is wrought I knowe þe plasmacion.   60
                                                                ij^us doctor

      Of þi wurdys I haue skorne and deryson [3]
      how schulde a chylde þat nevyr lettyr dude [4] lere
      Com to þe wytt of so hyȝ cognysion
      Of þo grete wurkys þat so wundyrfull [5] were.

                                                                    Jhesus
    ¶ All thynge is brought to informacion   65
      be thre personys · oo · god in trynite
      and on of þo thre hath take in-carnacion
      bothe flesch and blood of a mayd ffre
      And be þat myght of þo personys thre
      hevyn and erth and all thynge is wrought   70
      and as it plesyth þat hyȝ mageste
      all thynge xal leste and lenger nowght.

                                                                i^us doctor
    ¶ I grawnt weyl all thynge þat god dyde make
      and with-owtyn hym no thynge may be
      But o thynge þou seydyst and þat I for-sake   75
      þat oo god alone was personys thre
      Ryght on-possyble þat is to me
      that on is thre I kan not thynke
      If þou canst preve it a-non lett se
      Ffor in oure hertys it may nevyr synke.   80
                                                                    Jhesus
Fo. 107^v ¶ In þe sunne · consydyr ȝe thyngys thre
      The splendure þe hete and þe lyght
      as þo thre partys but oo sunne be
      Ryght so thre personys be oo god of myght.

---

[1] The *s* in the word has a double stroke and might be read as *ss*.
[2] The word *can* has been written above the word *telle* by another hand. The tailed *n* (ŋ) of this *can* is found again in a later correction on Fo. 135. It is not found in Ff. 95, 96, or 112.
[3] So in MS. for *son* (= sion).
[4] *dude* has been changed to *dyde* in darker ink and another hand.
[5] The *u* of *full* written by the scribe over the beginning of an *o*.

                                                    ij^us doctor
In very feyth þis reson is ryght   85
but ȝitt fayr babe oo thynge we pray ȝow
what do aℓℓ þo thre personys hyght
Vs to enforme · ȝe sey to me now.
                                                    Jhesus
¶ The fyrst is calde þe fadyr of myght
þe secunde þe sone of wysdam and wytt   90
þe holy gost þe iij^de. of grace he is hyght
and in oo substauns aℓℓ these iij. be knyt.
                                                    i^us doctor
A-nother questyon[1] I Aske ȝow ȝitt
ȝe. seyd on of þese iij. toke flesch and blood
and sche a clene mayde I kan not be-leue it   95
clene mayde and modyr nevyr ȝit in oo persone stood.
                                                    Jhesus
¶ Lyke as þe sunne doth perysch[2] þe glas
þe glas not hurte of his nature
ryght so þe godhed entryd has
þe virgynes wombe and sche mayd pure   100
That maydonys childe xal do grett cure
convicte þe devyl in þe opyn felde
and with his bolde[3] berst fecch hom his creature
mankende to saue his brest xal be þe shelde.
                                                    ij^us doctor
¶ This childys doctryne doth passe oure wytt   105
Sum Aungel of hevyn I trowe þat he be
But blyssyd babe of oo dowte ȝitt
We pray ȝow enforme us for charyte
Which toke flesch of þe personys thre
Ageyn þe fende to holde such batayle.   110
                                                    Jhesus
The secunde persone for sothe is he
xal fray þe fende with-owte fayle.
                                                    i^us doctor
¶ Why rather he than Any of þat other[4]
The fyrst or þe thyrde why come they nowth.

---
[1] ȝitt first written after questyon and crossed through.
[2] perysch crossed through and pers written above in another hand not certainly identifiable with that of Ff. 95, 96.
[3] bolde similarly replaced by bluddy.
[4] other first written and corrected by scribe.

	Jhesus
this is þe cawse why sertys and non other	115
Ageyn þe secunde þe trespas was wrought	
Whan þe serpent adam to synne browth	
He temptyd hym nowght be þe faderys myght	
Of þe gostys goodnes spak he ryght nowght	
but in connynge he temptyd hym ryght.	120

¶ Myght is þe faderys owyn propyrte
   to þe gost apperyd is goodnes
   in none of these tweyn temptyd he
   mankende to synne whan he dede dresse
   To þe sone connynge doth longe expres   125
   ther with þe serpent dyd Adam A-say
   Ete of þis Appyl he seyd no lesse
   and þou xalt haue connynge as god verray.

¶ Þus þe secunde person Attrybute
   Was only towchyd be temptacion   130
   Wherfore hym self wyl hold þe sewte
   And kepe his propyrte fro maculacion.

ij[us] doctor

This is An hevynly declaracion
oure naturall wytt it doth excede
so ȝonge a childe of such informacion   135
in al þis werld neuyr er non ȝede.

i[us] doctor

¶ We be not worthy to kepe þis sete
   Whyll þat oure mayster is in presens
   þe maystry of us þis childe doth gete
   We must hym wurchep with hyȝ reverens   140
   Come forth swete babe of grett excellens
   þe whysest clerke þat evyr ȝett was born [1]
   to ȝow we ȝeve þe hyȝ resydens
   Vs more to teche as ȝe haue done be-forn.

hic adducunt ihesum inter ipsos et in scanno altiori ipsum sedere faciunt ipsis in inferioribus scannis sedentibus et ait ij[us] doctor

2[us] doctor

¶ So ȝonge A chylde suche clergye to reche   145
   and so sadly to sey it we woundyr sore

---

[1] *bar* miswritten before *born* and crossed through.

Christ and the Doctors

Who was ȝowre mayster who dede ȝow teche
of what man had ȝe þis wurthy lore.

Jhesus

My wytt and my lernynge is no ȝonge store
or þis worde was wrought all þinge dede I knowe   150
Ffyrst or ȝe wore borne ȝerys many score
thorwe þe myght of my fadyr my wytt in me dede flowe[1].

i^{us} doctor

¶ Or þat we weryn born nay þat may nat be
þe ȝongest of us tweyn is iij. score ȝere of Age
and þi-selfe art but a chylde al men may wel se   155
late camst out of cradyl as it semyth be þi vesage.

Jhesus

I am of dobyl byrth and of dobyl lenage
Ffyrst be my fadyr I am without gynnynge
And lyke as he is hendeles in his hyȝ stage
So xal I also neuyr mor haue endynge.   160

¶ Ffor be my ffadyr kynge celestyall
Without begynnyng I am endles
but be my modyr þat is carnall
I am but xij ȝere of age þat is expres
My body of ȝoughȝ[2] doth shewe wyttnes   165
Which of my modyr here I dude take
but myn hyȝ godhede þis is no lesse
all thynge in þis world for sothe dude I make.

ij^{us} doctor

¶ Be ȝour ffadyr þat endles is
Who is ȝour modyr telle us we pray.   170

Jhesus

Be my fadyr þe hyȝ kynge of blys
A modyrles chylde I am veray.

i^{us} doctor

Who was ȝour ffadyr to us than say
be ȝour modyr A woman þat was.

Jhesus

I am fadyrles as for þat may   175
of fleschly luste she dude nevyr trespas

---

[1] The words *dede flowe* are written above the line by the scribe, some miswritten words (*dede flawe* ?) being crossed through.
[2] The *h* appears to have been written over some other letter.

              ij^us doctor
 ¶ Telle us I pray ʒow what is ʒour name
  what hyght ʒoure modyr telle us Also.
              Jhesu
  Jhesu of Nazareth I am þe same
  born of a clene mayd prophetys seyd so   180
  Ysaye seyd þus · Ecce virgo
  A mayd xal conceyve in clennes a chylde
  ʒitt ageyn nature and alkende loo
  ffrom all wem of synne pure and vndefylde.

Fo. 109^v ¶ Mary þe chylde of Joachym And anne   185
  ys þat clene mayd and here childe am I
  þe frute of here wombe xal saue euery manne[1]
  Ffrom þe grett dowte of þe ffyndys tormentry.
             i^us doctor
  All þe clerkys of þis worlde trewly
  can not brynge this to declaracion   190
  lesse þan þei haue of god Almyghty
  Sum influens of informacion
             ij^us doctor
 ¶ No jentyl jhesu we ʒow pray
  Whyl þat we stodye a whyle to dwelle
  In cas mo dowtys þat we fynde may   195
  þe trewth of hem ʒe may us telle.
             Jhesu
  Goo take ʒour stodye and avyse ʒow well
  And all ʒour leysere I xal a-byde
  If Any dowtys to me ʒe mell
  þe trewth þerof I xall vn-hyde.   200
             Maria
 ¶ Alas Alas myn hert is wo
  My blyssyd babe a-wey is went
  I wott nevyr whedyr þat he is go
  Alas for sorwe myn hert is rent
  Jentyl husbond haue[2] hym sent   205
  Out on herrande to Any place

---

[1] MS. mane. (with circumflex and dot over *a*). The *a* and the *n* are run together and the *e* appears an after thought.

[2] *yow* is written above the line after *haue*, probably by the later hand of Ff. 95, 96.

but yf ȝe knowe were her¹ ys bent
myn hert for woo A-sondyr wyl race.

           Joseph

¶ On my massage I hym not sent
fotsothe good² wyff in no degre   210
how longe is it þat he hens went
What tyme dude ȝe ȝour childe last se.

           Maria

Trewly gode spowse not þese days thre
þerfore myn herte is cast in care
hym for to seke wher so he be   215
in hast good husbonde lete us forth fare.

           Joseph

¶ Than to hjerusalem³ lete us streyte wende
Ffor kynred gladly to-gedyr wole gon
I hope he þer⁴ with sum good ffrende
þer he hath cosynys ryght many on.   220

           Maria

I am aferde þat he hath fon
Ffor his grett wyttys and werkys good
lyke hym of wytt ffor-soth is non
Euery childe with hym is wroth and wood.

¶ Also my babe my blys my blood   225
Whedyr art þou þus gon fro me.
my sowle my swetyng my frute myn ffood
Send me ssum wurd where þat þou be
Telle me good serys for charyte
Jhesu my childe þat babe of blysse   230
Amonge þis companye dude ȝe hym se
Ffor godys hyȝ⁵ loue telle where he is.

         iᵘˢ doctor

¶ Of oo qwestyon I am be-thought
All of ȝour modyr þat blyssyd may
In what governauns is she brought   235
How is sche rewlyd be nyght and day.

---

¹ So in MS. for *he*.   ² *god* first written and crossed through.
³ The *j* written over an *e*.
⁴ *is* is written above the line before *þer* in another ink.
⁵ This word is crossed out in different ink.

                   Jhesu

    An old man joseph · as I ʒow say
    here weddyd be meracle on to his Wyff
    Here for to fede *and* kepe Alway
    and bothyn in clennesse be maydonys Olyff.   240
                ij^us doctor

Fo. 110^v ¶ What nede was it here to be wedde
[marked   On to A man of so grett Age
110]   lesse þan þei myght bothe ago to bedde
    and kept þe lawe of maryage.

                Jhesus
    To blynde þe devyl of his knowlache   245
    and my byrth from hym to hyde
    þat holy wedlok was grett stopage
    þe devyl in dowte to do A-byde.

¶ Also whan sche xulde to egypte gon [1]
  and fle from herowde for dowte of me   250
  be-cawse she xulde nat go Alon
  Joseph was ordeyned here make to be
  my ffadyr of his hyʒ mageste
  here for to comforte in þe way
  these be þe cawsys as ʒe may se   255
  why joseph weddyd þat holy may.
                Maria

¶ A dere childe dere chylde why hast þou þus done
  Ffor þe we haue had grett sorwe *and* care
  thy ffadyr *and* I thre days haue gone
  Wyde þe to seke of blysse [2] ful bare.   260
                Jhesus
  Why haue ʒe sought me wi*th* evy [3] fare
  Wete ʒe not wele I muste been [4]

---

[1] The words *Jhesus adhuc* first written in the margin against this line with red loop as the name of speaker, and the line separating one speech from another drawn above this stanza; words and line cancelled in red ink.

[2] Crossed out and *comfort* roughly written above by a (? that of Ff. 95, 96) later hand.

[3] Altered to *hery* by the later hand of Ff. 95, 96.

[4] Or *ben*. Two *e*s seem to be run together. A final *e* is marked with a dot for deletion.

A-monge hem þat is my faderys ware
his gostly catel for to ovyrsen.

            Maria

¶ ȝour ffaderys wyl must nedys be wrought   265
It is most wurthy þat it so be
ȝitt on ȝour modyr haue ȝe sum thought
And be nevyr more so longe fro me

o. 111 As to my thynkynge these days thre
þat ȝe Absente haue ben A-way   270
be more lengere in þer degre
þan All þe space of xij ȝere day.

            Jhesus

¶ Now ffor to plese my modyr mylde
I xal ȝow folwe with obedyence
I am ȝour sone and subjecte childe   275
and Owe to do ȝow hyȝ reverence
home with ȝow I wyl go hens
Of ȝow clerkys my leve I take
Euery childe xulde with good dyligens
his modyr to plese his owyn wyl forsake [1].   280

          i^{us} doctor

¶ O blyssyd jhesu with ȝow we wende
of ȝow to haue more informacion
Fful blyssyd is ȝour modyr hende
of whom ȝe toke ȝour incarnacion
we pray ȝow jhesu of consolacion   285
At oure most nede of ȝow to haue
all þat hath herd þis consummacion
of þis pagent ȝour grace þem saue.

            Amen.

[Remainder of Fo. 111—2⅔ inches—and Fo. 111ᵛ—except for title of next play added by reviser and a scribbled name, He. Kinge the yownger,—left blank.]

[1] *to* added above the line between *wyl* and *forsake* in darker ink.

Hic Incipyt Johannes Baptysta.

Johannes[2]

Ecce vox clamantes in deserto
I am þe voys of wyldirnese
þat her spekyth *and* prechych yow to
loke ȝe for-sake all wrecchidnesse
fforsake all synne þat werkyth woo   5
And turne to[3] vertu and holynese
Beth clene of levyng in your sowle also
Thaṅ xall ȝe be savyd from peynfulnese
Of fyere brynnyng in hell
If þat ȝe for-sak synne   10
hevyṅ blysse xall ȝe wyne
Drede ȝe not þe devyllys gynne
with Angellys xall yow dwell.

Penitenciam nunc agite
Appropinquabit regnum celorum   15
ffor your trespas penaunce do ȝe
and ȝe xall wyṅ hevyṅ dei deorum
In hevyn blyse ye xall wyn to be
Among þe blyssyd company omnium supernorum
per as is all merth joye and glee   20
Inter agmina angelorum
In blyse to a-byde
Baptyme I cowncell yow[4] for to take
And do penaunce for your synnys sake
and for your offens amendys ȝe make   25
your synnys for to hyde.

I gyff baptyṁ in water puere
Þat is callyd flomjordoṅ
My baptyṁ is but sygnyfure

---

[1] This folio is an interpolated leaf of different paper, the wire lines of which correspond with those of Ff. 95 and 96. The handwriting is that found on Ff. 95 and 96. There is no rubrication.

[2] This name stands at the top of the page.

[3] *to* omitted and written over the line.

[4] *take* first written before *for* and crossed through.

The Baptism

  Of his baptyṁ þat his lyke hath non   30
112ᵛ He is a lord of gret valour
  I am not worthy to¹ oñ-bokyh his schoñ
  ffor he xaH baptyze as² seyth scryptour
  þat comyth of hem aH euery-chone
  In þe holy goost   35
  he may dampne and he may save
  aH goodnese of hem we haue
  þer may no³ man his werkys deprave
  ffor he is lord of myghtys most.

hic accedit Jhesus ad Johannem quem intuens Johannes dicat
digito demonstrans Jhesum

  Ecce Agnus dei qui tollit peccata mundi   40
  Be-holde þe lombe of god is this
  þat comyth now here be-forne
  þe wich xaH wasch þe worldys mys
  and saue aH þat that was for-lorne
  This same lombe for soth it is
  þat of a mayd fuH clene was borne   45
  Shamfuh deth þis lambe⁴ i-wys
  xaH suffer for us and be aH to-torne
  And rent oñ a roode
  he xaH suffer for mannys sake
  lytyh rest and⁵ moch gret sorow and⁵ wrake   50
  hys bake xaH be bowndyn to a stake
  And betyñ owt aH his bloode.
              Jhesus

113⁶ ¶ Johan Baptyste myn owyn good ffrende
  þat ffeythffully doth prech my wylle
  I the thanke with aH my mende   55
  Ffor þat good servyse þou dost me tylle
  thy desyre is synne to shende
  AH synful lyff⁷ þou woldyst spylle.
  thyn entente hath a good hende

---

¹ *v* first written before *oñ* and crossed through.
² *scrypture* first written before *seyth* and crossed through.
³ *no* omitted and written above the line.
⁴ The *b* is written over some other letter.
⁵ The words from *and* to *and* are written over an erasure.
⁶ The original hand of the MS. begins again here.   ⁷ MS. *synfullyft*.

þe lawe of god þou dost fful-fylle 60
  þis tyde
Baptym̄ to take I come to the
and conferme þat sacrement þat newe xal be
In flomjordon þou baptyze me
In water þat is wyde. 65
                              Johannes

¶ My lorde god þis be-hovyth me nought
  with my*n* hondys to baptyze the
  I xulde rather of the haue sought
  holy baptym þan þou of me.
                              Jhesus

Suffyr now Johan my wyl were wrought 70
all ryght-ffullnes[1] þus ffulfyll we
me to baptyze take þou no dowth
þe vertu of mekenes here tawth xal be
Euery man to lere
And take ensawmple here by me 75
How mekely þat I come to þe
Baptym confermyd now xal be
me to baptyze take þou no dwere[2].
                              Johannes

¶ All men may take exaunple[3] lo
  of lowly mekenes evyn ryght here 80
  be oure lorde god þat comyth me to
  hese pore servaunt *and* his su .. tere[4]

Fo. 113ᵛ Euery man[5] lere to worke ryght so
Bothe Kynge *and* Caysere *and* gret Empere[6]
be meke *and* lowe þe pore man to 85
And put out pryde in all manere
God doth here þe same
To þi byddynge my lord so dere
I me obey *with* gladsu*m* chere
And baptyze the *with* watyr clere 90
Euer halwyd be þi name.

---

[1] *ffl* first written instead of *fful* and crossed through.
[2] *dwere* crossed through and *fere* written after it by the later hand.
[3] So in MS. [4] Some letter, possibly a *w*, has been erased between *u* and *t*.
[5] *man* omitted and written above the line.
[6] The contraction mark might be expanded Emper*er*—the same being used for (r) *e* and *er*—but the rhyme demands *Empere*.

The Baptism

Spiritus sanctus hic descendat super ipsum et deus ¶ pater celestis dicet in celo.

¶ This is my welbelovyd chylde
  Ouer whom my spryte doth ouer sprede
  Clene and pure And vndefylyd
  of body of sowle ffor thought for[1] dede   95
  That he is buxhum meke and mylde
  I am wel plesyd with-owtyn drede
  Wysly to wysse ȝow ffrom weys wylde
  to lysten his lore all men I rede
  And ȝoure erys to herke   100
  Take good hede what he doth preche
  and ffolwyth þe lawys þat he doth teche
  Ffor he xal be ȝour Alther-is leche
  to saue ȝow from deuelys derke.

                            Johannes bap[tista
¶ Here I se with opyn syght   105
  The sone of God þat þou erte[2]
  the hooly goost ouer the doth lyght
  þi faderys voys I here fful smerte
  The childe of god as I þe plyght
  þat þou be whilys I am qwerte   110
  I xall wyttnes to every whyght
  and teche it trewly with all myn hert
  To sese it were grett synne
  Ffor goddys sone I wurchypp the
  Ffrom hevyn þin hyȝ mageste   115
  thu comyst hedyr ffrom dygnite
  mannys sowle to wynne.

                            Jhesus
¶ Johan Baptyste þou be wyttnes
  the trewth loke þat þou nat hyde
  Ffor now I passe forth in to wyldernes   120
  the holy gost xal be my gyde.
hic ihesus transit in desertum dicens et cetera.
  In whylsum place of desertnes

---

[1] *of* first written and crossed through; and *for* written above the line.
[2] *arte* first written and corrected by scribe to *erte*.

xl^ti days a terme¹ ful wyde
and ffourty nyght*ys* both more *and* lesse
With-owtyn bodyly ffode þer to a-byde 125
Ffor man þus do I swynke
In to deserte I² passe my way
ffor mannys sake as I ȝow say
xl^ti nyght*ys and* xl^ti day
I xal nowther ete nor drynke. 130
          Johan baptyst

¶ In place where I passe · wyttnes I bere
the trewth xal I telle where-so-evyr I go
þat cryst þe sone of god is be-come oure fere
clad in oure clothynge to sofer for us wo
I baptyzid with myn owyn handys cryst jhesu ryght here 135
and now he is to wyldyrnes penawns þer to do
In-formyng so all us þat lord þat hath no pere
to do for oure trespace penawnce here also

Fo. 114ᵛ Of penawnce do I preche
In wyttnes ryght be this 140
þat what man for his mys
doth penawns here i-wys
his sowle he doth wel leche.

¶ All men on ground þat be ȝitt on lyue
Ffor ȝour grett offens loke ȝe be repentaunt 145
Of all ȝour venym synne I rede þat ȝe ȝow shryve
Ffor god is ful redy mercy for to graunt
Be contryte³ for ȝour trespas *and* penauns do be-lyve
Reconsyle ȝour-self *and* be to god plesaunt
With contryscion schryffte *and* penauns þe devyl may ȝe dryve 150
Ffor fro ȝour Ffelachep he xal not be erraunt
ȝow for to meve
To penauns⁴ *and* synne forsake
Shryfte of mowth loke þat ȝe make

¹ The MS. has been rubbed here, and this word has been written in darker ink over the original word, which may, however, have been the same. The rubbing and subsequent touching up extends down seven lines, but in no other case is the original letter or word unclear.
² Inked over by the later hand.
³ *conty* first written for *contry[te]* and crossed through.
⁴ *per* first written for *pen[auns]* and crossed through.

The Baptism

And þan þe fende[1] in helle so blake 155
he xal ȝow nevyr more greve.

¶ A tre þat is bareyn and wyl bere no frute
þe ownere wyl hewe it down and cast it on þe fyre
Ryght so it be man þat folwyth þe fowle sute
of þe devyl of helle and werkyth his desyre 160
God wyl be vengyd on man þat is both dum and mute
þat wyl nevyr be shrevyn but evyr more doth delyre
Clothe the in clennes with vertu be indute
And god with his grace he wyl þe sone inspyre
to Amendynge of þi mys 165
Schryfte of mowthe may best þe saue
Penauns for synne what man wyl haue
whan þat his body is leyd in grave
His sowle xal[2] go to blys.

¶ Corne þat is good man kepe it ful clene 170
Chaff þat is sympyl is sett wul nere at nought
So good men of levynge to god chosyn bene
Whan synful men be lyke chaff and to helle xul be brought
Good penauns ȝow to preche ful hertyly do I mene
Shryfft and satysfaccion evyr more to haue in thought 175
What man in good penauns and schryfte of mowth be sene
of god he is welbelovyd þat all þis worlde hath wrought
and all þinge of nowth dede make
Now haue I tawght ȝow good penauns
god graunt ȝow grace at his plesauns 180
to haue of synne delyverauns
Ffor now my leve I take.

Sathan

¶ Now belyard and belzabub ȝe derwurthy devel of helle
And wysest of councel amonges all þe rowte
herke now what I sey a tale I xall ȝow telle
þat trobelyth sore my stomak þer of I haue grett dowte.

---
[1] *felle* first written and corrected to *fende*.
[2] *xal* and *go*, and *good* in the line below, are obscured by a stain on the MS.

                                            Belyall
          Syr sathanas oure souereyn syre *with* þe wol we dwelle   5
          All redy at þi byddynge to þe do we lowte
          If þou haue Any nede of oure wyse counselle
          telle us now þi qwestyon All out *and* oute
                  sey al þi dowte be-den[e.
                                            Belsabub
          ȝa sere telle us þi dowte by *and* by   10
          and we xul telle þe so sekyrly
          þat þou xalt knowe verryly
                  What þi dowte doth mene.
                                            Sathan
        ¶ The dowte þat I haue it is of cryst i-wys
          born he was in bedleem as it is seyd   15
          And many a man wenyth þat goddys sone he is
          born of a woman *and* she a clene mayd
          And all þat evyr he prechyth it is of hevyn blys
          he wyl lese oure lawe I am ryght sore afrayd
          Ffayn wold I knowe who were ffadyr his   20
          Ffor of þis grett dowte I am sore dysmayd
                  in dede
          If þat he be goddys childe
          and born of a mayd mylde
          than be we rygh¹ sore begylde   25
                  and short xal ben oure² spede.

        ¶ Therfore ser*ys* sum what þat ȝe shewe
          In þis grett dowth what is best to do
          If he be goddys sone he wyl brede a shrewe
          and werke us mech wrake both wrech *and* woo   30
          Sorwe *and* care he wyl sone strewe
Fo. 116ᵛ All oure gode days þan xulde sone be goo
          and all oure lore *and* all oure lawe he wyl down hewe
          and þan be we all lorn if þat it be soo
          he wyll don us all tene   35
          he wyll be lorde ouer hevyn *and* helle
          and ffeche awey all oure catelle
          þerfore shewe now sum good counselle
          What comfort may best bene.

              ¹ So in MS.   ² Some word before *oure* crossed through.

The Temptation

                Belyall

¶ The best wytt þat I kan say   40
hym to tempte forsoth it is
with sotyl whylys if þat þou may
A-say to make hym to don A-mys
If þat he synne þis is no nay
he may nat be kynge of blys   45
hym to tempte go walke þi way
Ffor best counsell I trowe be this
Go forth now and assay.

                Belsabub

The best wytt I hold it be
hym to tempte in synnys thre   50
the whiche mankende is frelte
doth ffalle sonest Alway.

                Sathan

¶ So Afftyr ȝour wytt now wyll I werke
I wyll no lengere here a-byde
be he nevyr so wyse a clerke   55
I xal apposyn hym with-inne A tyde.

                Belsabub

now louely lucyfer in helle so derke
Kynge and lorde of synne and pryde
with sum myst his wyttys to merke
He send þe grace to be þi gyde   60
and evyr more be þi spede.

                Belyall

p. 117 All þe deuelys þat ben in helle
shul pray to Mahound as I þe telle
þat þou mayst spede þis jurney well
and comforte the in þis dede[1].   65

                Jhesus

¶ xl^{ti} days and xl^{ti} nyght
now haue I fastyd for mannys sake
A more grett hungyr had neuyr no wyght
than I myself be-gynne to take
Ffor hungyr in peyn stronge am I pyght   70
and bred haue I non myn hungyr for to slake
A lytel of a loof relese myn hungyr myght
but mursele haue I non my comforte for to make

    [1] *caas* first written and crossed through.

This suffyr I man for the
Ffor þi glotenye and metys wrong 75
I suffyr for þe þis hungyr stronge
I am afferde it wyl be longe
      Or þou do þus for me.

                                                          Sathan

¶ The sone of god if þat þou be
be þe grett myght of þi godhede 80
turne these flyntys Anon lett se
ffrom Arde stonys to tendyr brede
more bettyr it is as I telle the
wysely to werke aftyr my reed
and shewe þi myght of grett majeste 85
than thorwe grett hungyr ffor to be dede
these stonys now bred þou make
goddys sone if þat þou be
make these stonys bred lett se
þan mayste þou ete ryght good plente 90
      thyn hungyr for to slake.

                                                         Jhesus

Fo. 117ᵛ ¶ Nott only be bred mannys lyff ȝitt stood
but in þe wurde of god as I þe say
to mannys sowle is neuyr mete so good
As is þe wurd of god þat prechid is Alway 95
bred materyal doth norch blood
but to mannys sowle þis is no nay
nevyr more may be a betyr food
þan þe wurd of god þat lestyth ay
* to here goddys wurde þerfore man loue 100
Thi body doth loue materal brede
With-oute þe wurde of god þi soule is but dede
to loue prechynge þerfore I rede
If þou wylt duellyn in blysse a-bove.

                                                         Sathan

¶ Ffor no grett hungyr þat I kan se 105
In glotony þou wylt not synne
now to þe temple com forth with me
and þer xal I shewe þe a praty gynne
Vp to þis pynnacle now go we

      * Paragraph mark erased here.

The Temptation

I xal þe sett on þe hyʒest pynne 110
ther I preue what þat þou be
Or þat we tweyn part a-twynne
I xal knowe what myght þou haue.

hic ascendit deus pinnaculum templi dum diabolus dicit quod sequitur.

Whan þou art sett upon þe pynnacle
þou xalt þer pleyn a qweynt steracle 115
Or ellys shewe a grett meracle
thy-sself[1] ffrom hurte þou saue.

hic satanas ponit ihesum super pinnaculum dicens

¶ Now If þou be goddys ssone of myght
Ryght down to þe erth anon þou ffalle
And saue þi-sylf in every plyght 120
Ffrom harm and hurte and Scappys alle
Ffor it is wretyn[2] with aungelys bryght
þat ben in hevyn þi faderys halle
the to kepe boþ day and nyght
xul be ful redy as þi thralle[3] 125
hurt þat þou non haue
þat þou stomele not a-geyn þe ston
and hurt þi fote as þou dost gon
Aungell be redy att everychon
  in weys þe to saue. 130

Jhesus

¶ It is wretyn in holy book
þi lorde god þou xalt not tempte
aH thynge must obeye to goddys look
out of his myght is[4] non exempt
Out of þi cursydnes and cruel crook 135
by godys grace man xal be redempt
whan þou to helle þi brennynge brook
to endles peyne xal evyr be dempt

---

[1] A dot by the side of the first *s* may be intended to indicate deletion, but cf *ssone* beneath.
[2] The final *n* is in darker ink and may have been added later, but there is the usual space between it and the next word.
[3] MS. *tharalle*, with deleting dot beneath first *a*.
[4] Some word (? *in*) first written.

ther in Alwey to a-byde
Thi lorde god þou tempt no more 140
It is nott syttenge to þi lore
I bydde þe sese a-non þerfore
And tempte god in no tyde.

Sathan

¶ Ow in gloteny nor in veynglory it doth ryght nott a-vayl
Cryst for to tempt it profyteth me ryght nought 145
I must now be-gynne to haue a newe travayl¹
In covetyse to tempt hym it comyth now in my thought
Ffor If I went þus A-way and shrynkyd as a snayle
lorn were þe labore all þat I haue wrought
þerfore in covetyse oure syre I xal a-sayle 150
And assay into þat synne yf he may be brought
anon forth ryght
Syr ȝitt onys I pray to the
to þis hyȝ hyl com forth with me
I xal þe shewe many a cete 155
And many a wurthy syght.

Tunc ihesus transit cum diabolo super montem et diabolus dicit

¶ In to þe northe loke fforth evyn pleyn
the towre of babolony þer mayst þou se
the cete of Jerusalem stondyth þer ageyn
and evyn ffast þer by stondyth Galyle 160
Nazareth naverne and þe kyngdom of Spayn
zabulon and neptalym þat is a rych countre
bothe zebee and salmana þou mayst se serteyn
Itayl and Archage þat wurthy remys be
bothe januense and jurye 165
Rome doth stonde be-fore þe ryght
the temple of Salamon as sylver bryght
And here mayst þou se opynly with syght
both Ffraunce and normandye.

¶ Turne þe now on þis syde and se here lumbardye 170
of spycery þer growyth many An C. balys
Archas and Aragon and grett Almonye

¹ *trav* first written for *trav[ayl]* and crossed through.

The Temptation

Parys *and* portyngale *and* þe towñ of Galys
Pownteys *and* poperynge *and* also pycardye
Erlonde scottlonde *and* þe londe of walys *   175
Grete pylis *and* castellys þou mayst se wit*h* eye
ȝa *and* aƚƚ þe wyd werde wit*h*-oute mo talys
Aƚƚ þis longygh to me
If þ*ou* wylt knele down to þe grownde
and wurchepp me now in þis stownde¹   180
aƚƚ þis world² þ*at* is so rownd
I xal it gyve to the.

                               Jh*es*us

¶ Go Abak þou fowle sathanas
   in holy scrypt*ur*e wretyn it is
   thi lorde god to wurchipp in every plas   185
   As for his thraƚƚ *and* þou servaunt his.

                               Sathañ

Out out harrow Alas Alas
I woundyr sore what is he this
I can not brynge hym to no³ trespas
nere be no synne to don a-mys   190
he byddyth me gon a-bakke
What þ*at* he is I kan not se
Whethyr god or man what þ*at* he be
I kan not telle in no degre
Ffor sorwe I lete a crakke.   195

*hic venient Angeli cantantes et ministrantes ei · Gloria tibi domine ·
dicens* ⁴

                               Jh*es*us

¶ Now Aƚƚ mankende exaumple⁵ take
   by these grete werkys þ*at* þou dost se
   how þ*at* þe devyƚƚ of helle so blake
   in synne was besy to tempte me
   Ffor aƚƚ hise maystryes þ*at* he dyd make   200

---

* Cf. similar lists in the *Castle of Perseverance* and in the Croxton *Play of the Sacrament*.

¹ The final *e* is blotted, possibly meant to be obliterated.

² The *l* is corrected from another letter; possibly the scribe first intended *word*.

³ Some word written before *no* and crossed through.

⁴ *dicens* is enclosed in a red loop.

⁵ Possibly *exaunple* corrected to *exaumple*.

The Temptation

Fo. 119ᵛ

he is ouercom̄ and now doth ffle
aƚƚ þis I suffyr ffor mannys sake
to teche þe how þou xalt rewle the
Whan þe devylle dothe the Assayle
loke þou concente nevyr to synne 205
For no sleytys ne for no gynne
and þan þe victory xalt þou wynne
þe devyl xal lesyn aƚƚ his travayl.

¶ To Suffyr temptacion it is grett peyn̄
If þou with-stonde it þou wynnyst grett mede 210
Of god þe more grace þou hast serteyn̄
If þou with-sett þe devyl in his dede
thow þat þe fende tempt þe Ageyn
of his power take þou no drede
Ffor god hath the ȝovyn botℏ myght and mayn 215
hym for to with-sytt evyr at nede
þou hast more myght than he
Whan þe devyl doth tempte the ,' thoo
Shewe þi myght aȝens þi ffoo 219
whan þi sowle partyth the froo   Amen
In blysse þan xal it be.

[Remainder of Fo. 119ᵛ—3⅝ inches—and Ff. 120, 120ᵛ blank (except for scribblings).]

Fo. 121 Hic de muliere in adulterio deprehensa.   ¶ Nolo mortem peccatoris[1].

¶ Jhesus

MAn for þi synne take repentaunce
If þou amende þat is amys
Than hevyn xal be þin herytaunce
Thow þou haue don Aȝens god grevauns
ȝett mercy to haske loke þou be bolde 5
his mercy doth passe in trewe balauns
Aƚƚ cruel jugement be many folde.

[1] This is written in larger but not liturgical script.
A note written very small in the hand of the scribe stands above it: gyn at nolo morte[m.

¶ thow þat ȝour synnys be nevyr so grett
Ffor hem be sad *and* aske mercy   10
sone of my ffadyr grace ȝe may gett
wi*th* þe leste teer wepynge owte of ȝo*ur* ey
My ffadyr me sent the man to bye
All þ*i* Raunsom my-sylfe must pay
Ffor loue of þe my-sylfe wyl dye   15
Iff þ*ou* aske mercy I sey nevyr nay.

¶ In to þe erth ffrom hevyn A-bove
þ*i* sorwe to sese *and* joye to restore
man I cam down all ffor þ*i* loue
Loue me ageyn I aske no mor*e*   20
þow þ*ou* mys-happe *and* synne ful sore
ȝit turne Aȝen *and* mercy craue
it is þ*i* fawte *and* þ*ou* be lore
haske þ*ou* mercy *and* þ*ou* xalt haue.

¶ Vppon þ*i* neybor*e* be not vengabyl   25
Ageyn þe lawe if he offende
lyke as he is þ*ou* art vnstabyl
thyn owyn frelte evyr þ*ou* attende
Eu*er* more þ*i* neybor*e* helpe to Amende
evyn as þ*ou* woldyst he xulde þe   30
Ageyn hym wrath if þ*ou* accende
the same in happ wyll falle on the.

¶ Eche man to othyr be mercyable
And mercy he xal haue at nede
What man of mercy is not tretable   35
Whan he Askyth mercy he xal not spede
Mercy to graunt I com in dede
Who so Aske mercy he xal haue grace
Lett no man dowte for his mysdede
But evyr Aske mercy whyl he hath space.   40
                                                        Scriba

¶ Alas Alas oure Lawe is lorn
A fals Ypocryte jh*es*u be name
þ*at* of a sheppherdis dowtyr was born
Wyl breke our*e* lawe *and* make it lame
he wyl us werke ryght mekyl shame   45

his fals purpos if he up-holde
all oure lawys he doth defame
þat Stynkynge beggere is woundyr bolde.

Phariseus

¶ Sere scrybe in feyth þat ypocryte
wyl turne þis londe al to his lore 50
þerfore I councell · hym to indyte
And chastyse hym ryght wel þerfore.

Scriba

On hym be-leve many A score
In his prechynge he is so gay
Ech man hym ffolwygh ever more and more 55
Aȝens þat he seyth no man seyth nay.

Phariseus

Fo. 122 ¶ A Ffals qwarel if we cowde feyne
(K quire) þat ypocrite to puttyn in blame
All his prechynge xulde sone disteyne
and than his wurchepp xuld turne to shame 60
with sum falshede to spyllyn his name
lett vs assay his lore to spylle
þe pepyl with hym yff we cowde grame
than xulde we sone haue al oure wyll.

Accusator
65

¶ Herke sere pharysew and sere scrybe
A ryght good sporte I kan ȝow telle
I vndyr-take þat ryght a good brybe
We all xul haue to kepe councell
A fayre ȝonge qwene here-by doth dwelle
both fflesch and gay upon to loke 70
And a tall man with here doth melle
the wey in to hyre chawmere ryght evyn he toke.

¶ lett us thre now go streyte thedyr
the wey fful evyn I xall ȝow lede
and we xul take them both to-gedyr 75
Whyll þat þei do þat synful dede.

Scriba

Art þou sekyr þat we xal spede
Shall we hym fynde whan we cum there.

Accusator

be my trowth I haue no drede
þe hare fro þe fforme we xal a-rere. 80

The Woman taken in Adultery

              Phariseus

¶ We xal haue game *and* þis be trewe
lete us thre werke by on Assent
We wyl here brynge evy*n* be-for*n* jh*es*u
and of here lyff þe truth present
How in advowtrye hyr*e* lyff is lent   85
Than hym be-for*n* whan she is browth
We xul hym Aske þ*e* trew jugement
What lawfull deth to her*e* is wrouth.

¶ Of grace *and* mercy hevyr he doth preche
And þat no man xulde be vengeable   90
Ageyn þ*e* woman if he sey wrech
Than of his prechynge he is vnstable¹
And if we fynde hym varyable
of his prechynge þat he hath tawth
than haue we cawse bothe juste *and* Able   95
Ffor a fals man þat he be cawth.

             Scriba

¶ Now be² grete god ȝe sey fful well
If we hym fyndyn in varyaunce
We haue good reson as ȝe do tell
Hym for to brynge to foule myschauns   100
If he holde stylle his dalyauns
and preche of mercy hi*r*e for to saue
than haue we mater of gret substauns
hym for to kylle *and* putt in graue.

¶ Grett reson why I xal ȝow telle   105
Ffor moyses doth bydde in our*e* lawe
that Eu*er*y Advowterere we xuld qwelle
and ȝitt w*ith* stonys thei xulde be slawe
Ageyn moyses³ if þat he drawe
þat Synful woman w*ith* grace to helpe   110
he xal nevyr skape out of our*e* Awe
but he xal dye lyke a dogge whelpe.

---

¹ *unstabyl* first written and corrected to *vnstable*.
² The words *Now be* and *we* in the line below are obscured by a blot.
³ The words *Ageyn moyses* are written in a slightly larger form.

                                                    Accusator
Fo. 123 ¶ ȝe tary ovyr longe ser*ys* I sey ȝow
         they wyl sone parte as þat I gesse
         þerfore if ȝe wyl haue ȝo*ur* pray now   115
         lete us go take them i*n* here whantownnesse.
                                                    Phariseus
         Goo þou be-forn þe wey to dresse
         We xal þe ffolwe w*ith*-in short whyle
         Iff þat we may þat quene dystresse
         I hope we xal jh*es*u be-gyle.   120
                                                    Scriba
      ¶ Breke up þe dore *and* go we inne
         Sett to þe shuldyr w*ith* a*ll* þi myght
         We xal hem take evyn in here synne
         here owyn trespas shal þem indite.

hic juuenis quidam extra currit indeploydo calligis non ligatis et braccas in manu tenens et dicit accusator
                                                    Accusator
         Stow þat harlot sum erthely wyght   125
         that in advowtrye here is ffownde.
                                                    Juuenis
         ȝiff Any man stow me þis nyth
         I xal hym ȝeve a dedly wownde.

      ¶ I[1] Any man my wey doth stoppe
         or we departe ded xal I be   130
         I xal þis daggare putt in his croppe
         I xal hym kylle or he xal me.
                                                    Phariseus
         Grett goddys curse mut go w*ith* the
         W*ith* suche a shrewe wy*ll* I not melle.
                                                    Juuenes[2]
         That same blyssynge I ȝyff ȝow thre   135
         And qwheth ȝow alle to þe devyl of helle.

       * In feyth I was so sore affrayd
Fo. 123ᵛ Of ȝone thre shrewys þe sothe to say
         my breche be nott ȝett we*ll* up-teyd
         I had such hast to renne A-way   140

                    [1] So in MS. for *If*.   [2] So in MS.
                    * No paragraph sign here in MS.

Thei xal nevyr cacche me in such affray
I am full glad þat I am gon
Adewe Adewe a xx^{ti} devyl way
and goddys curse haue ȝe every-chon.

Scriba 145

¶ Come forth þou stotte com forth þou scowte¹
com forth þou bysmare *and* brothel bolde
com forth þou hore *and* stynkynge bych clowte
how longe hast þou such harlotry holde.

Phariseus

Com forth þou quene com forth þou scolde
com forth þou sloveyn com forth þou slutte 150
we xal the tecche *with* carys colde
A lytyl bettyr to kepe þi kutte.

Mulyer

¶ A mercy mercy ser*ys* I ȝow pray
Ffor goddys loue haue mercy on me
of my mys-levynge me not be-wray 155
haue mercy on me for charyte.

Accusator

Aske us no mercy it xal not be
We xul so ordeyn ffor þi lott
þat þou xalt dye for þin Advowtrye
Þerfore com forth þou stynkynge stott. 160

Mulier

¶ Ser*ys* my wurchepp if ȝe wyl saue
And helpe I haue non opyn shame
bothe gold *and* sylvyr ȝe xul haue
So þat in clennes ȝe kepe my name.

Scriba 165

Mede ffor to take we were to blame
to save suche stottys it xal not be
We xal brynge the to suche a game
þat all advowterer*ys* xul lern be the.

Mulier

¶ Stondynge ȝe wyl not graunt me grace
but for my synne þat I xal dye 170
I pray ȝow kylle me here in þis place
and lete not þe pepyl up-on me crye
If I be sclaundryd opynly

---

¹ MS. *scowtte*, with deleting dot beneath the first *t*.

to all my frendys it xul be shame
I pray ȝow kylle me prevyly 175
lete not þe pepyl know my defame.

Phariseus

¶ Ffy on þe scowte þe devyl þe qwelle
Ageyn þe lawe xul we þe kyll
Ffyrst xal hange þe · þe devyl of helle
or we such folyes xulde ffulfyll 180
thow it lyke þe nevyr so ill
be-fforn þe prophete þou xalt haue lawe
lyke as moyse doth charge ut[1] tyll
With grett stonys þou xalt be slawe.

Accusator

¶ Com forth a-pase þou stynkynge scowte 185
be-fore þe prophete þou were þis day
or I xal ȝeue þe such a clowte
þat þou xalt fall down evyn in þe way.

Scriba

now be grett god and I þe pay
Such a buffett I xal þe take 190
þat all þe teth[2] I dare wel say
with-inne þin heed ffor who xul shake.

Phariseus

Fo. 124ᵛ ¶ Herke sere prophete we all ȝow pray
to gyff trewe dom and just sentence
Vpon þis woman which þis same day 195
In synfull advowtery hath don offense.

hic ihesus dum isti accusant mulierem continue debet digito suo scribere in terra.

Accusator

Se we haue brought here to ȝour presens
be-cawse ȝe ben a wyse prophete
þat ȝe xal telle be consyens
What deth to hyre ȝe thynke most mete. 200

Scriba

¶ In moyses lawe ryght þus we fynde
þat such fals louers xul be slayn
streyte to a stake we xul hem bynde
and with grett stonys brest out þer brayn

---

[1] So in MS. for *us*.
[2] This word first miswritten in some way and crossed through.

Of ȝour concyens telle us þe playn 205
with þis woman what xal be wrought
Shall we lete here go qwyte agayn
or to hire deth xal she be brought.

Jhesus nichil respondit sed semper scrybyt in terra.

¶ Now holy prophete be mercyable   Mulier [1]
vpon me wrecch take no vengeaunce 210
Ffor my synnys Abhomynable
In hert I haue grett repentaunce
I am wel wurthy to haue myschaunce
Both bodyly deth and werdly shame
but gracyous prophete of socurraunce 215
þis tyme pray ȝow for goddys name.

    Phariseus

¶ Ageyn þe lawe þou dedyst offens
þerfore of grace speke [2] þou no more
As moyses gevyth in law sentens
þou xalt be stonyd to deth þer-fore. 220

    Accusator

Ha don sere prophete telle us ȝoure lore
xul we þis woman with stonys kyll
or to hire hous hire home restore
In þis mater tell us ȝour wyll.

    Scriba

¶ In a colde stodye me thynkyth ȝe sytt 225
good sere awake telle us ȝour thought
xal she be stonyd telle us ȝour wytt
or in what rewle xal sche [3] be brought.

    Jhesus

Loke which of ȝow þat nevyr synne wrought
but is of lyff clennere þan she 230
Cast at here stonys and spare here nowght
Clene out of synne if þat ȝe be.

hic ihesus iterum se inclinans scribet in terra et omnes accusatores quasi confusi separatim in tribus locis se disiungent,'

---

[1] This word is written in the left-hand margin in another hand, the name of the speaker having been omitted.

[2] *spe* first written, *ke* inserted by the scribe over the line.

[3] Or *scho*—the letter is blotted.

Phariseus

¶ Alas Alas I am ashamyd
I am a-fferde¹ þat I xal deye
all myn synnys evyn propyrly namyd   235
ȝon prophyte dede wryte be-for myn eye
Iff þat my felawys þat dude Aspye
they wyll telle it bothe ffer and wyde
my synfull levynge if þei out crye
I wot nevyr wher myn heed to hyde.   240

Accusator

¶ Alas for sorwe myn herte doth blede
All my synnes ȝon man dude wryte
Fo. 125ᵛ If þat my felawys to them toke hede
I kan not me ffrom deth Acquyte
I wolde I wore hyd sum-where out of syght   245
þat men xuld me no-where se ne knowe
Iff I be take I am afflyght
In mekyl shame I xal be throwe.

Scriba

¶ Alas þe tyme þat þis be-tyd
Ryght byttyr care doth me enbrace²   250
All my synnys be now vnhyd
ȝon man be-for me hem all doth trace
If I were onys out of þis place
to suffyr deth gret ‖ and vengeauns Able³
I wyl nevyr come be-for his face   255
þow I xuld dye in a stable.

Mulier

¶ Thow I be wurthy ffor my trespas
to suffyr deth ab-homynable
ȝitt holy prophete of ȝour hyȝ grace
In ȝour jugement be mercyable   260
I wyl nevyr more be so vnstable
O holy prophete graunt me mercy
of myn synnys vnresonable
With all myn hert I am sory.

Jhesus

¶ Where be þi fomen þat dude þe Accuse   265
Why haue þei left us to Alone.

¹ *fed* first written for *fferde* and crossed through.
² So in MS.   ³ Line thus divided at *gret* in MS.

The Woman taken in Adultery

                Mulier

By-cawse they cowde nat hemself excuse
With shame they ffled hens Euery-chone
But gracyous prophete lyst to my mone
of my sorwe take compassyon   270
now all myn enmyes hens be gone
Sey me sum wurde of consolacion.

                Jhesus

¶ Ffor þo synnys þat þou hast wrought
hath Any man condempnyd the.

                Mulier

Nay for-soth þat hath þer nought   275
but in ȝour grace I putt me.

                Jhesus

Ffor me þou xalt nat condempnyd be
go hom A-geyn and walk at large
loke þat þou leve in honeste
and wyl no more to synne I þe charge.   280

                Mulier

¶ I thanke ȝow hyȝly holy prophete
Of þis grett grace ȝe haue me graunt
all my lewde lyff I xal doun lete
and ffonde to be goddys trewe servaunt.

                Jhesus[1]

What man of synne be repentaunt   285
of god if he wyl mercy craue
God of mercy is so habundawnt
þat what man haske it he xal it haue.

¶ Whan man is contrite and hath wonne grace
God wele not kepe olde wreth in mynde   290
but bettyr loue to hem he has
Very contryte whan he them fynde
Now god þat dyed ffor all mankende
saue all þese pepyl both nyght and day
and of oure synnys he us vnbynde   295
hyȝe lorde of hevyn þat best may.

                Amen.

[Ff. 126ᵛ and 127 left blank.]

---

[1] The word *Jhesus* has been crossed out and the word *Doctor* written below in the same hand as the word *Mulier* earlier in the play.

Fo. 127ᵛ   hic incipit de suscitacione lazari.

¶ Lazarus

    God þat all thynge dede make of nowth
    And puttyst eche creature to his fenaunce
Saue thyn handwerke þat þou hast wrought
As þou art lord of hyȝ substauns
O gracyous god att þi plesauns   5
of my dysese now comforte me
Which þurowe syknes hath such penawnce
On-ethys ffor heed Ache may I now se.

¶ Systyr Martha *and* Mawdelyn eke
With hast helpe me in bedde to dresse   10
Ffor trewly I am so woundyrly seke
I may nevyr schape¹ þis grett seknes
My deth is com̄ now I gesse
help in to chawmere þat I be led
my grett desesse I hope xal lesse   15
If I were leyd upon a bed.

                                Martha

¶ Lazarus brother be of good cher
I hope ȝour syknes ryght wel xal slake
Vpon þis bed rest ȝow rygh² her
And a good slep assay to take.   20

                                Magdalyn

Now jentyl brothyr ffor goddys sake
lyfte up ȝowre hert *and* be not feynt
An hevy housholde with vs ȝe make
If dedly syknes haue ȝow Ateynt.

                                Lazarus

¶ Ffor sothe dere systeryn I may not slepe   25
my seknes so sore doth evyr encrese
of me I pray ȝow take ryght good kepe
tyll þat my peyne be-gynne relese.

                                Martha

Fo. 128³   God graunt grace þat it may sese
of syknes god make ȝow sownde   30

---

¹ The *h* is corrected from some other letter.   ² So in MS.
³ The red loops of the names of the speakers are omitted on Fo. 128.

The Raising of Lazarus

or ellys oure joy wyll sone dyscres
In so grett peynes if ȝe ly bownde.

Magdalyn

¶ A brothir brothir lyfte up ȝoure herte
ȝour hevy cher doth us grevaunce
If deth from us ȝow xulde de-parte   35
than were we brought in comberaunce
ȝe be oure brothyr syb of Alyaunce
If ȝe wore deed þan had we none
ȝe do us brynge in distemperaunce
Whan ȝe us telle ȝe xal hens gone.   40

1us consolator

¶ Dame martha *and* magdalyne
how faryth ȝour brothire lete us hym se.

Martha

he is ryght seke *and* hath grett pyne
I am aferde deed he xal be.

Magdalyn

A man may haue ryght grett pete   45
the fervent hete of hym to fele.

ij^{us} consolator

Take ȝe no thought in no degre
I hope þat he xal ffare fful wele.

Martha

¶ he may nat leve his colowre doth chaunge
com to his bed ȝe xal hym se.   50

Magdalyn

Iff he longe leve it wyl be straunge
but as god wole so mut it be
Chere hym gode frendys ffor charyte
Comforte of hym we kan non gete
128^v Alas Alas what eylight me   55
myne herte for wo is wundyr grete.

3us consolator

¶ Al heyl syr lazarus how do ȝe fare
how do ȝe ffele ȝow in ȝour herte.

lazarus

I am with syknes all woundyn in care
And loke whan deth me xulde departe.   60

4us consolator et nu*n*cius

ȝe xal haue hele *and* leue in qwart
If ȝe wol take to ȝow good chere.

              Lazarus

  Whan deth[1] on me hath shet his dart
  I xal haue hele *and* ly on bere.

             1ᵘˢ consolator

¶ Be of good comforte *and* thynke not so   65
  put out of herte þat idyl thought
  ȝoure owyn mys-demynge may werke ȝow wo
  and cause ȝow sonere to deth be brought.

             ijᵘˢ consolator

  W*i*t*h* gret syknes þow ȝe be sought
  Vpon ȝoure-sylf haue no mystruste   70
  If þat ȝe haue I wundyr ryght nought
  Þow ȝe be deed *and* cast in duste.

             3ᵘˢ consolator

¶ Many on hath had ryght grett syknesse
  And aftyr hath had his hele agayn[2]
  and many a man þis is no lesse   75
  W*i*t*h* his wantruste hym-sylf hath slayn[3]
  ȝe be a man of ryght sad brayn
  þow þat ȝo*ur* syknes greve ȝow ryght ill
  Pluk up ȝo*ur* herte w*i*t*h* myght *and* mayn
  and chere ȝo*ur*-sylf with all ȝo*ur* wyll.   80

             Lazarus

Fo. 129 ¶ Ageyn my syknes þer is non ese
  but jh*e*su cryst my maystyr dere
  If þat he wyst of my dyssese
  Ryght sone I trust he wolde ben here.

             4ᵘˢ Consolator

  I xal go to hym w*i*t*h*-outyn dwere   85
  And of ȝo*ur* syknes telle hym serteyn
  loke þat ȝe be of ryght good chere
  Whyll þat I go *and* com ageyn.

             Martha

¶ Now jentyl ffrend telle hym ryght thus
  he þat he lovyth hath grett syknes   90
  hedyr to come *and* comforte vs
  Say þat we prayd hym of his goodnes.

---

[1] Some correction has been made by the scribe in this word.
[2] *ageyn* first written and corrected to *agayn*.
[3] The *a* here seems also to be written over another letter.

The Raising of Lazarus

              Magdalyn

Recomende us on to his hyʒnes
And telle hym all oure hertys wo
but he comforte oure hevynes   95
oure werdly[1] joy awey wyl go.
          4us consolator et nuncius

¶ The trewth ffor soth all every dele
 as ʒe haue told so xal I say
 go to ʒour brothyr and cheryse hym wele
 Ffor I walke fforth streyte in my way. *here goth he his way*[2] 100
             Martha

What chere good brothyr telle me I pray
What wele ʒe ete what wele ʒe drynk
loke what is plesynge to ʒour pay
ʒe xal haue what ʒe wole thynke.
           Lazarus

¶ My wynde is stoppyd gon is my breth   105
 And deth is come to make myn ende
 to god in hevyn my sowle I qweth
 Ffarwell systeryn ffor hens I wende.
hic lazarus moritur et cetera.
          Magdalyn

129ᵛ Alas ffor wo myn here I rende
Myn owyn dere brothyr lyth here now ded   110
now haue we lost a trusty ffrende
þe sybbest blood of oure kynreed.
          Martha

¶ Alas alas and wele way
 now be we tweyn bothe brotherles
 Ffor who my hert is colde as clay   115
 A hoo xal comforte oure carefulnes
 Ther had nevyr woman more doolfulnes
 A systyr Magdalyn[3] what is ʒour reed
 What whith may helpe oure hevynes
 now þat oure brother is gon and deed.   120
          Magdalyn

¶ Alas dere systyr I cannot telle
 þe best comforte þat I can sey[4]

---

[1] Some letter, possibly *s*, is erased at the end of this word.
[2] Written roughly in the hand of the scribe of Ff. 95, 96.
[3] *Mad* (?) first written for *Mag[dalyn]* and crossed through.
[4] *say* first written and corrected to *sey*.

but sum man do us sle *and* qwelle
lete us ly down by hy*m and* dey
Alas why went he alone awey 125
If we had deyd wit*h* hy*m* Also
than had our*e* car*e* all turnyd to pley
ther now all joye is turnyd to woo.

i*us* consolator

¶ Be of good comforte *and* thank god of al
Ffor det*h* is dew to every man 130
What tyme þat deth on us xal ffal
non erthely wyght þe our*e* telle can.

Martha

We all xul dye þat is sertan͞ ¹
but ȝit þe blood of kynde natur*e*
Whan det*h* þe brothyr Awey hath tan 135
must nedys murne þat sepultur*e*.

ij*us* consolator

Fo. 130 ¶ Good ffrendys I pray ȝow holde ȝo*u*r pes
All ȝo*u*r wepynge may not amende itt
of ȝo*u*r sorwynge þer for*e* now ses
And helpe he were buryed in a cley pitt. 140

Magdalyn͞

Alas þ*at* wurde my*n* herte doth slytt
þ*at* he must now in cley be graue
I wolde su*m* man my throte wulde kytt
þat I wit*h* hym myght lyne in caue.

iij*us* consolator

¶ Bothe heed a*n*d ffoot now he is wounde 145
in a chete ² bothe ffayr *and* clene
lete us ber*e* hym streȝte to þat grounde
Wher*e* þat ȝe thynke his graue xal bene.

Martha

We be ffull lothe þat pytt to sen
but stondynge it may no bettyr be 150
The coors take up ȝow thre betwen
wit*h* carefull herte ȝow ffolwe xal we.

*hic portauit corpus ad sepelliendu*m.

---

¹ The scribe began to write *e* after the *t* (? for *serteyn*).
² An initial *s* has been added to this word in a different ink—apparently by a later corrector.

The Raising of Lazarus   215

                                             Magdaleyn

¶ Alas comforte I se non othyr
but all of sorwe *and* care *and* woo
We dulfull women must burry our*e* brothir   155
Alas þat deth me wyl not slo
If I to pitt w*ith* hym myght go
þ*er*-in evyr mor*e* w*ith* hym to Abyde
than wer*e* my care all went me fro
þer now grett sorwe doth wounde me wyde.   160
                          i*us* consolator

¶ This coors we burry here in þis pytte
all myghty god þe sowle mut haue
and w*ith* þis ston þis graue we shytte
ffro ravenous[1] best*ys* þe body to saue.
                                     Magdalyn
He is now brought in to his cave[2]   165
my*n* hert ffor wo þis syght doth kylle
Lete us sytt down her*e* by þe grave
or we go hens wepe all our*e* fylle.

                                        Martha
¶ Vs for to wepe no man may lett
be-forn our*e* face to se þis syght   170
Alas qwy doth deth us not fett
Vs for to brynge to þis same plyght.
                          ij*us* consolator
Arys for shame ȝe do not ryght
streyth from þis grave ȝe xul go hens
þus for to grugge ageyns godys myght   175
Aȝens hyȝ god ȝe do offens.
                                     Magdalen

¶ Syth I must nedys w*ith* ȝow hens gon
my brother*ys* graue lete me fyrst kys
Alas no whith may helpe my mon
Ffare wel my brothyr far*e* wel my blys.   180
                          iij*us* consolator
hom to ȝo*ur* place we xal ȝow wysse
Ffor goddys loue be of good cher*e*
In dede ȝe do ryght sor*e* amys
so sore to wepe as ȝe do here.

[1] *rauenous* first written, the *u* corrected to *v* by scribe.
[2] This *v* also appears to have been touched up, but there is no trace of an original *u*.

                                                    Martha
¶ Lete us go hom than to oure place   185
  we pray ȝow aʟʟ with us to Abyde
  Vs to comforte with sum solace
  tyl þat oure sorwe doth slake and sclyde.
                                            1ᵘˢ consolator
  ȝow for to comforte at every tyde
  We xaʟʟ dwelle here bothe nyght and day   190
  and god þat made þis werd so wyde
  be ȝowre comforte þat best may.

hic iiij^{us} consolator et nuncius loquitur jhesu dicens
                                            4ᵘˢ consolator
Fo. 131 ¶ Heyl holy prophete jhesu be name
  Martha and Mawdelyn þo systeryn too
  Recomende hem to ȝour hyȝ fame   195
  And bad me sey to ȝow þus loo
  How þat Lazare qwhich þat ȝe lovyd so
  with grett syknes is sore dyssesyd
  to hym they prayd ȝow þat ȝe wolde goo
  If þat ȝour hyȝnes þerwith were plesyd.   200
                                                    Jhesus
¶ Dedly syknes lazare hath non
  but for to shewe goddys grete glorye
  Ffor þat syknes is ordeynyd a-lon
  þe sone of god to gloryfie.
                                                    Nuncius
  They be in dowte þat he xal deye   205
  grett syknes hym sore doth holde
  Ffor veruent hete his blood doth dreye
  his colore chaungyth as they me tolde.
                                                    Jhesus
¶ Goo hom ageyn And telle hem thus
  I xal come to hem whan þat I may.   210
                                                    Nuncius
  At ȝour comaundement · O prophete jhesus
  I xal hem telle as ȝe do say.
                                                    Jhesus
  Come forth bretheryn walke we oure way
  In to jurye go we A-non
  I cam not there ful many a day   215
  þerfore thedyr now wyl I gon.

              Omnes discipuli

¶ The jewys ageyn the were grym and grylle
 Whan þou were there they wolde þe aslayn
 With stonys they sowte þe ffor to kyll
 And wylt þou now go thedyr ageyn.   220
              Jhesus

 xij owrys þe day hath in certeyn
 In them to walke both clere and bryght
 he xal not stomble ageyn hyll nor pleyn
 þat goth þe wey whyl it is day lyght.

¶ but if men walke whan it is nyght[1]   225
 Sone they offende in þat dyrknes
 be-cawse they may haue no cler syght
 they hurte there ffete ofte in suche myrkenes
 but as ffor this ȝitt nevyr-þe-lesse
 the cawse þerfore I thedyr wyl wende   230
 is ffor to reyse ffrom bedde expresse
 lazare þat slepyth · oure Althere ffrende.
             Omnes discipuli

¶ Of his syknes[2] he xal be save
 If þat he slepe good sygne it is.
              Jhesus

 lazare is deed and leyd in grave   235
 Of his slepynge ȝe deme amys
 I was not there ȝe know weyl[3] this
 to strengthe ȝoure feyth I am ful glad
 þerfore I telle ȝow þe trewthe i-wys
 oure ffrende is deed and vndyr erth clad.   240
              Thomas

¶ Than goo we all ryght evyn streyth thedyr
 there as oure ffrende lazare is deed
 and lete us deye with hym to-gedyr
 þer as he lyth in þe same stede.
              Jhesus

 The ffor[4] to deye haue þou no drede   245
 the wey streyth thedyr in hast we take

---

[1] The colour of the ink becomes darker with this line and the writing slightly more angular. This continues more or less to the bottom of Fo. 134.
[2] The k is corrected from some other letter.
[3] The e omitted and written above the line.   [4] So in MS.

be þe grett myght of myn godhede
oute of his slepe he xal awake.

Nuncius

¶ All heyl Martha and mawdelyn eke
to jhesu I haue ȝour massage seyd¹ 250
I tolde hym how þat ȝour brothyr was seke
and with grett peyn in his bed leyd
He bad ȝe xulde not be dysmayde
all his syknes he xal Askape
he wyll byn here with-in a brayde 255
as he me tolde he comyth in rape.

Mawdelyn²

¶ That holy prophete doth come to late
oure brothyr is beryed iij days or this
A grett [stone]³ stoppyth þe pyttys gate
there as oure brothere beryde is. 260

Nuncius

Is lazare deed now god his sowle blys
ȝit loke ȝe take non hevynes
So longe to wepe ȝe don A-mys
It may not helpe ȝour sorynes.

Martha

¶ Oute of myn herte all care to lete 265
all sorwe and wo to caste Away
I xal go forth in þe strete
to mete with jhesu if þat I may.

ij^us consolator

God be ȝour spede bothe evyr and Ay
Ffor with ȝour sustyr we wyl abyde 270
here to comforte we xal a-say
and all here care to caste a-syde.

3us consolator

Fo. 132^v ¶ Mary Mawdelyn be of good herte
And wel be-thynke ȝow in ȝour mynde
Eche creature hens must depart 275
þer is no man but hens must wende.

---

¹ *Sayd* first written and corrected to *seyd*.
² The *w* written over a *d*.
³ This word is written above the line in a different ink, possibly by the scribe of Ff. 95, 96.

Deth to no wyht can be a frende
Aḧ þinge to erth he wyl down cast
Whan þat god wol aḧ thynge hath ende
lengere than hym lyst no thynge may last. 280

Magdalyn

¶ I thanke ȝow frendys ffor ȝour good chere
myn hed doth ake as it xulde brest
I pray ȝow therfore while¹ ȝe ben here
A lytil whyle þat I may rest.

4ᵘˢ consolator nunc[ius

þat lord þat made bothe est and west 285
graunt ȝow good grace suche rest to take
þat on to hym xulde plese most best
As he þis worlde of nought dyd make.

Martha

¶ A gracyous lord had ȝe ben here
My brother lazare þis tyme had lyved² 290
but iiij days gon upon a bere
We dede hym berye whan he was ded
ȝitt now I knowe with-owtyn drede
What thynge of god þat þou do craue
þou xalt spede of þe hyȝ godhede 295
what so þou aske · þou xalt it haue.

Jhesus

¶ Thy brothyr lazare Aȝen xal ryse
A levynge man aȝen to be.

Martha

I woot wel þat at ·⁄ þe grett last syse
he xal a ryse | and also we³. 300

Jhesus

Resurreccion þou mast me se
And hendeles lyff I am also
What man þat deyth and levyth in me
Ffrom deth to lyve he xal ageyn go.

¶ Eche man in me þat feythful is 305
and ledyth his lyff aftere my lore
of hendeles lyff may he nevyr mys

---

¹ *wyl* first written and crossed through.
² There are traces of correction in this word—possibly *leved* first written.
³ These line divisions seem intended to mark the extra rhyme.

euere he xal leve *and* deye nevyr more
'The body *and* sowle I xal restore
to endeles joye dost þou trowe this.   310
          Martha

I hope in the ‖ O cryst ful sore[1]
þou art þe sone of god in blys.

¶ Thy Ffadyr is god of lyff endeles
þ*i* self is sone of lyff *and* gras
to sese these wordlys wrecchydnes   315
Ffrom hefne to erth þou toke þe pas.
          Jhesus

Of hevynly myght ryght grett solas
to all þis world me[2] xul sone se
go calle þ*i* systyr in to þis plas
byd mary mawdelyn come hedyr to me.   320
          Martha

¶ At þ*i* byddyng I xal here calle
In hast we were here ȝow be-forn.
          Mawdelyn

Alas my mowth is bytter as galle
grett sorwyn my herte on tweyn hath scorn
Now þ*at* my brothyr from syth is lorn   325
þer may no myrth my care releve
Alas þe tyme þat I was born
þe swerde of sorwe my*n* hert doth cleve.
         i*us* consolator

Fo. 133ᵛ ¶ Ffor his dere loue þ*at* all hath wrought
Ses su*m* tyme of ȝo*ur* wepynge   330
and put all thynge out of thought
in-to þis care þat ȝow doth brynge.
         2*us* consolator

ȝe do ȝo*ur*-self ryght grett hyndrynge
and short ȝoure lyff or ȝe be ware
Ffor goddys loue ses of ȝour sorwynge   335
and wi*th* good wysdam refreyn ȝo*ur* care.
         Martha

¶ Sustyr magdalen com out of halle
oure maystyr is com as I ȝow say

---

[1] The reason for the dividing strokes in this line is not apparent.
[2] So in MS. for *men*.

he sent me hedyr ȝow for to calle
come forth in hast as I ȝow pray.

Magdalen 340

Ha · where hath he ben many a longe day
Alas why cam he no sonere hedyr
In hast I folwe ȝow anon þe way
me thynkyth longe or I come thedyr.

3*us* consolator 345

¶ Herke gode ffrendys I ȝow pray
Aftyr þis woman in hast we wende
I am a-ferde ryght in good fay
here-self for sorwe þat she wyl shende.

Nuncius

Here brothyr so sore is in hire mende
She may not ete drynke nor slepe 350
streyte to his graue she goth on ende
As a mad woman þer for to wepe.

Magdalen

¶ A Souereyn lord *and* mayster dere
had ȝe with us ben in presens
Than had my brother on lyue ben here 355
nat ded but qwyk þat now is hens
Ageyn deth is no resystens
Alas myn hert is woundyrly wo
Whan þat I thynke of his Absens
þat ȝe ȝour-self in herte lovyd so. 360

1*us* consolator

whan we haue mynde of his sore deth
he was to us so gentyl[1] *and* good
þat mend of hym oure hertys sleth
þe losse of hym doth marre oure mood.

ij*us* [2] consolator

Be bettyr neybore nevyr man stood 365
to euery man he was ryght hende
Vs he dede refresch with drynk *and* ffood
now he is gon | gon is oure frende.

Jhesus

¶ ȝowre grett wepynge doth me constreyne
Ffor my good ffrend to[3] wepe also 370

[1] This initial *g* has been corrected from an original *j* by the scribe.
[2] 3 first written for ii and crossed through.
[3] Some letter (? *w*) before *to* crossed through.

I can not me for wo restreyn
but I must wepe lyke as ȝe do.

hic ihesus fingit se lacrimari.   3ᵘˢ consolator

Be-holde þis prophete how he doth wepe lo
he louyd lazare ryght woundyrly sore
he wolde not ellys for hym þus wepe so   375
but if þat his loue on hym were þe more.

                                           Nuncius

¶ A straw for þi tale what nedyth hym to wepe
A man born blynde · dede¹ he nat ȝeue syght
myght he nat thanne his frende on lyve kepe
be the uertu of þat same hyȝ myght.   380

                                           Jhesus

Where is he put telle me anon ryght
brynge me þe weye streyth to his grave.

                                           Martha

lord at ȝour wylle we xal brynge ȝow tyght
evyn to þat place þer he doth lyne in caue.

                                         Magdalyn

Whan þat we had þe massangere sent   385
Or he had fullych half a myle gon
deyd² my brother and up we hym hent
here in þis graue we beryed hym anon.

                                           Jhesus

þe myght of þe godhed xal glathe³ ȝow every-chon
suche syght xal ȝe se hens or ȝe wende   390
Sett to ȝour handys take of þe ston
A syght lete me haue of lazare my ffrende.

                                           Martha

¶ he stynkygh ryght fowle longe tyme or this
iiij days gon for sothe he was dede
lete hym ly stylle ryght evyn as he is   395
þe stynke of his careyn myght hurte⁴ us I drede.

                                           Jhesus

As I haue þe tolde syght of þe god-hede
thy-self xuldyst haue · feythful if þou be

---

¹ *dede* has been altered to *dyde* in different ink and by a later hand, the *y* being written over the *e*.

² *deyd* has been altered to *dyyd* in a different ink and (probably) by a later hand—the *y* written over the *e*.

³ *glathe* has been altered to *gladd* in a different ink, the *dd* written over the *the*.

⁴ *hurf* first written and crossed through.

The Raising of Lazarus

  take of þe ston do Aftyr my rede
  þe glorye of þe godhede a-non 3e xal se.   400

                1us consolator

¶ 3oure byddynge xal be don a[non]¹ ful swyfte
  Sett to 3our handys and helpe echoñ
  I pray 3ow serys help me to lyfte
  I may not reyse it my-self² a-lon.

                2us consolator

In feyth it is An hevy ston   405
Ryth sad of weyth and hevyof peys.

                3us consolator

Thow it were twyes so evy³ as on⁴
Vndyr vs foure we xal it reyse.

                Nuncius

¶ Now is þe ston take ffrom þe caue
  here may men se A rewly sygth   410
  of þis ded body þat lyth here graue⁵
  Wrappyd in a petefful plyght.

Jhesus eleuatis ad celum oculis dicit

                Jhesus

  I thanke þe fadyr of þin hy3 myglit⁶
  þat þou hast herd my prayour þis day
  I know ful wel bothe day and nyght   415
  Euer þou dost graunt þat I do say.

¶ But for þis pepyl þat stondyth about
  And be-leue not þe power of þe and me
  them for to brynge clene out of dowt
  this day oure myght thei⁷ all xul se.   420

hic Jhesus clamat voce magna dicens

  Lazare Lazare my frende so fre
  Ffrom þat depe pitt come out a-noñ

---

¹ MS. *añ* with *non* written in the left-hand margin.
² *seff* first written, imperfectly corrected to *self*.
³ An initial *h* has been added in different ink by a later corrector.
⁴ Some letter has been corrected to *o* in this word and to make it clear an *o* is written above.
⁵ The word *in* has been written above the line before *graue* by the later hand.
⁶ This line as far as *fadyr* written and crossed through in red ink to make room for the stage direction, then written again beneath.
⁷ *thei* has been altered to *they* in a different ink by the later corrector.

        be þe grett myght of þe hyȝ mageste
        A-lyve þou xalt on erth ageyn gon.

lazarus

¶ At ȝoure comaundement I ryse up ful ryght   425
   hevyn helle And erth ȝoure byddyng must obeye
   Ffor ȝe be god *and* man *and* lord of most myght
   Of lyff *and* of deth ȝe haue both lok *and* keye.

hic resurget lazarus ligatis manibus et pedibus ad modum sepulti[1] *et dicit ihesus*

Jhesus

Fo. 135ᵛ   Goo forthe bretheryn[2] *and* lazare ȝe vntey
   And all his bondys losyth hem asundyr   430
   Late hym walke hom *with* ȝow in þe wey
   Ageyn god*ys* myght þis meracle is no wu*n*dyr.

Petrus

¶ At ȝour byddynge his bondys we vnbynde
   all thynge muste lowte[3] ȝour mageste
   be þis grett meracle opynly we fynde   435
   Very god *and* man in trewth þat ȝe be.

Johannes

   þat þou art very god every man may se
   be this meracle so grett *and* so meruayll
   all thynge vndyr hevyn must nedys obey þe
   whan Aȝens þe þowh deth be · he may not preuayll.   440

Omnes consolatores

¶ We All *with* o voys ffor god do þe knowe
   and for oure sauyour we do þe reverens
   All oure hool loue now in þe doth growe
   O sovereyn lord of most excellens
   helpe vs of ȝour grace whan þat we go hens   445
   Ffor aȝens deth us helpyht not to stryve
   but aȝen ȝoure myght is no resistens
   oure deth ȝe may A-slake *and* kepe vs[4] stylle on lyve.

Jhesus

¶ Now I haue shewyd in opyn syght
   of my godhed þe gret glorye   450

---

[1] MS. sepult.
[2] *breryn* first written and crossed through.
[3] The word *vnto* has been written above the line by the later hand between *lowte* and *ȝour*.
[4] *ust* written before *vs* and crossed through.

to-ward my passyon I wyl me dyght
the tyme is nere þat I must deye
Ffor all mankynde his sowle to bye
A crowne of thorn xal perchyn myn brayn
and on þe mont of caluarye 455
Vpon a cros I xal be slayn.

The Passion Play. I

¶ Demon

¶² I am ȝour lord lucifer þat out of helle cam
Prince of þis werd · and gret duke of helle
Wherefore my name is clepyd sere satan
Whech Aperyth among ȝow · A matere to spelle.

¶ I am Norsshere of synne · to þe confusyon of man 5
To bryng hym to my dongeon · þer in fyre to dwelle
Ho so evyr serve me so reward hym I kan
þat he xal syng wellaway · ever in peynes ffelle.

¶ Lo þus bountevous A lord þan · now am I
To reward so synners · as my kend is 10
Who so wole folwe my lore · and serve me dayly
Of sorwe and peyne A-now · he xal nevyr mys.

¶ Ffor I began in hefne synne for to sowe *
Among all þe Angellys · þat weryn þere so bryth
And þer fore was I cast out · in to helle ful lowe 15
Not withstandyng I was þe fayrest and berere of lyth.

¶ ȝet I drowe in my tayle · of þo Angelys bryth
With me in to helle · takyth good hed what I say
I lefte but tweyn A-ȝens on · to Abyde þere in lyth
But þe iij^de part come with me · þis may not be seyd nay. 20

---

¹ The writing is less regular in this quire and the rubrication of the initial letters of the line more abundant and less careful. On Ff. 136 and 136ᵛ the writing varies noticeably in size from line to line.
² The paragraph marking of this Prologue is irregular, as is the scheme of linked quatrains.
* Wylliam Dere scribbled in the margin.

¶ Takyth hed to ȝour prince þan · my pepyl euery-chon
and seyth what maystryes in hefne · I gan þer do play
To gete A thowsand sowlys in an houre · me thynkyth it but skorn
Syth I wan Adam And Eve · on þe fyrst day.

¶ But now mervelous mendys rennyn in myn rememberawns 25
Of on cryst wiche is clepyd Joseph · and maryes sone
Thryes I tempte hym be ryth sotylle instawnce
Aftyr he fast fourty days · ageyns sensual myth or reson
Fo. 136ᵛ Ffor of þe stonys to a mad bred · but sone I had conclusyon
Þan upon a pynnacle but Angelys were to hym Assystent 30
His Answerys were mervelous · I knew not his intencion
and at þe last to veyn glory · but nevyr I had myn intent.

¶ And now hath he xij dysypulys · to his Attendauns
To eche town and cety · he sendyth hem as bedellys
in dyverce place to make ffor hym puruyauns 35
The pepyl of hese werkys ful grettly merveyllys
To þe crokyd blynd and down · his werkys provaylys [1]
Lazare þat foure days lay ded · his lyff recuryd
and where I purpose me to tempt · A-non he me Asaylys
Mawdelyn playne remyssyon · Also he hath ensuryd. 40

¶ Goddys son he pretendyth · and to be born of A mayde
and seyth he xal dey · for mannys saluacion
þan xal þe trewth be tryed · and no fordere be delayd
Whan þe soule fro þe body · xal make separacion
And as for hem þat be vndre · my grett domynacion 45
He xal fayle of hese intent and purpose [2] Also
Be þis tyxt of holde · remembryd to myn intencion
Quia in inferno nulla est redempcio.

¶ But whan þe tyme xal neyth of his persecucion
I xal Arere new Engynes · of malycious conspiracy 50
Plente of reprevys · I xal provide · to his confusyon
þus xal I false þe wordys · þat his pepyl doth testefy
His discipulis xal for-sake hym · and here mayster denye
In-novmberabyl xal his woundys be · of woful grevauns
A tretowre xal countyrfe · his deth · to fortyfye 55
Þe rebukys þat he gyf me · xal turne to his displesauns.

[1] So (pvaylys) in MS.   [2] MS. p̄pose.

Prologue of Demon

fo. 137
¶ Some of hese dyscypulys · xal be chef · of þis Ordenawns
Þat xal fortefye þis term͞ · þat in trost is treson͞
Þus xal I venge · be sotylte al my malycious grevauns
For no thyng may excede · my prudens and dyscrecion.   60

¶ Gyff me ȝour love · grawnt me myn Affeccion
And I wyl vnclose · þe tresour of lovys Alyawns
And gyff ȝow ȝoure desyrys afftere ȝoure intencion
no poverte xal aproche ȝow · fro plentevous Abundauns.

¶ By-holde þe dyvercyte · of my dysgysyd varyauns   65
Eche thyng sett ꝓ of dewe naterall, dysposycion
and eche parte Acordynge · to his resemblauns
Ffro þe sool of þe ffoot, to þe hyest Asencion.

¶ Off ffyne cordewan͞, A goodly peyre of long pekyd schon
hosyn enclosyd ꝓ of þe most costyous cloth · of Crenseyn͞ 70
þus a bey to a jentylman · to make comparycion
With two doseyn poyntys of cheverelle ꝓ þe Aglottys of syluer¹
        feyn͞.

¶ A shert of feyn holond² ꝓ but care not for þe payment
A stomachere of clere reynes · þe best may be bowth
Þow poverte be chef ꝓ lete pride þer be present   75
And all þo þat repreff pride, þou sette hem at nowth.

¶ Cadace · wolle · or flokkys · where it may be sowth
to stuffe with-al þi dobbelet, and make þe of proporcyon
two smale legges · And a gret body · þow it ryme nowth
ȝet loke þat þou desyre · to An þe newe faccion.   80

¶ A gowne of thre ȝerdys loke þou make comparison͞
Vn-to all degrees dayly · þat passe þin astat
A purse with-outyn mony ꝓ a daggere for devoscyon͞
And þere repref is of synne ꝓ loke þu make debat.

. 137ᵛ ¶ With syde lokkys I schrewe þin here³ · to þi colere hangyng
        down͞   85
to herborwe qweke bestys þat tekele men onyth

---

¹ The initial *s* is written over some other letter (? þ).
² The word *cloth* first written after *holond* but crossed through.
³ ouer þin eyn *and* þin herys is written by the same hand above the line, as if for an alternate reading for *I schrewe þin here.*

An hey smal bonet · for curyng of þe crowne
And all beggerys and pore pepyll · haue hem on¹ dyspyte
On to þe grete Othys · And lycherye gyf þi delyte
to maynteyn þin astate lete brybory be present   90
And yf þe lawe repreve² þe · say þou wylt ffyth
And gadere þe A felachep after þin entent.

¶ Loke þou sett not be precept · nor be comawndement
Both sevyle and Canone³ · sett þou at nowth
Lette no membre of god · but with othys be rent   95
Lo þus þis werd at þis tyme · to myn intent is browth
I Sathan with my felawus · þis werd hath Sowth
And now we han it · at houre plesawns
Ffor synne is not shamfast · but boldnes hath bowth
Þat xal cause hem in helle to han inerytawns.   100

¶ A beggerys dowtere to make gret purvyauns⁴
To cownterfete a jentyl woman · dysgeysyd as she can
And yf mony lakke · þis is þe newe chevesauns
With here prevy plesawns to gett it of sum man
Here colere splayed / and furryd with Ermyn calabere or satan 105
A seyn to selle lechory · to hem þat wyl bey
And þei þat wyl not by it, yet i-now xal þei han
And telle hem it is for love · she may it not deney.

¶ I haue browth ȝow newe namys, and wyl ȝe se why
Ffor synne is so plesaunt, to ech Mannys intent   110
ȝe xal kalle pride · oneste · and naterall kend' lechory
And covetyse wysdam · there tresure is present.

¶ Wreth manhod, and envye callyd chastement
Seyse nere sessyon · lete perjery be chef
Glotonye · rest · let Abstynawnce · beyn Absent   115
And he þat wole exorte þe to vertu · put hem to repreff.

¶ To rehers al my servauntys · my matere is to breff
But all þese xal eneryth · þe dyvicion eternal

---

¹ Or *en*—the letter is blotted.
² *ll* written before *repreve* and crossed through.   ³ MS. Canoñ.
⁴ An α (mark of stage direction) crossed through in red ink in the left-hand margin here, and the beginning of the red underlining of stage directions beneath the first word of the line.

Þow cryst by his sotylte · many materys meef
In evyr-lastynge peyne · with me dwellyn þei xal. 120

¶ Remembre oure seruauntys · whoys sowlys ben mortall
Ffor I must remeffe · for more materys to provyde
I am with ȝow at all tymes · whan ȝe to councel me call
But for A short · tyme · my-self I devoyde.

           Johannes baptis[ta

¶ I johan baptyst ·/ to ȝow þus prophesye
Þat on xal come aftyr me · and not tary longe
In many folde more · strengere þan I
Of whose shon ·/ I am not worthy to lose þe thonge
Where-fore I councel þe ȝe reforme all wronge   5
in ȝour concyens of þe mortall dedys · sevyn
And for to do penawns · loke þat ȝe ffonge
Ffor now xal come þe kyngdham of hevyn.

¶ Þe weys of oure lord cast ȝow to Aray
And þer-in to walk loke ȝe be Applyande   10
And make his pathys · as ryth as ȝe may
Kepyng ryth forth · and be not declinande
138ᵛ   Neyther to fele · on ryth · nor on lefte hande [1]
But in þe myddys · purpose ȝow · to holde
For þat in all wyse is most plesande   15
As ȝe xal here · whan I have tolde.

¶ Of þis wey for to make · moralysacyon [2]
Be þe ryth syde ȝe xal vndyrstonde mercy
And on þe lefte syde · lykkenyd dysperacion
And þe patthe be-twyn bothyn · þat may not wry   20
Schal be hope and drede · to walke in perfectly
Declynyng not to fele · for no maner nede
grete cawsys I xal shove ȝow why
þat ȝe xal sowe [3] þe patthe of hope and drede.

[1] *syde* first written and crossed through.
[2] Two lines following this are crossed through: *Be þe ryth syde lyknyd disperacion* and *and þe pathe betwyn bothyn*.
[3] Or *sewe*—the vowel is not clear.

¶ On þe mercy of god, to meche ȝe xal not holde 25
as in þis wyse · be-hold¹ · what I mene
Ffor to do synne · be þou no more bolde
Iu trost þat god wole · mercyful bene
And yf be sensualyte · as it is ofte sene
Synnyst dedly · þou xalt not perfore dyspeyre 30
but perfore do penawns and confesse þe clene
And of hevyn · þou mayst trost to ben eyre.

¶ Þe pathe þat lyth · to þis blyssyd · enherytawns
Is hope and drede · copelyd be conjunccyon
Be-twyx þese tweyn · may be no dysseuerawns 35
Ffor hope with-outyn drede · is maner of presumpcion
And drede · with-owtyn hope · is maner of dysperacion
So these tweyn must be knyt be on Acorde
How ȝe xal aray þe wey · I haue made declararacion²
Also þe ryth patthis · Aȝens þe comyng of oure lord. 40

Here xal annas shewyn hym-self in his stage be-seyn after a busshop of þe hoold lawe in a skarlet gowne · and ouer þat a blew tabbard furryd with whyte and a mytere on his hed after þe hoold lawe · ij doctorys stondyng by hym in furryd hodys and on be-forn hem with his staff of A-stat and eche of hem on here hedys a furryd cappe with a gret knop in þe crowne and on stondyng be-forn as a sarazyn þe wich xal be his masangere · Annas þus seyng

Annas

¶ As a prelat am I properyd, to provyde pes
And of jewys · jewge · þe lawe to fortefye
I Annas be my powere · xal comawnde dowteles
Þe lawys of moyses · no man xal denye
Hoo excede my comawndement · Anon ȝe certefye
Yf Any eretyk here reyn · to me ȝe compleyn
For in me lyth þe powere · all trewthis to trye
And pryncypaly oure lawys · þo must I susteyn.

¶ Ȝef I may aspey · þe contrary³ no wheyle xal þei reyn

---

¹ Or *held*. ² So in MS.
³ *contrary* apparently first written, corrected to *contraly*, and again, imperfectly but possibly by the scribe, to *contrary*.

But a-non to me be browth · *and* stonde *p*resent   10
Be-fore her*e* jewge · wich xal not feyṅ
But after*e* her*e* trespace · to gef hem jugement
Now serys for A prose · heryth my*n* intent
Ther*e* is on jhesus of nazareth · þ*at* our*e* lawys doth excede
Yf he *p*rocede · thus · we xal us aƚƚ repent   15
For our*e* lawys · he dystroyt · dayly w*ith* his dede.

o. 139ᵛ ¶ Ther*e* fore be ȝcur cowncel we must take hede
What is be¹ to *p*rovyde or do in þis case
Ffor yf we let hym þus go · *and* ferdere *p*rosede
Ageyn sesar*e* *and* our*e* lawe we do trespace.   20

                                      prim*us* doctor
¶ Sere þis is my*n* Avyse · þ*at* ȝe xal do   Annas²
Send to cayphas for cowncel knowe his intent
For yf jhesu proce³ *and* þus forth go
Our*e* lawys xal be dystroyd thes se we *p*resent.

                                      ij*us* doctor
¶ Sere remembre þe gret charge · þ*at* on ȝow is leyd   annas⁴ 25
Þe lawe to ke⁵ which may not ffayle⁶
Yf any defawth *p*revyd of ȝow be seyd
Þe jewys w*ith* trewth · wyl ȝow a-sayl
Tak hed whath cownsayl · may best *p*rovayl⁷
After rewfyn *and* leyoṅ · I rede⁸ þ*at* ȝe sende   30
They arn temp*er*al jewgys, þ*at* knowyth þe p*a*rayl
W*ith* ȝoure cosyn cayphas · þis mater*e* to Amende.

                                        Annas
¶ Now surely þis cowncel · revyfe my*n* herte
ȝoure cowncel is best as I can se
Arfexe in hast loke þ*at* þou styıte   35
And pray cayphas my cosyn come speke w*ith* me.

¶ To rewfyn *and* leon þu go Also
And pray hem þei speke w*ith* me in hast

---

¹ So in MS. for *best*.
² *Annas* written in a different hand, possibly that of Ff. 95, 96.
³ So in MS. for *procede*.
⁴ *annas* written as before.   ⁵ So in MS. for *kepe*.
⁶ The first three letters in this word have been written over others.
⁷ So (pvayl) in MS.
⁸ With this word the ink becomes darker and the writing is perhaps slightly more angular from here to the bottom of Fo. 141ᵛ, where the darker ink ceases.

   For A pryncipal matere · þat haue to do
   Wich must be knowe or þis day be past.   40
                Arfexe
 ¶ My souereyn at ȝour intent · I xal goṅ
  In al þe hast þat I kan hy
  On to Cayphas · rewfyn and lyoṅ
  And charge ȝoure intent þat þei xal ply.

Fo. 140 *here goth þe masangere forth* and *in þe mene tyme cayphas shewyth him-self in his skafhald · Arayd lych to Annas savyng his tabbard xal be red furryd with white ij doctorys with hym arayd with pellys aftyr þe old' gyse and furryd cappys on here hedys · Cayphas þus seyng·*

               Cayphas
 ¶ As A primat most preudent · I present here sensyble   45
  buschopys of þe lawe with al þe cyrcumstawns
  I Cayphas am jewge · with powerys possyble
  To distroye all errouris · þat in oure lawys make varyawns
  All thyngys I convey be reson and temperawnce
  And all materis possyble · to me ben palpable   50
  Of þe lawe of moyses I haue A chef governawns
  To seuere [1] ryth and wrong in me is termynable.

 ¶ but þer is on Cryst þat oure lawys is varyable
  he perverte þe pepyl with his prechyng ill
  We must seke A mene · on to hym reprevable   55
  ffor yf he procede oure lawys he wyl spyll.

 ¶ We must take good cowncel in þis case
  Of þe wysest of þe lawe [2] þat kan þe trewthe telle
  of þe jewgys of pharasy · and of my cosyn Annas
  for yf he procede be prossesse · oure lawys he wyl felle. 60
             primus doct[or
 ¶ Myn lord plesyt ȝow to pardon me for to say   Cayfas [3]
  þe blame in ȝow · is as we fynde [4]
  to lete cryst contenue þus day be day
  With his fals wichcraft þe pepyl to blynde

---

    [1] *deuere* first written and crossed through.
    [2] *ll* written before *lawe* and crossed through.
    [3] *Cayfas* written as *Annas* above in the Fo. 95, &c., hand.
    [4] *haue fow* first written and crossed through.

|140ᵛ| He werkyth fals meraclis Ageyns¹ all kende | 65
	And makyth oure pepyl to leve hem in
	It is ȝour part to take hym and do hym bynde
	And gyf hym jugement for his gret syn.

ijᵘˢ doctor
¶ For-sothe sere of trewth this is þe case   Cayphas²
On to our lawe ȝe don oppressyon   70
Þat ȝe let cryst from ȝou pace
and wyl not don on hym correxion
Let Annas knowe ȝour intencion
With prestys and jewgys · of þe lawe
And do cryst fforsake his fals oppynyon   75
Or in to A preson lete hem be thrawe.

                             Cayphas
¶ Wel serys ȝe sal se with-inne short whyle
I xal correcte hym · for his trespas
He xal no lenger oure pepyl be-gyle
Out of myn dawngere he xal not pas.   80

*here comyth þe masangere to cayphas and in þe mene tyme revfyn
and lyon schewyn hem in þe place in ray tabardys furryd and ray hodys
a-bouth here neckys furryd · þe masangere seyng·*

                             Masangere
¶ Myn reverent souereyn · and it do ȝow plese
Sere·annas my lord hath to ȝou sent
He prayt ȝou þat ȝe xal not sese
Tyl þat ȝe ben with hym present.

                             Cayphas
¶ Sere telle myn cosyn I xal not fayl   85
it was my purpose hym for to se
For serteyn materys þat wyl³ provayle⁴
Þow he had notwth a sent to me.

                             Masager
|141| ¶ I recomende me to ȝour hey degre
on more massagys I must wende.   90
                             Cayphas

Ffare wel sere and wel ȝe be
gret wel my cosyn and my ffrede⁵

---

[1] *all kende* first written before *Ageyns* and crossed through.
[2] *Cayphas* writtten as *Annas* above in the Fo. 95, &c., hand.
[3] The *l* seems to have been omitted and inserted later.
[4] So in MS.   [5] So in MS. for *ffrende*.

here þe masager metyth with þe jewgys sayng

Masager

¶ Heyl jewgys of jewry · of reson most prudent
of my massage to ȝou · I make relacion
my lord sere Annas hath for ȝou sent
to se his presens with-owth delacion.   95

Rewfyn

¶ Sere we Are redy At his comawndement
To se sere Annas in his place
it was oure purpose and oure intent
to A be with hym with-inne short space.   100

leyon

¶ We are ful glad his presence to se
sere telle hym · we xal come in hast
no declaracion þer-in xal be
but to his presens hye us fast.

masager

¶ I xal telle my lord seris as ȝe say   105
ȝe wyl ful-fylle al his plesawns.

Rewfyn

Sere telle hym · we xal make no delay
but come in hast · at his instawns.

here þe masangere comyth to Annas þus seyng

masan[ger

¶ My lord and it plese ȝou to haue intellygens
Ser Cayphas comyth to ȝou in hast   110
Rewfyn and lyon · wyl se ȝour presens
And se ȝow here or þis day be past.

Annas

Fo. 141ᵛ
(marked 140)

¶ Sere I kan þe thank of þi dyligens
Now ageyn my cosyn I wole walk
Serys folwyth me on to his presens   115
Ffor of these materys we must talk.

a here Annas goth down to mete with cayphas and in þe mene tyme þus seyng

Cayphas

¶ Now on to annas let us wende
ech of vs to knowe otherys intent
Many materys I haue in mende
þe wich to hym I xal present.   120

The Council of the Jews

              i^us doctor C.
¶ Sere of all othere thyng remembre þis case
Loke þat ihesus be put to schame.

             ij^us doctor C.
Whan we come present beforn annas
Whe xal rehers all his gret blame.

⁂ here þe buschopys with here clerkys and þe Pharaseus mett and [1]
þe myd place and þer xal be a lytil oratory with stolys and cusshonys
clenly be-seyn [2] lych as it were a cownsel hous · Annas þus seyng

                Annas
¶ Wel come ser Cayphas · and ȝe jewgys [3] alle   125
now xal ȝe knowe all myn entent
A wondyr case serys · here is be-falle
On wich we must · gyf jewgement [4]
Lyst þat we aftyre þe case repent
Of on cryst þat goddys sone · som doth hym calle   130
He shewyth meraclys · and sythe present
þat he is prynce of pryncys alle.

¶ The pepyl so fast to hym doth falle
Be prevy menys as we a-spye
ȝyf he procede · son sen ȝe xalle   135
þat oure lawys he wyl dystrye [5].

o. 142 ¶ It is oure part þus [6] to deny
What is ȝour cowncell in þis cas.

              Cayphas
Be reson þe trewth here may we try
I cannot dem hym with-outh trespace   140
Be-cause he seyth in every A place
þat he kyng of jewys in every degre
þerfore he is fals knowe wel þe case
Sesar is kyng and non but he.

              Rewfyn
¶ He is An eretyk [7] and a tretour bolde   145
To sesare and to oure lawe sertayn

---

[1] So in MS. for *at*.   [2] A mark over *be*, as for *ben*, is probably accidental.
[3] The *g* omitted and written over the line.
[4] *alle* first written after *jewgement* and crossed through.
[5] *þat oure lawys* scribbled beneath in a slightly later hand (?).
[6] *þus* perhaps corrected to *þis*. The MS. is rubbed.
[7] The *t* in this word is blotted and may have been corrected in some way.

Bothe in word *and* in werke *and* ȝe be-holde
He is worthy to dey wi*th* mekyl peyn.

                   leon̄

¶ þe cawse þat we been here *p*resent
To fortefye þe lawe *and* trewth to say   150
Jh*es*us ful nere oure lawys hath shent
*p*erfore he is worthy for to day.

               i*us* doctor An[nas

¶ Ser*y*s ȝe *þat* ben rewelerys of þe lawe
On jh*es*u ȝe must gyf jugement
Let hy*m* fyrst ben hangyn *and* drawe   155
and þanne his body in fyre be brent.

               ij*us* doctor An[nas

¶ Now xal ȝe here þe intent of me
Take jh*es*u þat werke us all gret schame
Put hym to deth · let hym not fle
For þan þe comownys[1] þei wyl ȝow blame.   160

               i*us* doct*or* Cayp[has

Fo. 142ᵛ ¶ He werke wi*th* weche[2]-crafte · in eche place
And drawyth þe pepyl to hese intent
Be whare ȝe jewgys let hym not passe
þan be my trewthe[3] ȝe xal repent[4].

               ij*us* doctor Cayphas

¶ Ṣerys takyth hede on to þis case   165
and in ȝo*ur* jewgement be not slawe
þ*er* was nevyr man dyd so gret trespace
As jh*es*u hath don Ageyn oure lawe.

               Annas

¶ Now bretheryn þan wyl ȝe here my*n* intent
These ix days let us A-byde   170
We may not gyf so hasty jugement
but eche man inqwere on his syde
Send spyes A-bouth þe cou*n*tre wyde
tó se *and* recorde *and* testymonye
And þan hese werkys he xal not hyde   175
nor haue no power hem to denye.

               Cayphas

This cowncell A-cordyth to my reson.

---

[1] *Cowm* first written.   [2] Some miswritten letter— *lf*—after *weche*.
[3] Or *trowthe*—letter blotted.   [4] *repepent* first written.

The Council of the Jews

                                                                                               Annas

    And we All to þe same.

va-*cat* / *here Enteryth þe Apostyl petyr and johan þe euangelyst with hym.*

                                                                    petyr sey*ng*

    O 3e pepyl · dyspeyryng be glad
    A gret cause 3e haue · *and* 3e kan se
    þe lord of all þing · of nowth mad
    is comyng 3o*ur* comfort to be
    All 3o*ur* langorys · salvy*n* xal he

143[2] -*cat*. \\ 3o*ur* helthe is more þan kan wete.[1]

                                                                    ¶ jhe*s*us

¶ Ffrendys be-holde þe tyme of mercy
The whiche is come now w*ith*-owt dowth   180
ma*n*nys sowle in blys now xal edyfy
and þe prynce of þe werd is cast owth.

¶ Go to 3on castel þat standyth 3ow Ageyn
su*m* of my*n* dyscyplis · go forth 3e to
þer*e* xul 3e ffyndyn bestys tweyn   185
An Asse tyed *and* her*e* fole Also
Vn-losne þat Asse · and brynge it to me pleyn
Iff any mas[3] Aske why þat 3e do so
Sey þat I haue nede to þis best certeyn
and he xal not lett 3ow · 3o*ur* weys for to go   190
þat best brynge 3e to me.

                                                           1us Ap*os*tolus

holy p*r*ophete we gon oure way
we wyl not 3our*e* wourd de-lay
Also sone as þat we may
        We xal it brynge to the.   195

here þei ffecch þe Asse with þe ffole and þe burgeys seyth

                                                           Burgensis

¶ Herke 3e men who 3aff 3ow leve
    Thus þis[4] best ffor to take Away

---

[1] This passage is cancelled in the MS. by a red stroke through the stage direction, black strokes through the speech, and a red bracket and the word *vacat*, as above, in black.

[2] This folio is interpolated. It is of different paper and the writing (though this is the same as that of the body of the MS.) and rubrication have not the features that distinguish the rest of quire N.

[3] So in MS.   [4] *ff* crossed through after *þis*.

but only ffor pore men to releve
this Asse¹ is ordayned as I ʒow say.

Philippus

Good sere take this At no greff²
Oure mayster us sent hedyr þis day
he hath grett nede with-owt repreff
þerfore not lett us I þe pray
   þis best for to lede.

Burgensis

Fo. 143ᵛ Sethyn þat it is so ? þat he hath ʒow sent
werkyth his wyll *and* his intent
take þe beste as ʒe be bent
and evyr wel mote ʒe spede.

jacobus minor

¶ This best is brought ryght now here lo
holy prophete At þin owyn wylle
And with þis cloth Anon Also
þis bestys bak we xal sone hylle.

Philippus

Now mayst þou ryde whedyr þou wylt go
Thyn holy purpos to ffulfylle
thy best fful redy is dyth þe to
bothe meke *and* tame þe best is stylle
And we be redy Also
Iff it be plesynge to þi ssyght
The to helpe A-non forth ryght
Vpon þis best þat þou were dyght
þi jurney ffor to do.

here cryst rydyth out of þe place And he wyl · and Petyr and Johan Abydyn stylle · And at þe last whan þei haue don þer prechyng þei mete with jhesu,

Petrus

¶ O ʒe pepyl dyspeyryng³ · be glad
A grett cawse ʒe haue *and* ʒe kan se
þe lord þat all thynge · of nought mad
is comynge ʒour comforte to be
All ʒour langoris · salvyn xal he

---

¹ *assa* first written.
² *Thus this best to take* first written as the following line and crossed through.
³ The last *y* in this word is altered from an *e*.

ȝour helthe is more than ȝe kan wete
He xal cawse þe blynde · þat þei xal se
þe def to here · þe dome · for to speke.

¶ þei þat be crokyd · he xal cause hem to goo   230
In þe wey · þat johan baptyst · of prophecyed'
Sweche A leche · kam ȝow nevyr non too
Wher-fore what he comawndyth · loke ȝe Applyed'
þat som of ȝow · be blynd · it[2] may not be[3] denyid[4]
Ffor hym þat is ȝour makere · with ȝour gostly ey ȝe xal not
    knowe   235
of his comaundement · in ȝow gret necglygens is Aspyed'
Where-fore def · fro gostly heryng · clepe ȝow I howe.

¶ And some of ȝow · may not go · ȝe be so crokyd
For of good werkyng · in ȝow · is lytyl habundawns
Tweyn[5] fete · heuery man xuld haue · and it were lokyd   240
Wyche xuld bere · þe body gostly · most of substawns
Ffyrst is to love god Above all other plesawns
þe secunde · is to love · þi neybore · as þin owyn persone
and yf þese tweyn · be kepte in perseverawns
Into þe celestyal habytacion · ȝe Arn habyl to gone.   245

¶ Many of ȝow be dome · why · for ȝe wole not redresse
Be Mowthe · ȝour dedys mortal · but þer-in don perdure
Of þe wych · but ȝe haue contrycyon · and ȝow confesse
ȝe may not in-heryte hevyn · þis I ȝow ensure
And of All þese maladyes · ȝe may haue gostly cure   250
For þe hevynly leche · is comyng ȝow · for to vicyte
And as for payment · he wole shewe ȝow. · no redrure
Ffor with þe love of ȝowre hertys · he wole be Aqwhyte.
                                       Johannes apostolus
                Apostolus johannes

¶ On to my brotherys for-seyd rehersall
þat ȝe xuld ȝeve þe more · veray confydens   255

---

[1] The less regular writing and less discriminating rubrication of the N quire resumed.

[2] *it* has been written over an erasure in darker ink.

[3] *dey* first written instead of *be* and crossed through and *be* written above the line.

[4] The *id* seems to have been added as a correction.

[5] The contraction mark is a rough blotted stroke.

I come with hym as testymonyall[1]
Ffor to conferme · and fortefye his sentens
þis lord xal come with-out resystens
On to þe Cety-ward he is now comyng
Where-fore dresse ȝow with all dew dylygens   260
To honowre hym as ȝour makere and kyng.

¶ And to fulfylle þe prophetys prophese
Vp-on An Asse he wole hedyr ryde
Shewyng ȝow exawmple of humylyte
Devoydyng þe Abhomynable synne of pryde   265
Whech hath ny conqweryd all þe werd wyde
Grettest cause of All ȝour trybulacyon
Vse it ho so wole · for it is þe best gyde
þat ȝe may haue to þe place of dampnacyon.

¶ Now brothyr in god syth we have intellygens   270
þat oure lord is ny come to þis cete
To Attend up-on his precyous presens
it syttyth to us · as semyth me
Wherfore to mete whit hym now go we
I wold fore no thyng we where[2] to late   275
to þe cete-ward fast drawyth he
me semyth he is ny at þe gate[3]

a here spekyth þe iiij ceteseynys · þe fyrst þus seyng[4]

                               1us ciues de jherusalem

Fo. 145[5] ¶ Neyborys gret joye in oure herte we may make
þat þis hefly kyng · wole vycyte þis cyte.

                              ij^{us} ciues
Yf oure eerly kyng · swech a jorne xuld take   280
to don hym honour and worchepe · besy xuld we be.

                              iij^{us} ciues
Meche more þan to þe hevynly kyng · bownd Are we
Ffor to do þat xuld be to his persone reuerens.

---

[1] The second *l* is written over some blotted letter (? *e*).
[2] The *h* in this word is written over an *e*.
[3] The *g* has been altered from some other letter.
[4] A note in a later hand stands in the left-hand margin at the bottom of this folio : *here entrith þe fyrst prophete.*
[5] Two numbers, 16 and 40, are here crossed out in paler ink.

iiij ci[ues

late vs þan welcome hym with flowrys and brawnchis of þe tre
Ffor he wole take þat to plesawns · becawse of redolens.   285

⁂ here þe iiij ceteseynys makyn hem redy for to mete with oure lord
goyng barfot and barelegged and in here shyrtys savyng þei xal
haue here gownys cast A-bouth theme and qwan þei seen oure lorde þei
xal sprede þer clothis be-forn hyme and he xal lyth and go þer upone
and þei xal falle downe up-on þer knes alle atonys þe fyrst þus seyng
                                                                    i^{us} ciues

¶ Now blyssyd he be þat in oure lordys name
to us in Any wyse wole resorte
And we be-leve · veryly · þat þou dost þe same
For be þi mercy xal spryng mannys comforte.

⁂ here cryst passyth forth · þer metyth with hym a serteyn of
chylderyn with flowrys and cast be-forn hyme and They synggyn
Gloria laus and be-forn on seyt[. . .

Thow sone of davyd' þou be oure supporte   290
At oure last day whan we xal dye
Where-fore we Alle Atonys to þe exorte
Cryeng mercy mercy mercye.¹
                                                              Jhesu ²

f. 145ᵛ ¶ Ffrendys be-holde þe tyme of mercy
þe wich is come now · with-owtyn dowth   295
Mannys sowle in blysse now xal edyfy
And þe prynce of þe werd · is cast owth
As I haue prechyd in placys A-bowth
And shewyd experyence · to man and wyf
In to þis werd goddys sone hath sowth   300
Ffor veray loue · man to revyfe.

¶ The trewthe of trewthis xal now be tryede
and A perfyth of corde be-twyx god and man
Wich trewth xal nevyr be dyvide
Confusyon on to þe fynd sathan.   305
                                              i^{us} pauper homo
¶ þou sone of davyd on vs haue mercye
As we must stedfast be-levyn in þe

¹ A similar note in the same later hand stands at the bottom of Fo. 145 with a reference mark to the close of this speech : *here entreth þe parte of þe ij^{de} prophete*.
² The name of the speaker stands as usual at the bottom of the last page, but in this case is written again at the top of Fo. 145ᵛ.

R

   þi goodnesse lord lete us be nye
   Whech lyth blynd here and may not se.

              ij^{us} pauper homo

¶ Lord lete þi mercy to us be sewre   310
 and restore to us oure bodyly syth
 We know þou may us wel recure
 With þe lest poynt · of þi gret myth.

                Jhesu

¶ ȝowre be-leve hath mad ȝou for to se
 And delyveryd ȝou fro All mortal peyn   315
 blyssyd be All þo þat be-leve on me
 And se me not with here bodyly eyn.

 here Cryst blyssyth here eyn and þei may se þe fryst seyng

Fo. 146   i^{us} pauper homo

 *Gromercy¹ lord of þi gret grace
 I þat was blynd · now may se.

              ij^{us} pauper homo

 Here I for-sake al my trespace   320
 And stedfastly wyl be-levyn on þe.

 here cryst procedyth on fote with his dyscipulys After hym cryst wepyng up-on þe cyte sayng þus †

                Jhesu

¶ O Jherusalem woful is þe ordenawnce
 Of þe day of þi gret persecucyon
 þou xalt be dystroy² with woful grevans
 and þi ryalte browth · to trew confusyon   325
 ȝe þat in þe cete · han habytacyon
 þei xal course þe tyme þat þei were born
 So gret advercyte and trybulacion
 Xal falle on hem both evyn and morwyn.

             27

¶ þei þat han most chylderyn sonest xal wayle   330
 and seyn Alas · what may þis meen
 Both mete and drynk · sodeynly xal fayle
 þe vengeance of god þer xal be seen

---

 * No paragraph sign here in MS.
 ¹ Somewhat larger capital than usual.
 † A large sign in darker coloured ink stands in the left-hand margin against this stage direction and the first lines of the next speech.
 ² So in MS.

The Last Supper

    þe tyme is comyng · hes woo xal ben
    þe day of trobyl and gret grevauns   335
    Bothe templys and towrys they xal down cleen
    O cete fful woful · is þin Ordenawns.

Petrus

¶ Lord where wolte þou kepe þi maunde
    I pray þe now lete us haue knowyng
    þat we may make redy for þe   340
    þe to serve with-owte · latyng.

Johannes

f. 146   To provyde lord · for þi comyng
    With all þe obedyens · we kan A-tende
    And make redy for þe in Al thyng
    In to what place þou wytl[1] us send.   345

Jhesu

¶ Serys goth to syon and ȝe xal mete
    A pore man in sympyl A-ray
    Beryng watyr in þe strete
    Telle hym I xal come þat way
    On-to hym mekely · loke þat ȝe say   350
    þat hese house I wele come tylle
    he wele not onys to ȝow sey nay
    But sofre to haue all ȝour wylle.

Petrus

¶ At þi wyl lord it xal be don
    To seke þat place we xal us hye.   355

Johannes

    In All þe hast · þat we may go
    þin comawdement[2] nevyr to denye.

*α here petyr and johan gon forth metyng with symon leprows beryng
A kan with watyr petyr þus seyng*

Petrus

¶ Good man þe prophete · oure lord jhesus
    þis nyth wyl · rest wyth-in þin halle
    On massage to þe he hath sent vs   360
    þat ffor his supere ordeyn þou xalle[3].

---

[1] So apparently in MS. for *wylt*—Halliwell reads *wyth*; *wyll* is possible; the letters are not clear.
[2] So in MS.   [3] The *e* is written over another letter (? *t*).

  Johannes

    ʒa for hym *and* his dyscipulys Alle
    Ordeyn þu for his maunde
    A paschall lomb what so be-falle
    Ffor he wyl kepe his pasch wi*th* the.   365
                                          Symon

Fo. 147 ¶ What wyl my lord · vesyte my plase
    Blyssyd be þe tyme of his comyng
    I xal ordeyn wi*th*-inne short space
    Ffor my good lordys wel comyng
    Serys walkyth in at þe begynny*n*g   370
    And se what vetaylys · þat I xal take
    I am so glad of þis tydyng
    I wot nevyr what joye þat I may make.

⊄ here þe dyscypulys gon in with Symone to se þe ordenawns and cryst comyng thedyr-ward þus seyng

                                             Jhesus

    ¶ Þis path is Calsydon¹ · be goostly ordenawns
    Wech xal co*n*uey us wher we xal be   375
    I knowe ful redy is þe p*ur*vyaunce
    Of my frendys þat lovyn me
    Contew*n*yng² in pees · now p*ro*cede we
    For mannys love þis wey I take
    Wi*th* gostly ey I veryly se   380
    Þ*a*t man ffor man · an hende must make.

⊄ here þe dyscipulys come A-geyn to cryst petyr þus seyng

                                             Petrus

    ¶ All redy lord is our*e* ordenawns
    As I hope to ʒow plesyng xal be
    Seymon hath don at ʒoure instawns
    He is ful glad ʒo*ur* prese*n*s to se.   385
                                           Johannes

    ¶ All thyng we haue lord at our*e* plesyng
    Þat longyth to ʒoure mawnde wi*th* ful glad chere
    Whan he herd telle of ʒo*ur* comyng
    Gret joye in hy*m* þan dyd Appere.

Fo. 147ᵛ *here comyth symon owt of his hous to welcome cryst.*

---

¹ So in MS. Halliwell prints *cal Sydon*.   ² So in MS.

Symon

¶ Gracyous lord · wel come þu be 390
Reverens be to þe both god *and* man
My poer hous · þat þou wylt se
Weche am þi servau*n*t as I kan.

Jhesu

¶ There joye of Aȴ joyis to þe is sewr*e*
Symon I knowe þ*i* trewe intent 395
þe blysse of hefne · þou xalt recur*e*
Þis rewarde I xal þe grawnt p*re*sent.

a here Crist enteryth in-to þe hous with his disciplis and *ete þe paschal lomb* and *in þe mene tyme þe counsel hous beforn-seyd xal sodeynly onclose schewyng þe buschopys prestys* and *jewgys syttyng in here Astat lyche as it were A convocacyone · Annas seyng þus*

Annas

¶ Be-hold [1] it is nowth al þat we do
In alle houre materys we *p*rophete nowth
Wole [2] ȝe se wech peusawns of pepyl drawyth hym to 400
Ffor þe mervaylys þat he hath wrowth.

¶ Some othyr sotylte · must be sowth [3]
Ffor in no wyse we may not þus hy*m* leve
Than to A schrewde co*n*clusyoñ · we xal be browth
Ffor þe romaynes þan wyl us myscheve. 405

¶ And take our*e* Astat *and* put us to repreve
and convey aȴ þe pepyl at here owyn request
and þus aȴ þe pepyl in hym xal be-leve
Þerfore I pray ȝow cosyn · say what is þe best.

Cayphas

o. 148 Attende now serys to þat I xal seye 410
on-to us aȴ it is most expedyent
þat o man ffor þe pepyl xuld deye
þan aȴ þe pepyl xuld perysch *and* be shent.

¶ Þerfor late us werk wysely þat we us not repent
We must nedys put on hym som fals dede 415
I sey for me I had levyr he were brent
þan he xuld us alle þus ouyr-lede

---

[1] Larger capital than usual. [2] Or *Wele*.
[3] *wrowth* first written.

þer-fore every man on his party help at þis nede
and cowntyrfete aƚƚ þe sotyltes þat ȝe kan
now late se ho kan ȝeve best rede   420
to ordeyn sum dystruccion ffor þis man.

                                                  Gamalyel

¶ Late us no lenger make delacion
but do jhesu be takyn in hondys fast
And aƚƚ here ffolwerys to here confusyon
And in-to a preson do hem be cast   425
Ley on hem yron þat wol last
Ffor he hath wrouth a-ȝens þe ryth
And sythyn Aftyr we xal iń hast
Jewge hym to deth with gret dyspyth.

                                                   Rewfyn

¶ Ffor he hath trespacyd A-ȝens oure lawe   430
me semyth þis were best jewgement
With wyld hors lete hym be drawe
And afftyr in fyre he xal be brent.

                                                   leyon

¶ Serys o thyng my self herd hym sey
Þat he was kyng of jewys Alle   435
Þat is a-now to do hym dey
Ffor treson to sezar we must it calle.

Fo. 148ᵛ ¶ He seyd Also to personys þat I know
Þat he xuld and myth serteyn
Þe gret tempyl mythtyly ovyr-throw   440
and þe thrydde day reysynt Ageyn.

¶ Seche materys þe pepyl doth conseyve [1]
To ȝeve credens to his werkys Alle
In hefne he seyth xal be his reyn
Bothe god and man he doth hym calle.   445

                                                   Rewfyn

And Aƚƚ þis day we xuld contryve
What shameful deth jhesu xuld haue
We may not do hym to meche myscheve
Þe worchep of oure lawe to save.

                                                   leyon

¶ Vp on A jebet lete hym hongyn be   450
Þis jugement me semyth it is reson

---

[1] So in MS. for some word (? *constreyn*) to rhyme with *reyn*.

  þat all þe countre may hym se
  and be ware be his gret treson.
              Rewfyn
¶ ȝet o thyng serys ȝe must A-spye
 And make A ryth sotyl ordenawns   455
 Be what menys ȝe may come hym bye
 For he hath many folwerys at his instawns.
              Annas
¶ Serys þer of we must have avysement
 and ben Acordyd or þan we go
 How we xal han hym at oure entent   460
 Som wey we xal fynd þerto.¹
            Mawdelyn
¶ As a cursyd creature closyd all in care Maria Magdalen
 and as a wyckyd wrecche all wrappyd in wo
 Of blysse was nevyr no berde so bare
 as I my-sylf þat here now go   465
 Alas Alas I xal for fare
 ffor þo grete synnys þat I haue do
 lesse than my lord god sum-del spare
 and his grett mercy receyve me to
    Mary mavdelyn is my name   470
 Now wyl I go to cryst jhesu
 ffor he is lord of all vertu
 and for sum grace I thynke to sew
    ffor of my-self I haue grett shame.

¶ A mercy lord and salve my synne   475
 Maydenys ffloure þou wasch me fre
 þer was nevyr woman of mannys kynne
 so ful of synne in no countre
 I haue be ffowlyd be fryth and ffenne

---

¹ At the bottom of the folio below this line stands the stage direction : *a . here judas caryoth comyth in-to þe place*, followed by the word *Jhesus* as name of the next speaker. These are crossed through in black and red ink. Beneath stand three alternative catchwords: *now cownter fedyd*, in the same ink as the text (cf. Fo. 152) crossed through in red ; *myn hert is ryth*, in fainter ink (cf. Fo. 150) crossed through in black ; and *as a cursyd* (cf. Fo. 149), in the same ink as *Mawdelyn*, written as name of speaker under the cancelled *Jhesus*.

² Quire O is of different paper from quire N and the following quires P, Q, R ; and the writing and rubrication have not the characteristic features found in quire N and quires P, Q, R.

and sowght synne in many A cete 480
but þou me borwe lord I xal brenne
with blake ffendys Ay bowne to be
   where fore kynge of grace
With þis oynement þat is so sote
lete me A-noynte[1] þin holy fote 485
and for my balys þus wyn sum bote
   and mercy lord ffor my trespace.

          Jhesus

¶ Woman ffor þi wepynge wylle
Sum socowre god xal þe sende
þe to saue I haue grett skylle 490
ffor sorwefful hert may synne Amende
All þi prayour I[2] xal fulfylle
to þi good hert I wul attende
and saue þe fro þi synne so hylle[3]
and fro vij develys I xal[4] þe ffende 495
   ffendys fleth ʒour weye
Wyckyd spyritys I ʒow conjowre
Ffleth out of hire bodyly bowre
In my grace she[5] xal evyr fflowre
   tyl deth doth here to deye. 500

        Maria Magdalene

¶ I thanke þe lorde of þis grett grace
Now þese vij ffendys be fro me fflytt
I xal nevyr fforfett nor do trespace
Fo. 149ᵛ in wurd nor dede ne wyl nor wytt[6]
Now I am brought from þe fendys brace 505
In þi grett mercy closyd and shytt
I xal nevyr returne to synful trace
þat xulde me dampne to helle pytt
I wurchep the on knes bare
blyssyd be þe tyme þat I hedyr sowth 510

---

[1] *noyy* first written and crossed through.   [2] A miswritten letter crossed through before *I*.   [3] *hende* first written and crossed through.
[4] This *x* is written over another letter.   [5] An *x* crossed through before *she*.
[6] Spaces are left in the MS. between lines 504 and 505 and between 508 and 509, also between lines 516 and 517 and between 518 and 519, as if the structure of these stanzas was not understood. The tail-verses are not written to the right on fo. 149ᵛ, and the last quatrain, ll. 522-5, is more roughly written than the rest.

And þis oynement þat I heydr brought
Ffor now myn hert is clensyd from thought
þat Ffyrst was combryd with care.

Judas

¶ Lord me thynkyth þou dost ryght ylle
To lete þis oynement so spylle   515
To selle it | yt were more skylle
and bye mete to poer men
The box was worth of good mone
iij C. pens fayr and fre
þis myght a bowht mete plente   520
to Ffede oure power ken.

Jhesus

¶ Pore men xul abyde
A-geyn þe woman þou spekyst wronge
And I passe forth in A tyde
Off mercy is here mornyng songe.   525

*here cryst restyth and etyth A lytyl and seyth syttyng to his disciplis ·
and mary mawdelyn*[1]

¶ Jhesus

¶ Myn herte is ryght sory And no wondyr is
Too[2] deth I xal go and nevyr dyd trespas
But ȝitt most grevyth[3] myn hert evyr of this
On of my bretheryn xal werke þis manas
On of ȝow here syttynge my treson xal tras   530
On of ȝow is besy my deth here to dyth
and ȝitt was I nevyr in no synful[4] plas
Where-fore my deth xuld so shamfully be pyght.

Petrus

¶ My dere lord I pray the þe trewth for to telle
Whiche of vs ys he þat treson xal do   535
Whatt traytour is he þat his lord þat wold selle
expresse his name lord þat xal werke þis woo.

Johannes

If þat þer be on þat wolde selle so
Good mayster telle us now opynly his name

---

[1] The words *gohth here outh* follow, but have been crossed out in black and red ink.
[2] *Thoo* first written and the *h* marked with deleting dot.
[3] The *g* has been written over some correction.
[4] Some miswritten letters (? *syv*) crossed out before *synful*.

What traytour is hym þat · from þe þat wolde go 540
And with ffals treson ffullfylle his grett shame.
                                                    Andreas

¶ It is right dredfull such tresson to thynke
and wel more dredfful to werk þat bad dede
Ffor þat ffals treson to helle he xal synke
In endles peynes grett myscheff to lede. 545
                                                    Jacobus maior

It is not I lord ffor dowte I haue drede
þis synne to fulfylle cam nevyr in my mende
Iff þat I solde þe · thy blood ffor to blede
In doyng þat treson my sowle xulde I shende.
                                                    Matheus

¶ Alas my dere lord what man is so wood 550
Ffor gold or for sylvyr hym-self so to spylle
he þat þe doth selle ffor gold or for other good
with his grett Covetyse hym-self he doth kylle.
                                                    Bartholomeus

What man so evyr he be of so wyckyd wylle
dere lord among vs · tell vs his name all owt 555
He þat to hym tendyth · þis dede to fulffille
Ffor his grett treson his sowle stondyth in dowt.
                                                    Philippus

¶ Golde sylver and tresoour sone doth passe away
but with-owtyn ende evyr doth laste þi grace
A lorde who is that · wyll chaffare þe for monay 560
Ffor he þat sellyth his lord to grett is þe trespace.
                                                    Jacobus minor

That traytour þat doth þis orryble manace
bothe body and sowle I holde he be lorn
Dampnyd to helle pytt fer from þi face
Amonge All ffowle fyndys to be rent and torn. 565
                                                    Symon

¶ To bad A marchawnt þat traytour he is
and ffor þat monye¹ he may mornyng make
Alas what cawsyth hym to selle þe kyng of blys
Ffor his fals wynnynge þe devyl hym xal take.
                                                    Thomas

Ffor his ffals treson þe feudys so blake 570
xal bere his sowle depe down into helle pytt

¹ *may* first written before *monye* and crossed through.

The Last Supper

resste xal he non haue · but evyr-more wake
brennyng in hoot fyre in preson evyr shytt.

Thadeus

¶ I woundyr ryght sore who þat he xuld be
Amonges vs all bretheryn þat xulde do þis synne   575
Alas he is lorn þer may no grace be
In depe helle donjeon his sowle he doth pynne.

Jhesus

In my dysche he Etyht þis treson xal be-gynne
Wo xal be-tydyn hym for his werke of dred
he may be ryght sory swych ryches to wynne   580
ad¹ whysshe hymself vn-born ffor þat synful ded.

Judas

¶ The trewth wolde I knowe as leff as ȝe
And þerfore good ssere þe trewth þou me telle
whiche of vs All here þat traytour may be
Am I þat person þat þe now xal selle.   585

Jhesus

So seyst þi-selff Take hed att þi spelle
þou Askyst me now here if þou xalt do þat treson
Remembyr þi-self A-vyse þe ryght welle
þou art of grett Age and wotysst what is reson.

*here judas rysyth prevely and goth in þe place and seyt now
counter[fetyd.*

[Remainder of Fo. 151—2¼ inches—and Fo. 151ᵛ left blank.*]

Judas

¶ Now cowntyrfetyd I haue A prevy treson   590
My Maysterys power for to felle
I judas xal A-say be some³ encheson
On-to þe jewys hym for to selle
Som mony for hym ȝet wold I telle
Be prevy menys I xal a-say   595
Myn intent I xal fulfylle
No lenger I wole make delay.

---

¹ So in MS.

*[*John Holand* is scribbled on Fo. 151ᵛ (marked 150), and two and half a third black tapering marks extend from near the top to near the bottom of the page; cf. Fo. 164.]

² In this quire the writing and rubrication have the characteristics found in quire N.

³ A *j* (*i*) first written and crossed through after *some*.

¶ Þe princys of prestys now be present
Vn-to hem now my way I take
I wyl go tellyn hem myn entent 600
I trow ful mery I xal hem make
Mony I wyl non for-sake
And þei profyr to my plesyng
For covetyse I wyl with hem wake
and on-to my maystyr I xal hem bryng. 605

¶ Heyl prynsesse and prestys þat ben present
New tydyngys to ȝow I come to telle
ȝyf ȝe wole folwe myn intent
my mayster jhesu I wele ȝow selle
Hese intent and purpose for to felle 610
For I wole no lenger folwyn his lawe
Late sen what mony þat I xal telle
and late jhesu my maystyr ben hangyn and drawe.

Gamalye[l]

¶ Now welcome judas oure owyn frende
Take hym in serys be þe honde 615
We xal þe both geve and lende
And in every qwarel by þe stonde.

Rewfyn

Fo. 152ᵛ ¶ Judas what xal we ffor þi mayster pay*
Þi sylver is redy and we A-corde
Þe payment xal haue no delay 620
But be leyde down here At a worde.

Judas

¶ Late þe mony here down be layde
And I xal telle ȝow as I kan
In old termys I haue herd seyde
þat mony makyth schapman. 625

Rewfyn

¶ Here is thretty platys of sylver bryth
Fast knyth withinne þis glove
And we may haue þi mayster þis nyth
Þis xalt þou haue and all oure love.

* The name John Holand is scribbled again in the margin of this page and of Fo. 153ᵛ; on Fo. 155ᵛ Holland Hary; and on Fo. 154ᵛ there are indecipherable scribblings in similar ink and hand.

Conspiracy of the Jews and Judas

                      Judas

¶ ʒe Are resonable chapmen to bye *and* selle   630
þis bargany *with* ʒow now xal I make
Smyth up ʒe xal haue al ʒo*ur* wylle
Ffor mony wyl I non for-sake.

                      leyon͡

¶ Now þis bargany is mad ful *and* fast
Noyther part may it for-sake   635
But judas þ*ou* must telle us in hast
Be what menys we xal hym take.

                      Rewfyn

¶ ʒa þ*er* be many þat hym nevyr sowe
Weche we wyl sende to hym in fer*e*
Þ*er*-for be A tokyn we must hy*m* knowe   640
Þ*at* must be *p*revy be-twyx us here.

                      leyon͡

¶ ʒa be ware of þat for ony thynge
For o dyscypil is lyche þ*i* mayst*er* in al parayl
153 And ʒe go lyche in All clothyng
So myth we of our*e* purpose fayl.   645

                      Judas

¶ As for þat serys haue ʒe no dowth
I xal ordeyn so ʒe xal not mysse
Whan þat ʒe cvm[1] hy*m* All A-bowth
Take þe man þat I xal kysse.

¶ I must go to my maystyr A-geyn   650
Dowth not serys þis mater*e* is sur*e* i-now.

                      Gamalyel

Fare wel judas our*e* frend serteyn
þ*i* labo*ur* we xal ryth wel A-low.

                      Judas

¶ Now wyl I sotely go seke my mast*er* Ageyn
and make good face as I nowth knew   655
I haue hym solde to wo *and* peyn
I trowe ful sore he xal it rew.

α here judas goth in sotylly wher-as he cam fro.

                      Annas

¶ Lo serys a part we haue of our*e* entent
For to take jh*es*u now we must *p*rovyde

---

[1] The *v* seems to be written over some other letter (? *u*).

A sotyl meny to be present 660
þat dare fyth and wele A-byde.

Gamalye[l]

¶ Ordeyn eche man on his party
Cressetys lanternys and torchys lyth [1]
And þis nyth to be þer redy
With exys gleyvis [2] and swerdys bryth. 665

Cayphas

¶ No lenger þan make we teryeng
But eche man to his place hym dyth
And ordeyn preuely for þis thyng
þat it be don þis same nyth.

Fo. 153ᵛ *Here The Buschopys partyn in þe pLace And eche of [3] hem takyn here leve þe contenawns resortyng eche man to his place with here meny to make redy to take cryst and þan xal þe place þer cryst is in xal sodeynly vn-close rownd Abowtyn shewyng cryst syttyng at þe table and his dyscypulys eche in ere degre cryst þus seyng*

Jhesu

¶ Brederyn þis lambe þat was set us beforñ 670
þat we Alle haue etyn in þis nyth
it was comawndyd be my fadyr to moyses and Aaroñ
Whan þei weryn with þe chylderyn of israel in egythp.

¶ And as we with swete bredys haue it ete
And Also with þe byttyr sokelyng 675
And as we take þe hed with þe fete
So dede þei in all maner thyng.

¶ And as we stodyn so dede þei stond
and here reynes þei gyrdyn veryly
With schon on here fete and stavys in here hond 680
And as we ete it so dede þei hastyly
þis fygure xal sesse A-nothyr xal folwe þer-by
Weche xal be of my body þat am ȝour hed
weche xal be shewyd to ȝow be A mystery
Of my fflesch and blood in forme of bred. 685

¶ And with fervent desyre of hertys Affeccioñ
I have enterly desyryd to kepe my mawnde

---

[1] *bryth* first written.   [2] The *le* is written over other letters.
[3] The capital letters in this line are unusually tall, and the initial B of the following speech is elaborated.

The Last Supper

<small>p. 154
quire)</small>

  A-mong ȝow er þan I suffre my passyon
  For of þis no more to-gedyr suppe xal we
  And as þe pascHal Lomb etyn Haue we [1]   690
  In þe old lawe was vsyd for A sacryfyce
  So þe newe lomb þat xal be sacryd be me
  Xal be vsyd for A sacryfyce most of price.

☾ *here xal jhesus take An oble in his hand lokyng vpward in to hefne to þe fadyr þus seyng*

 ¶ Where fore to þe fadyr of hefne þat art eternall
  Thankyng and honor I ȝeld on to þe   695
  To whom be þe godhed I am eqwall
  But be my manhod I am of lesse degre
  Wherefore I as man · worchep þe deyte
  Thankyng þe fadyr þat þou wylt shew þis mystery
  And þus þurwe þi myth fadyr and blyssyng of me   700
  Of þis þat was bred is mad my body.

☾ *here xal he spekyn ageyn to his dyscipulys þus seyng*

 ¶ bretheryn be þe [vertu] of þese wordys þat [re]hercyd be [2]
  Þis þat shewyth as bred to ȝour Apparens [3]
  Is mad þe very flesche and blod of me
  To þe weche þei þat wole be savyd must ȝeve credens.   705

 ¶ And as in þe olde lawe it was comawndyd and precepte
  To ete þis lomb to þe dystruccyon of pharao vn-kende
  So to dystroy ȝour gostly [4] enmye · þis xal be kepte
  ffor ȝour paschal lombe in-to þe werdys ende.

 ¶ Ffor þis is þe very lombe with-owte spot of synne   710
  Of weche Johan þe baptyst dede prophesy
  Whan þis prophesye he dede be-gynne
  Seyng · Ecce agnus dey.

 ¶ And how ȝe xal ete þis lombe I xal ȝeve infformacion
  In þe same forme as þe eld lawe doth specyfye   715

---

[1] The capital letters in this top line are again unusually tall.
[2] This line is omitted and written in the margin, where it is partly cut away, and again in large form in red ink at the bottom of the page.
[3] *apperens* first written and corrected to *apparens*.
[4] Perhaps corrected from *goostly*.

As I shewe be gostly interpretacyon
þer-fore to þat I xal sey ȝour wyttys¹ loke ȝe replye.

Fo. 154ᵛ² ¶ With no byttyr bred þis bred ete xal³ be
þat is to say with no byttyrnesse of hate and envye
But with þe suete bred of loue and charyte 720
Weche ffortefyet þe soule gretlye.

¶ And it schuld ben etyn with þe byttyr sokelyng
þat is to mene ȝyf A man⁴ be of synful dyspocycion
Hath led his lyff here with mys-levyng
þerfore in his hert he xal haue byttyr contrycion. 725

¶ Also þe hed with þe feet ete xal ȝe
Be þe hed ȝe xal vndyr-stand my godhed
And be þe feet ȝe xal take myn humanyte
þese tweyn ȝe xal receyve to-gedyr in dede.

¶ This immaculat lombe þat I xal ȝow ȝeve 730
Is not only þe godhed A-lone
But bothe god and man þus must ȝe beleve
þus þe hed with þe feet ȝe xal receyve ech-on.

¶ Of þis lombe vn-ete · yf owth be levyth i-wys
Yt xuld be cast in þe clere fyre and brent 735
Weche is to mene yf þou vndyrstande nowth al þis
Put þi feyth in god and þan þou xalt not be shent.

¶ The gyrdyl þat was comawndyd here reynes to sprede
Xal be þe gyrdyl of clennes and chastyte
þat is to sayn to be contynent in word thought and dede 740
and all leccherous levyng · Cast ȝow for to fle.

¶ And þe schon þat xal be ȝour feet vp-on
Is not ellys but exawnpyl of vertuis levyng
Of ȝour form faderys ȝou be-forn
With þese schon my steppys ȝe xal be sewyng. 745

---

[1] The ttys of this word blotted—ȝe first written after it.
[2] On Ff. 154ᵛ and 155 each paragraph begins with a larger and more definitely rubricated capital than has been usual hitherto in the MS. and these larger capitals continue more or less for some pages. They appear sparsely in quires N and P.
[3] ȝe first written after xal.   [4] Some word before man crossed through.

The Last Supper

155 ¶ And þe staf þat in ȝour Handys ȝe xal holde
    Is not ellys · but þe exawmplys to other men teche
    Hold fast ȝour stauys in ȝour handys and beth bolde
    To every creature myn precepttys for to preche.

  ¶ Also ȝe must ete þis paschall lombe hastyly   750
    Of weche sentens þis is þe very entent
    At every oure and tyme ȝe xal be redy
    Ffor to fulfylle my cowmawndement.

  ¶ Ffor þow ȝe leve þis day · ȝe are not sure
    Whedyr ȝe xal leve to-morwe or nowth   755
    þer-for hastyly every oure do ȝoure besy cure
    To kepe my preceptys and þau þar ȝe not dowth.

  ¶ Now haue I lernyd ȝow how ȝe xal ete
    ȝour paschal lombe þat is my precyous body
    Now I wyl fede ȝow all with Awngellys mete   760
    Wherfore to reseyve it · come fforth seryattly [1].
                                               Petrus

  ¶ Lord ffor to receyve þis gostly sustenawns
    In dewe forme it excedyth myn intellygens
    Ffor no man of hym-self may have substawns
    To receyve it with to meche reverens.   765

  ¶ Ffor with more delycyous mete lord þou may us not fede
    Þan with þin owyn precyous body
    Wherfore what I haue trespacyd in word thought or dede
    With byttyr contrycion · lord I haske þe mercy.

*whan oure lord ȝyvyth his body to his dyscypulys he xal sey to eche
of hem · except to judas*

¶ This is my body Fflesch and blode   770
    Þat for þe xal dey up-on þe rode.

ᴂ And whan judas comyth last oure lord xal sey to hym

  ¶ Judas art þou Avysyd what þou xalt take.
                                               Judas
    Lord þi body I wyl not for-sake.

ᴂ And sythyn oure lord xal sey on-to judas

---

[1] *sey* miswritten before this word.

                                                    Jhesu

¶ My*n* body to þe I wole not denye
 Sythyn þo*u* wylt presume þ*er*-upon   775
 Yt xal be þi dampnacyo*n* verylye
 I ȝeve þe warnyng now be-forn.

a   *A*nd *aftyr þat judas hath reseyvyd he xal syt þ*er *he was cryst seyng*

¶ On of ȝow hath be-trayd me
 Þat at my borde wit*h* me hath ete
 Bettyr it hadde hym for to A be   780
 Bothe vn-born *and* vn-begete.

a   *Than eche dyscypyl xal loke on other* and *petyr xal sey*

            Petrus

¶ Lord it is not I.   and *so alle xul seyn tyl þei comyn at*
        judas weche xal sey

            Judas

 Is it owth I lord
        þan jhesus xal sey

            Jhesu

 Judas þo*u* seyst þat word'
 Me þo*u* ast solde þat was þ*i* ffrend   785
 Þat þo*u* hast be-gonne brenge to An ende.

*þan judas xal gon A-geyn to þe jewys · And yf men wolne xal mete
with hym and sey þis spech folwyng* | *or levynt whether þei wyl · þe
devyl þus seyng*

            Demon

Fo. 156 ¶ A · A · Judas Derlyng myn
(R quire) Þo*u* art þe best to me þat evyr was bor*e*
 Þo*u* xalt be crownyd in helle peyn
 and þ*er*-of þo*u* xalt be sekyr for evyr-more.   790

¶ Thow hast solde þi maystyr *and* etyn hym also
 I wolde þo*u* kowdyst bryngyn hy*m* to helle every del
 But ȝet I fer*e* he xuld do þ*er* sum sorwe *and* wo
 Þat aft helle xal crye out on me þat sel.

¶ Sped up þi matere þat þo*u* hast be-gonne   795
 I xal to helle for þe to mak redy
 Anon þo*u* xalt come wher þo*u* xalt wonne
 In fyr*e and* stynk þo*u* xalt sytt me by.

Jhesu

¶ Now þe sone of god claryfyed is
and god in hym is claryfyed also 800
I am sory þat judas hath lost his blysse
Weche xal turne hym to sorwe and wo.

¶ But now in þe memory of my passyon
To ben partabyl with me in my reyn above
ȝe xal drynk myn blood with gret devocyon 805
Wheche xal be xad ffor mannys love.

¶ Takyth þese chalys of þe newe testament
And kepyth þis evyr in ȝour mende
As oftyn as ȝe do þis with trewe intent
It xal defende ȝow fro þe ffende. 810

℀ *Than xal þe dysciplys com* and *take þe blod · Jhesus seyng*

156ᵛ Þis is my blood þat for mannys synne
Outh of myn herte it xal renne.

℀ *And þe dyscipulys xul sett þem Aȝen þer þei wore · and jhesus xal seyn*

¶ Takyth hed now bretheryn what I haue do
With my flesch and blood I haue ȝow fed
Ffor mannys love I may do no mo 815
Þan for love of man to be ded.

¶ Werfore petyr and ȝe every-chon
ȝyf ȝe loue me fede my schep
þat for fawth of techyng þei go not wrong
But evyr to hem takyth good kep. 820

¶ ȝevyth hem my body as I haue to ȝow
Qweche xal be sacryd be my worde
And evyr I xal þus A-byde with ȝow
In-to þe ende of þe werde.

¶ Ho so etyth my body and drynkyth my blood 825
Hol god and man he xal me take
It xal hym defende from þe deuyl wood
and at his deth I xal hym nowth for-sake.

¶ *and* ho so not ete my body nor drynke my blood
Lyf in hym is nevyr A dele 830

Kepe wel þis in mende for ȝour good
and every man save hym-self wele.

a here jhesus takyth a basyn with watyr and towaly gyrt abowtyn hym and fallyth beforn petyr on his o kne.

Fo. 157 ¶ Jhesus[1]

A-nother exawmpyl I xal ȝow showe
How ȝe xal leve in charyte
Syt here down at[2] wordys fewe 835
and qwat I do ȝe sofre me.

here he takyth þe basyn and þe towaly and doth as þe roberych seyth beforn.

                Petrus

¶ Lord what wylt þou with me do
Þis servyce of þe I wyl for-sake
To wassche my feet þou xal not so
I am not worthy it of þe to take. 840

                Jhesu

¶ Petyr and þou for-sake my seruyce[3] all
Þe weche to ȝow þat I xal do
No part with me haue þou xal
And nevyr com my blysse on-to.

                Petrus

Þat part lord we wyl not for-go 845
We xal A-bey his comawndement
Wasche hed and hond we pray þe so
We wyl don After þin entent.

a here jhesus wasshyth his dyscipulys feet by and by and whypyth hem
And kyssyth hem mekely and sythym settyth hym down þus seyng

                Jhesu

ffrendys þis wasshyng xal now prevayll
ȝoure lord and mayster ȝe do me calle 850
And so I am with-owtyn fayl
ȝet I haue wasschyd ȝow Alle

Fo. 157ᵛ A memory of þis haue ȝe xall
Þat eche of ȝow xal do to othyr
With vmbyl hert submyt egal 855
As eche of ȝow were otherys brother.

---

[1] The name of the next speaker (possibly *Petrus*) has been erased at the bottom of Fo. 156ᵛ and *Jhesus* thus written at the top of the page.
[2] *ſe* written before *wordys* and crossed through. [3] MS. *seruycē* (? *seruyces*).

The Last Supper

¶ No thyng serys so wele plesyth me
Nor no lyf þat man may lede
As þei þat levyn in charyte
In efne I xal reward here mede 860
Þe day is come I must procede
Ffor to fulfylle þe prophecy
Þis nyth for me ȝe xal han drede
Whan novmbyr of pepyl xal on me cry.

¶ Ffor þe prophetys spoke of me 865
And seydyn of deth þat I xuld take
Fro whech deth I wole not fle
But for mannys synne A-mendys make.

¶ This nyth fro ȝow be led I xal
and ȝe for fer fro me xal fle 870
Not onys dur speke whan I ȝow call
And some of ȝow for-sake me.

¶ Ffor ȝow xal I dey and ryse ageyn
Vn þe thrydde day ȝe xal me se
Be-forn ȝow all walkyng playn 875
In þe lond of Galyle.

Petrus

¶ Lord I wyl þe nevyr for-sake
Nor for no perellys fro þe fle
I wyl rather my deth take
Þan onys lord for-sake þe. 880

Jhesu

¶ Petyr yn[1] ferthere þan þu doyst[2] knowe
As for þat promese loke þou not make
For or þe cok hath twyes crowe
Thryes þou xal me for-sake.

¶ But all my frendys þat Arn me dere 885
Late us go þe tyme drawyth ny
We may no lengere A-bydyn here
Ffor I must walke to betany.

---

[1] Halliwell reads *þu*, the forms of *yn* and *þu* being indistinguishable in this MS.
[2] The *o* is written over another letter.

¶ Þe tyme is come þe day drawyth nere
On-to my deth I must in hast   890
Now petyr make haṫ þi felawys chere
My flesch for fere is qwakyng¹ fast.

α here jhesus goth to betany ward and his dyscipulys folwyng with sad contenawns jhesus seyng

¶ Now my dere frendys and bretheryn echoñ
Remembyr þe wordys þat I xal sey
Þe tyme is come þat I must gon   895
For to fulfylle þe prophesey.

Þat is seyd of me þat I xal dey
Þe fendys power fro ȝow to flem
Weche deth I wole not deney
Mannys sowle my spovse² for to redem.   900

¶ Þe oyle of mercy is grawntyd playn
Be þis jorne þat I xal take
Be my fadyr I am sent sertayn
Be-twyx god and man An ende to make.

Fo. 158ᵛ ¶ Man for my brother may I not for-sake   905
Nor shewe hym vn-kendenesse be no wey
In peynys for hym my body schal schake
And for love of man man xal dey.

α here jhesus and his discipulys go toward þe mount of olyvet and whan he comyth a lytyl þer be-syde in a place lych to A park he byddyt his dyscipulys A-byde hym þer and seyth to petyr or he goth

¶ Petyr with þi felawys · here xalt þou A-byde
and weche tyl I come A-geyn   910
I must make my prayere here ȝou be-syde
My flesch qwakyth sore for fere and peyn.

Petrus

Lord þi request doth me constreyn
In þis place I xal Abyde stylle
Not remeve tyl þat þou comyst Ageyn   915
In confermyng lord of þi wylle.

α here jhesu goth to olyvet and settyth hym down one his knes and prayth to his fadyr þus seyng

¹ The q is roughly written, as if over some other letter.
² The v is written over an original u.

The Betrayal

Jhesu

¶ O ffadyr fadyr for my sake
þis gret passyon þou take fro me
Wech arn ordeyned þat I xal take
ȝyf mannys sowle savyd may be   920
And ȝyf it be-hove fadyr for me
To save mannys sowle þat xuld spylle
I am redy in eche degre
þe vyl of þe for to fulfylle.

ᴄ here jhesus goth to his dyscipulis and fyndyth hem sclepyng jhesus þus seyng to petyr

¶ Petyr petyr þou slepyst fast   925
A-wake þi felawys and sclepe no more
Of my deth ȝe Are not Agast
ȝe take ȝour rest and I peyn sore.

ᴄ here cryst goth Ageyn þe second tyme to olyvet and seyth knelyng

¶ Ffadyr in hevyn I be-seche þe
Remeve my[1] peynes be þi gret grace   930
and lete me fro þis deth fle
As I dede nevyr no trespace
The Watyr and blood owth of my face
Dystyllyth for peynes þat I xal take
My flesche qwakyth in ferful case   935
As þow þe joyntys A-sondre xuld schake.

ᴄ here jhesus goth A-ȝen to his discipulis and fyndyth hem Asclepe Jhesus þus seyng latyng hem lyne

¶ Ffadyr þe thrydde tyme I come A-geyn
Ffulleche myn erdon for to spede
Delyuere me fadyr fro þis peyn
Weche is reducyd with ful gret dred   940
On to þi sone fadyr take hede
þou wotyst I dede nevyr dede but good
It is not for me þis peyn I lede
But for man I swete bothe watyr and blode.

ᴄ here An Aungel descendyth to jhesus and bryngyth to hym A chalys with An host þer in.

[1] þi first written before my.

The Betrayal

Angelus

Fo. 159ᵛ ¶ Heyl bothe god *and* man in dede 945
The ffadyr hath sent þe þis p*re*sent
He bad þ*at* þ*ou* xuldyst not drede
But fulfylle his intent
As þe parlement of hefne hath ment
Þ*at* mannys sowle xal now redemyd be 950
Ffrom hefne to herd lord þ*ou* wore sent
Þ*at* dede Appendyth on-to þe.

¶ Þis chalys ys þ*i* blood þis bred is þ*i* body
Ffor ma*n*nys synne evyr offeryd xal be
To þe fadyr of heffne þ*at* is al-mythty 955
Þ*i* dyscip*u*lis *and* aff presthood xal offere fore the.

a here þe Aungel Ascendyth A-ȝen sodeynly.

Jhesu

Ffadyr þ*i* wyl ffulfyllyd xal be
It is nowth to say A-ȝens þe case
I xal fulfylle þe p*r*ophesye
and sofre deth ffor mannys trespace. 960

*a here goth cryst A-geyn to his dysciɟulys and fyndyth he*m *sclepyng stylle.*

¶ A-wake petyr þ*i* rest is ful long
of sclep þu wylt make no delay
Judas is redy w*ith* pepyl strong
and doth his part me to be-tray
Ryse up serys I ȝ*ou* pray 965
On-close ȝ*ou*r eyne for my sake
We xal walke in to þe way
and sen hem com þat xul me take.

Fo. 160 ¶ Petyr whan þ*ou* seyst I Am for-sake
Amonge my*n* frendys *and* stond Alone 970
Aff þe cher þ*at* þ*ou* kanst make
Geve to þ*i* bretheryn every-chone.

*a here jhesus wit*h *his dyscipulis goth in-to þe place and þer xal co*m*e in A x personys weyl be-seen in white Arneys and bregandery*s *and some dysgysed in odyr garmentys with swerdys gleyvys and other* *straunge wepone*¹ *as cressettys with feyr and lanternys and torchis lyth* and *judas formest of Al conveyng hem to jhesu be contenawns*

Jhesus þus s[eyng

¶ Serys² in ȝour way ȝe haue gret hast
To seke hym þat wyl not fle
Of ȝow I am ryth nowth A-gast 975
Telle me serys whom seke ȝe.

leyon

¶ Whom we seke here I telle þe now
A tretour is worthy to suffer deth
We knowe he is here A-mong ȝow
His name is jhesus of nazareth. 980

Jhesu

¶ Serys I am here þat wyl not fle
Do to me all þat ȝe kan
For sothe I telle ȝow I am he
Jhesus of nazareth þat same man.

☉ *here alle þe jewys falle sodeynly to þe Erde whan þei here cryst speke and qwan byddyth hem rysyn þei rysyn Aȝen cryst þus seyng*

¶ A-ryse serys whom seke ȝe fast haue ȝe gon 985
Is howth ȝour comyng hedyr for me
I stond be-forn ȝow here echon
Þat ȝe may me bothe knowe *and* se.

Rufyne

160ᵛ ¶ Jhesus of Nazareth³ we seke⁴
And we myth hym here A-spye. 990

Jhesu

I told ȝow now with wordys meke
Be-forn ȝou Al þat it was I.

Judas

¶ Welcome jhesu my mayster dere
I haue þe sowth in many A place
I am ful glad I fynd þe here 995
For I wyst nevyr wher⁵ þou wace.

☉ *here judas kyssyth jhesus and A-noon Alle þe jewys come A-bowth hym and ley handys on hym and pullyn hym as þei were wode and makyn on hym A gret cry All At-onys and Aftyr þis Petyr seyth*

---

¹ MS. wepoñ, not the usual ñ (= ne).
² *What* written at the beginning of the line before *Serys* and crossed through.
³ A more elaborate capital than usual.   ⁴ *and we* first written after *seke*.
⁵ *h* omitted and written above the line—cf. *werfore*, Ff. 156ᵛ and 160ᵛ.

The Betrayal

Petrus

¶ I drawe my swerd now þis sel
Xal I smyte mayster fayn wolde I wete¹.

and forth-with he smytyth of Malcheus here and he cryeth help myn here myn here · and cryst blyssyth it and tys hol.

Jhesus

Put þi swerd in þe shede fayr *and* wel
Ffor he þat smyth *with* swerd · *with* swerd xal be smete.   1000

¶ A judas þis treson cowntyrfetyd hast þou
And þat þou xalt ful sore repent
Þou haddyst be bettyr A ben vn-born now
Þi body *and* sowle þou hast shent.

Gamalyel

¶ Lo jhesus þou mayst not þe cace refuse   1005
Bothe treson *and* eresye in þe is fownde
Stody now fast on þi*n* excuse
Whylys þat þou gost in cordys bownde
Þou kallyst þe kyng of þis werd rownde
Now lete me se þi gret powere   1010
And saue þi-self here hool *and* sownde
And brynge þe out of þis dawngere.

leyon

Fo. 161   Bryng forth þis tretoure spare hym nowth
On-to Cayphas þi jewge we xal þe ledde
In many A place we haue þe sowth   1015
And to þi werkys take good hede.

Rufyne

¶ Come on jhesus *and* folwe me
I am ful glad þ*at* I þe haue
Þou xalt ben hangyn up-on A tre
A² melyon of gold xal þe not save.   1020

leyon

¶ Lete me leyn hand on hym in heye
On to his deth I xal hym bryng
Shewe forth þi wyche-crafte *and* nygramansye
What helpyth þe now Al þi fals werkyng.

---

¹ *Petyr put þi s written as the beginning of the next line and crossed through.*
² *m written at the beginning of the line before A and crossed through.*

The Betrayal

Jhesu

¶ Ffrendys take hede ȝe don vn-ryth   1025
So vn-kendely with cordys to bynd me here
And þus to falle on me be nyth
As thow I were A thevys fere
Many tyme be-forn ȝow I dede A-pere
With-inne þe[1] temple sen me ȝe have   1030
Þe lawys of god to teche and lere
To hem þat wele here sowlys sawe[2].

¶ Why dede ȝe not me dysprave
And herd me preche bothe lowd and lowe
But now as woodmen ȝe gynne to Rave   1035
And do thyng þat ȝe notwth knove[3].

Gamaly[el

¶ Serys I charge ȝow not o word more þis nyth
But on-to Cayphas in hast loke ȝe hym lede
Have hym forth With[4] gret dyspyte
And to his wordys take ȝe non hede.   1040

α here þe jewys lede cryst outh of þe place with gret cry and noyse some drawyng cryst forward and some bakwarde and so ledyng forth with here weponys A-lofte and lytys brennyng and in þe mene tyme maryę magdalene xal rennyn to oure lady and telle here of oure lordys takyng þus seyng

Maria Magdelene

¶ O in-maculate modyr of all women most meke
O devowtest in holy medytacion evyr A-bydyng
Þe cawse lady þat I to ȝour person seke
Is to wetyn yf ȝe heryn ony tydyng.

¶ Of ȝour swete sone and my reverent lord jhesu   1045
Þat was ȝour dayly solas ȝour gostly consolacyon.

Marya

I wold ȝe xuld telle me mawdelyn and ȝe knew
Ffor to here of hym it is All myn Affeccyon.

Maria Magd[alen

¶ I wold fayn telle lady and I myth for wepyng
For sothe lady to þe jewys he is solde   1050

---

[1] þis first written, corrected to þe.   [2] So in MS.
[3] So in MS.   [4] W unusually tall.

With cordys þei haue hym bownde and haue hym in kepyng
Þei hym bety spetously¹ and haue hym fast in holde¹.
             Maria uirgo
A · A · A · how myn hert is colde
A hert hard as ston how mayst þou lest
Whan þese sorweful tydyngys Are þe told   1055
So wold to god hert þat þou mytyst brest.

¶ A jhesu · jhesu · jhesu · jhesu ·
Why xuld ȝe sofere þis trybulacyon and Advercyte
How may thei fynd in here hertys · ȝow to pursewe
Þat nevyr trespacyd in no maner degre   1060
For nevyr thyng but þat was good thowth ȝe

Fo. 162   Where-fore þan xuld ȝe sofer þis gret peyn
I suppoce veryly it is for þe tresspace of me
And I wyst þat myn hert xuld cleve on tweyn.

¶ Ffor þese langowrys may I susteyn²   1065
Þe swerd of sorwe hath so thyrlyd my meende
Alas what may I do · Alas what may I seyn
Þese prongys myn herte A-sondyr þei do rende.

¶ O ffadyr of hefne wher ben Al þi be-hestys
Þat þou promysyst me whan A modyr þou me made   1070
Þi blyssyd sone I bare be-twyx tweyn bestys
And now þe bryth colour of his face doth fade.

¶ A good fadyr why woldyst þat þin owyn dere sone xal sofre
  Al þis
And dede he nevyr Aȝens þi precept but³ evyr was obedyent
And to every creature most petyful most jentyl and benyng
  i-wys   1075
And now for all þese kendnessys is now most shameful schent.

¶ Why wolt þou gracyous fadyr þat it xal be so
May man not ellys be savyd be non other kende
ȝet lord fadyr þan · þat xal comforte myn⁴ wo   1079
Whan man is savyd be my chylde and browth to A good ende.

---

[1] The *sl* is written over some correction, as is also the *d* in *holde*.
[2] So in MS., *not* omitted (?).   [3] MS bᵗ.
[4] Omitted and written over the line.

¶ Now dere sone syn þou hast evyr be so ful of mercy
Þat wylt not spare þi-self · for þe love þou hast to man
On All man-kend now haue þou pety
And Also thynk on þi modyr þat hevy woman.

[Remainder of Fo. 162—2¼ inches—and Fo. 162ᵛ left blank.]

163

                                                                                           1 doctor

¹O Thou Altitude · of Al gostly Ryches
  O þou incomperhensibele · of grete excyllence
O þou luminarye · of pure lyghtnes
Shete out þi bemys · on-tyl þis Audyens.

                                                                                           2 doctor

O Ffily Altissimi · clepyd by eternalyte   5
Hele þis congregacion · with þe salve of þi passyon
And we prey þe · spiritus paraclyte
With þe ffyre of þi love · to slake All detraccion.

                                                                                           1 doctor

To þe pepyl not lernyd · I stonde as A techer
Of þis processyon · to ʒeve informacion   10
And to them þat be lernyd · As A gostly precher
that in my rehersayl · they may haue delectacion.

                                                                                               2

Welcome of þe apostelys · þe gloryous qwere
Ffyrst Petyr ʒour prynce · And eke ʒour presydent
And Andrewe ʒour half brother · to-gedyr in ffere   15
that Ffyrst ffolwyd Cryst · be on Assent.

                                                                                               1

O ʒe tweyn luminaryes · Jamys · and Jhon
Contynualy brennyng · As bryght As þe sonnbem
With þe chene of charyte · bothe knyt in on
And offeryd of ʒour modyr · to cryst in jherusalem.   20

                                                                                               2

163ᵛ   Welcome Phelypp · þat conuertyd samaryan
And conuertyd þe tresorere · of þe qwene cavdas
With Jamys · þe lesser · that Apud jherosolyman
Was mad fyrst patryarke · by þe ordenauns of Cephas².

---

¹ No number is attached to this section, which stands by itself in the MS. between blank folios. It is rubricated on a different plan. Each speech has an initial red capital, and the lines connecting the rhyme words are red instead of black as elsewhere in the MS. except four lines on Fo. 214ᵛ. No paragraph marks are used.

² † *Petyr* written over *Cephas*.

j^{us} 1

Heyl mathew the Apostel · *and* Also evangelyst   25
that was clepyd to þe fflok · of gostly conue*r*sacion
Ffrom thyrknes of concyens · þat ȝe were in ffest
with Bertylmew þat ffled · Aͷ carnaͷ temptaci*on*.

2us 1

Heyl Symeon zelotes · þus be ȝo*ur* name
And Judas þat bothe · wel lovyd oure lord   30
There ffore ȝe haue bothe joye *and* game
Wher nevyr is sstryff · but good A-corde.

j^{us}

Heyl poul grett docto*ur* · of þe ffeyth
And vessel chosyn be trewe Elecci*on*
heyl Thomas · of whom þe gospel seyth   35
in crystys wounde · was ȝo*ur* refecci*on*.

2us

Heyl Joh*a*n baptyst · most sovereyn creatur*e*
that evyr was born · be naturaͷ conseyvyng
And hyest of p*r*ophetys · As wytnessyth scryptur*e*
heyl joys [2] þat in desert · was Aͷ-wey cryeng.   40

1us 3

[Ff. 164 and 164^v originally left blank. Fo. 164 is the first of quire S, and is discoloured, having apparently been an outside leaf at some time. It has on it a tapering stain which corresponds with similar stains lying in an opposite direction on Fo. 151^v, the last leaf of quire O. On Fo. 164 stands the second signature of R. Hegge. At the top of the page in book-hand are the words:

In nom*j*ne Dei · Amen.

In the middle of the page in cursive hand:

ego R. H. Dunelmensis
possideo
οὐ κτησις* ἀλλὰ χρησις.]

---

[1] These contractions are added in paler ink.   [2] So in MS. for *voys*.
[3] This stands at the bottom of the folio—the last of quire R—and suggests that the Doctors' proclamation is incomplete. It is, however, a more roughly made figure than the others.
* This τ is inserted in different ink.

The Passion Play. II

p. 165 *What tyme þat processyon is enteryd* in to þe place and þe herowdys takyn his schaffalde · and pylat and annas and cayphas here schaffaldys Also þan come þer An exposytour in doctorys wede þus seyng

              Contemplacio

¶ Sofreynes and frendys · ȝe mut alle be gret with gode
 Grace love and charyte · evyr be ȝou A-mong
 þe maydenys sone preserve ȝou · þat for man deyd on rode
 he þat is o god · in personys thre · defende ȝou fro ȝour fon.

¶ Be þe leue and soferauns of all-mythty god   5
 We intendyn² to procede þe matere þat we lefte þe last ȝere
 Where fore we be-seche ȝow þat ȝour wyllys be good
 to kepe þe passyon in ȝour mende þat xal be shewyd here.

¶ The last³ ȝere we shewyd here how oure lord for love of man
 Cam to þe cety of jherusalem · mekely his deth to take   10
 And how he made his mawnde · his body ȝevyng þan
 to his Apostelys evyr with us · to A-bydyn for mannys sake.

¶ In þat mawnde he was betrayd · of judas þat hym solde
 to þe jewys for xxx^ti platys · to delyvyr hym þat nyth
 With swerdys and gleyvys · toke⁴ jhesu · they come with þe tretour bolde   15
 And toke hym amonges his Apostelys · About myd nyth.

¶ Now wold we procede⁵ how he was browth þan
 be-forn Annas and cayphas · and syth beforn pylate
 And so forth in his passyon · how mekely he toke it for man   19
 besekyng ȝou · for mede of ȝour soulys · to take good hede þer Atte.

α here þe herowndys xal shewe hymself and speke.

              herowdys

165ᵛ   ¶ herodes Rex⁶

¶ Now sees of ȝour talkyng · And gevyth lordly Audyence
 Not o word I charge ȝou þat ben here present

---

¹ This is the only number written at the head of the page and not in the margin.
² The *d* is written over an original *t*.
³ *lass* miswritten before *last* and crossed out.   ⁴ So in MS. for *to*.
⁵ Some word has been erased before *how*.   ⁶ Name in larger form.

noon so hardy to presume · in my hey presence
to on-lose hese lyppys Ageyn myn intent
I am herowde of jewys · kyng most reverent 5
þe lawys of Mahownde · my powere xal fortefye
reverens to þat lord of grace · moost excyllent
Ffor be his powere all þinge doth multyplye.

¶ ȝef ony crystyn be so hardy · his feyth to denye
Or onys to erre Ageyns · his lawe 10
On gebettys with cheynes · I xal hangyn hym heye
and with wylde hors þo traytorys · xal I drawe
To kylle A thowsand crystyn · I gyf not An hawe
to se hem hangyn or brent · to me is very plesauns
to dryvyn hem in to doongenys · dragonys to knawe 15
And to rend here flesche and bonys · on-to here sustenauns.

¶ Johan þe baptyst crystenyd cryst · and so he dede many on
þer-fore my-self · dede hym bryng o dawe
It is I · þat dede hym kylle · I telle ȝou every-chon
Ffor And he had go forth · he xuld A dystroyd our lawe 20
Where as crystyn Apperyth · to me is gret grevauns
It peynyth myn hert of tho tretowrys to here
Ffor þe lawys¹ of mahownde I have in governawns
þe which I wele kepe · þat lord hath no pere
Fo. 166 Ffor he is god most prudent 25
Now I charge ȝou my lordys þat ben here
Yf Any crystyn doggys · here² doth Apere
Bryng þo tretorys · to my hey powere
And þei xal haue sone jewgement.
          i^{us} miles

¶ My sovereyn lord heyest of excillens 30
In ȝou all jewgement · is termynabyle
all crystyn doggys · þat do not here dyligens
ȝe put hem to peynes þat ben inportable.
         ij^{us} miles

No þing in ȝou may be more comendable
as to dysstroye þo traytorys þat erre 35
ageyn oure lawys · þat ben most profytable
be rythwysnesse þat lawe ȝe must proferre.

---

¹ *aw* is written over some other letters.
² The *r* is written over some other letter and the word is not clear.

King Herod

Rex Herow[. . . .

¶ Now be gloryous mahownd · my sovereyn savyour
these promessys I make · as I am trewe knyth
þoo þat excede his lawys be ony errour 40
to þe most xamefullest deth I xal hem dyth
But o thyng is sore in my gret delyte
þere is on jhesus of nazareth · as men me tellyth
of þat man I desyre to han A sythte
ffor with many gret wondrys · oure lawe he fellyth. 45

¶ The son of god hym-self · he callyth
And kyng of jewys he seyth is he
166ᵛ   And many woundrys of hym be-fallyth
My hert desyryth hym for to se
Serys yf þat he come in þis cowntre 50
With oure jurresdyccion loke ȝe A-spye
And a-non þat he be brouth on-to me
And þe trewth myself þan xal trye.
                                                            iᵘˢ miles

¶ To-morwe my jorne · I xal begynne
to seke jhesus with my dew dilygens 55
ȝyf he come ȝour provynce with-inne
he xal not a-skape ȝour hey presens.
                                                            ijᵘˢ miles

¶ Myn sovereyn þis my councel þat ȝe xal take
A man þat is bothe wyse and stronge
thurwe all galyle A serge to make 60
yf jhesu be enteryd ȝour pepyl A-mong
Correcte hese dedys þat be do wronge
Ffor his body is vndyr ȝour bayle
as men talkyn hem among
þat he was born in galyle. 65
                                                            Rex

¶ Thanne of þese materys serys take hede
Ffor A whyle I wele me rest
Appetyde requyryth me so in dede
And ffesyk tellyth me it is þe best.

 *here xal A massanger com in-to þe place rennyng and criyng
 Tydyngys tydyngys · and so rownd Abowth þe place · jhesus of*

nazareth is take · Jhesus of nazareth is take · and *forth-with heylyng
þe prynces þus seyng*

                   Massanger

Fo. 167 ¶ Aḷḷ heyle my lordys · princys of prestys   70
    Sere cayphas *and* sere Annas lordys of þe lawe
    tydyng*ys* I brynge ȝou · reseyve þe*m* in ȝo*ur* brestys
    Jhesus of nazareth is take · þ*er*-of ȝe may be fawe.

   ¶ he xal be browth hedyr to ȝou A-non
    I telle ȝou trewly wi*th* A gret rowth   75
    whan he was take · I was hem Among
    a*nd* þer was I ner to kachyd a clowte.

   ¶ Malcus bar A lanterne · *and* put hym in pres
    A-noon he had A towche *and* of went his ere
    Jhesus bad his dyscyple put up his swerd *and* ces   80
    And sett malcus ere Ageyn · as hool as it was ere.

   ¶ So moty the methowut it was A strawnge syth
    Whan we cam fyrst to hy*m* he cam vs A-geyñ
    And haskyd whom we sowth þ*at* tyme of nyth
    We seyd jhesus of nazareth · we wolde haue hy*m* fayn.   85

   ¶ And he seyd it is I · þ*at* am here in ȝo*ur* syth
    Wi*th* þ*at* word we ovyr-throwyn · bakward every-choñ
    and some on here bakkys · lyeng up-ryth
    but standyng up on fote manly · þer was not oñ.

   ¶ Cryst stod on his fete · As meke as A lom   90
    and we loyn stylle lyche ded men · tyl he bad us ryse
    Whan we we*re* up · fast handys we leyd hy*m* up-oñ
    but ȝet me thought I was not plesyd wi*th* þe newe gyse.

   ¶ Ther-fore takyth now ȝo*ur* cowncel · *and* A-vyse ȝou ryth weyl
Fo. 167ᵛ And beth ryth ware · þ*at* he make ȝou not A-mat   95
    Ffor be my thryfte I dare sweryn at þis seyl
    ȝe xal fynde hym A strawnge watt.

  α here bryng þei jhesus be-forn Annas and C. and on xal seyn þus
    Lo · Lo · Lordys here is þe man
    þat ȝe sent us fore.

                    Annas
    þ*er*fore we cone ȝou thanke than   100
    and reward ȝe xal haue þe more.

The Trial before Annas and Caiaphas

¶ Jhesus þou Art welcome hedyr to oure presens
ful oftyn-tymes we han þe besyly do sowth
we payd to þi dyscyple for þe thretty pens
and as an ox or an hors · we trewly þe bowth.   105

¶ þer-fore now art oure · as þou standyst us be-fore
Sey why þou ast trobelyd us · and subuertyd oure lawe
þou hast ofte concludyd us · and so þou hast do more
Where-fore it were ful nedful to bryng þe A dawe.

          Cayphas
¶ What arn þi dysciplys · þat folwyn þe A-boute   110
and what is þi dottryne¹ þat þou dost preche
telle me now some-whath and bryng us out of doute
þat we may to othere men þi prechyng forth teche. .

          jhes[us]²
¶ Al tymes þat I haue prechyd · opyn it was doñ
in þe synagog or in þe temple · where þat all jewys com̄   115
Aske hem what I haue seyd · and Also what I haue doñ
þei con telle þe my wordys · Aske hem everychoñ.

         iᵘˢ judeus
o. 168 ¶ What þou fela to whom spekyst þou
 xalt þou so speke to A buschop
 þou xalt haue on þe cheke · I make A vow   120
 and ʒet þer-to A knok.

α *here he xal smyte jhesus on þe cheke.*

          jhesus
¶ yf I haue seyd Amys
þer-of wytnesse þou mayst bere
and yf I haue seyd but weyl in þis
þou dost Amys me to dere.   125

          Annas
¶ Serys takyth hed now to þis mañ
þat he dystroye not oure lawe
and brynge ʒe wytnesse · A-ʒens hym þat ʒe can
so þat he may be browt of dawe.

         iᵘˢ doctor
¶ Sere þis I herd hym · with his owyn mowth seyn   130
brekyth down þis temple · with-out delay
and I xal settynt up Ageyn
as hool as it was be þe thrydde day.

---

¹ Or *doctryne*; *correcte* on Fo. 166ᵛ is also doubtful.
² The last letters of the contraction are blotted.

                ij^us doctor
¶ ȝa ser and I herd hym seyn Also
þat he was þe sone of god   135
and ȝet many A fole wenyth so
I durst leyn þer-on myn hod¹.
                iij^us doctor
¶ ȝa . ȝa . and I herd hym preche meche þing
And A-ȝens oure lawe every del
Of wheche it were longe to make rekenyng   140
to tellyn all at þis seel.
                Cayphas
¶ What seyst now jhesus whi Answeryst not
Fo. 168ᵛ heryst not what is seyd A-ȝens þe
Spek man spek · spek þou fop
hast þou scorn to speke to me   145
Heryst not in how many thyngys þei þe Acuse²
α³ Now I charge þe and conjure be þe sonne and þe mone
þat þou telle us and þou be goddys sone.
                Jhesus
¶ Goddys sone I am I sey not nay to þe
And þat ȝe all xal se At domys-day   150
Whan þe sone xal come · in gret powere and majeste
And deme þe qweke and dede as I þe say.
                Cayphas
¶ A · out out Allas what is þis
heryth ȝe not how he blasfemyth god
what nedyth us to haue more wytness   155
here ȝe han herd All his owyn word
α Thynk ȝe not he is worthy to dey. *et clamabunt omnes*
ȝys · ȝys · ȝys · All we seye · he is worthy to dey · ȝa ȝa ȝa.
                Annas
α Takyth hym to ȝow · and betyth hym sōm del
ffor hese blasfemyng at þis sel.   160
α *here þei xal bete jhesus A-bout þe hed and þe body and spyttyn in his
face and pullyn hym down and settyn hym on A stol and castyn A
cloth ouyr his face · and þe fyrst xal seyn*

---

¹ *hed* first written, corrected to *hod*.   ² This line is extra-metrical.
³ This and the following similar signs affixed to the rhyming couplet on
Ff. 168ᵛ, 169, 169ᵛ, and again on Fo. 171ᵛ on Fo. 177 and 178 and 178ᵛ and
179 and 183ᵛ are in red. With a very few exceptions (Fo. 74), such red
signs only occur otherwise in the play of the Assumption, Ff. 215ᵛ to 222. Those
affixed to stage directions are, as heretofore in the MS., black.

i^{us} jude*us*

ᛣ   A felawys be war*e* what ʒe do to þis man
     ffor he *p*rophecye weyl kan.

ij^{us} jude*us*

ᛣ   Þat xal be A-sayd be þis batte
     what þ*ou* jhe*s*us ho ʒaff þe þat.

*et p*er*c*uci*et s*uper *c*aput¹
iij^{us} jude*us*   165

169*ᛣ*   Whar whar now wole I
     Wetyn how he can p*r*ophecy
     ·ho was þ*at*.

iiij^{us} jude*us*

ᛣ   A *and* now wole I a newe game begynne
     þat we mon pley at aĦ þat arn here-inne.

ᛣ   whele *and* pylle · whele *and* pylle   170
     comyth to halle ho so wylle
     ho was þ*at*.

ᛣ *here xal þe woman come to jewys and seyn*

i^a ancilla

ᛣ   What serys how take ʒe on wi*th* þis man
     se ʒe not on of hese dysciplys · how he be-heldyth ʒ*ou* þan.

ᛣ *here xal þe tother woman seyn to petyr*

ij^a ancille

ᛣ   A · good man me semyth be þe   175
     þat þ*ou* on of his dysciplys xulde be.

Pet*r*us

ᛣ   A woman I sey nevyr er þis man
     syn þat þis werd fyrst be-gan.

*et cant*ab*it gall*us.
i^a Ancilla

ᛣ   What þ*ou* mayst not sey nay þ*ou* art on of hese men
     be þi face wel we may þe ken.   180

Petrus

ᛣ   Woman þ*ou* seyst A-mys of me
     I knowe hym not so mote I the.

i^{us} jude*us*

ᛣ   A fela my*n* wel met
     for my cosynys er*e* þ*ou* of smet

---

¹ This stage direction is not underlined in red.

	Whan we þi mayster in þe ʒerd toke	185
α	þan aH þi ffelawys hym for-soke	
α	and now þou mayst not hym for-sake	
	for þou art of galyle I vndyr-take.	

Petrus

¶ Sere I knowe hym not be hym þat made me
and ʒe wole me be-leve ffor An oth   190
Fo. 169ᵛ I take record of aH þis companye
þat I sey to ʒow is soth   *et cantabit gallus.*

α *And þan jhesus xal lokyn on petyr · and petyr xal wepyn and þan he xal gon out and seyn*

¶ A weel a-way · weel away · fals hert why whylt þou not brest
Syn þi maystyr so cowardly · þou hast forsake
Alas qwher xal I now on erthe rest   195
tyl he of his mercy to grace wole me take.

¶ I haue for-sake my mayster and my lord jhesu
thre tymes as he tolde me · þat I xulde do þe same
Wherfore I may not haue sorwe A-now
I synful creature · am so mech to blame.   200

¶ Whan I herd þe cok crowyn he kest on me A loke
As who seyth be-thynke þe · what I seyd be-fore
Alas þe tyme þat I evyr hym for-soke
And so wyl I thynkyn from hens evyr more.

Cayphas

α Massangere Massangere.   205

Massangere

here lord here.

Cayphas

¶ Massanger to pylat · in hast þou xalt gon
and sey hym we comawnde us in word and in dede
and prey hym þat he be at þe mothalle A-noon
ffor we han A¹ gret matere · þat he must nedys spede.   210

¶ In hast now go þi way
And loke þou tery nowth.

---

¹ This *A* is in darker ink—perhaps written over another letter later.

The Death of Judas

                                                                         Massang*er*

      it xal be do lord be þis day
      I am as whyt as thought.

α here pylat syttyth in *his skaffald* and *þe massang*er *knelyth to hy*m
ºo. 170 *þ*us *seyng*

    ¶ Al heyl sere pylat þat semly is to se   215
    Prynce of al þis jure · *and* kepere of þe lawe
    My lord busshop cayphas comawndyd hy*m* to þe
    And prayd the to be At þe mothalle by þe day dawe.

                                                                           Pylat

    ¶ Go þ*i* way *p*raty masanger · *and* comawnde me Also
    I xal be þere in hast *and* so þou mayst say   220
    be þe oure of *p*rime I xal comy*n* hem to
    I tery no longer · no¹ make no delay.

α here þe massanger comith ² *A*ȝen and *bryngith* ² *An Ansuere þus seyng*

                                                                    Massanger

    ¶ Al heyl my*n* lordys · *and* buschoppys · *and* princys of þe lawe
    Ser pylat comawndyth hy*m* to ȝou · *and* bad me to ȝou say
    He wole be at þe mothalle i*n* hast sone After þe day dawe   225
    He wold ȝe xuld be þer be *p*rime · wi*th*-outh lenger de-lay.

                                                                     Cayphas

α   Now weyl mote þou fare my good page
    take þou þis for þi massage.

α here enteryth judas on to þe juwys þus seyng

                                                                         judas

    ¶ I judas haue synyd · *and* treson haue don
    ffor I haue be-trayd þis rythful blood   230
    here is ȝour mony A-ȝen Aȝ̄ And som̄
    Ffor sorwe *and* thowth · I am wax wood.

                                                                          Annas

    ¶ What is þat to us A-vyse þe now
    þ*ou* dedyst wi*th* us cou*n*awnt ³ make
    þou soldyst hy*m* us · as hors or kow   235
    þerfore þin owyn dedys · þou must take.

ºo. 170ᵛ *þan judas castyth down þe mony · and goth and hangyth hym-self.*

---

        ¹ *and* first written and crossed through before *nc*.
        ² MS. comᵗ and brȳgᵗ; cf. waxit, Fo. 204ᵛ, seyt (saith), Fo. 151.
        ³ MS. cõnawnt. Halliwell prints *covnawnt*.

Fo. 170ᵛ   Cayphas

¶ Now serys þe nyth is passyd þe day is come
  it were tyme þis man had his jewgement
  And pylat Abydyth in þe mothalle Alone
  tyl we xuld þis man present.   240

¶ And þer-fore go we now forth with hym in hast.
                                                                        iᵘˢ judeus
  it xal be don and þat in short spas.
                                                                        ijᵘˢ judeus
  ȝa but loke yf he be bownd ryth wel and fast.
                                                                        iijᵘˢ judeus
  he is saff A-now · go we ryth A good pas.

☞ here þei ledyn jhesu A-bowt þe place · tyl þei come to þe halle.
                                                                          Cayphas
¶ Sere pylat takyht hede to þis thyng   245
  jhesus we han be-forn þe browth
  wheche oure lawe doth down bryng
  and mekyl schame he hath us wrowth.
                                                                            Annas
¶ ffrom þis cetye¹ · in to þe lond of Galyle
  he hath browth oure lawys · neyr in-to confusyon   250
  with hese craftys wrowth be nygramancye
  shewyth to þe pepyl · be fals symulacyon.
                                                                        iᵘˢ doctor
¶ ȝa ȝet ser A-nother · and werst of Alle
  Aȝens sesare · oure emperour þat is so fre
  kyng of jewys · he doth hym calle   255
  So oure emperourys power · nowth xulde be.
                                                                       ijᵘˢ doctor
¶ Sere pylat we kan not telle half þe blame
  þat jhesus in oure countre · hath wrowth
  þerfore we charge þe in þe emperorys name
  þat he to þe deth · in hast be browth.   260
                                                                           Pylat
Fo. 171 ¶ What seyst to these compleyntys jhesu
  these pepyl hath þe sore acusyd
  be-cause þou bryngyst up lawys newe
  þat in oure days were not vsyd.

¹ *ceyt* first written and crossed through,

The Trial before Pilate

Jhesus
265

¶ Of here A-cusyng me rowth nowth
   so þat þei hurt not here soulys · ne non mo
   I haue nowth ȝet founde · þat I haue sowth
   ffor my faderys wyl · fforth must I go.

Pylat

¶ Jhesus be þis þan I trowe · þou art A kyng
   and þe sone of god · þou art Also
   lord of erth · and of All þing
   telle me þe trowth · if it be so.

270

Jhesus

¶ In hefne is knowyn · my faderys intent
   And in þis werlde · I was born
   be my fadyr · I was hedyr sent
   for to seke · þat was for-lorn.

275

¶ Alle þat me heryn · And in me belevyn
   And kepyn here feyth stedfastly
   þow þei weryn dede · I xal þem recuryn
   and xal þem bryng to blysse · endlesly.

280
Pilate

¶ Lo serys now ȝe An erde þis man · how thynk ȝe
   thynke ȝe not All be ȝoure resou
   but as he seyth it may wel be
   and þat xulde be be þis incheson.

¶ I fynde in hym · non obecyon
   of errour nor treson · ne of no maner gylt
   The lawe wele · in no conclusyon
   With-owte defawth he xuld be spylt.

285

i^{us} doctor

¶ Sere pylat þe law restyth in þe
   and we knowe veryly his gret trespas
   to þe emperour · þis mater told xal be
   yf þou lete jhesus · þus from þe pas.

290

Pylat

¶ Serys þan telle me o thyng
   What xal be his A-cusyng.

Annas
295

α Sere we telle þe al to-gedyr
   Ffor his evyl werkys · we browth hym hedyr
α And yf he had not An evyl doere be
   we xuld not Abrowth hym to þe.

		Pylat
ℂ	Takyth hym þan Aftyr ȝour sawe	
	and demyth hym Aftyr ȝour lawe.	300
		Cayphas
ℂ	it is not lefful to vs ȝe¹ seyn	
	no maner man for to slen.	
ℂ	þe cawse why · we bryng hym to þe	
	þat he xuld not oure kyng be.	
ℂ	weyl þou knowyst kyng we haue non	305
	but oure emperour Alon.	
ℂ	Jhesu · þou Art kyng of jure.	Pylat
		Jhesus
	So þou seyst now to me.	
		Pylat
ℂ	Tel me þan	310
	where is² þi kyngham.	
		Jhesus
ℂ	my kyngham is not in þis werld	
	I telle þe At o word.	
ℂ	yf my kyngham here had be	
	I xuld not A be delyveryd to þe³	
		Pylat
ℂ	Serys A-vyse ȝow as ȝe kan	315

Fo. 172   I can fynde no defawth in þis man.

                                                              Annas

¶ Sere here is a gret record · take hed þer to
   and knowyng gret myschef in þis man
   And not only in o day or to
   it is many ȝerys · syn he began   320
   We kan telle þe tyme where and whan
   þat many A thowsand turnyd hath he
   As Aĺĺ þis pepyĺĺ record weyl kan
   from hens in-to þe lond of galyle   et *clamabunt* ȝa. ȝa. ȝa.
                                                              Pilat
¶ Serys of o thyng than · gyf me relacyon   325
   if jhesus were out born in þe lond of galelye
   ffor we han no poer ne no jurediccyon

---

¹ *we* first written, crossed through and *ȝe* written over the line.
² *is* omitted and written over the line.
³ From line 309 to line 314 two lines are written in one.

The Trial before Pilate

    of no man of þat contre
    Ther-fore þe trewth · ȝe telle me
    And A-nother wey I xal provyde   330
    if jhesus were born · in þat countre
    þe jugement of herowdys he must A-byde.

Cayphas

¶ Sere as I am to þe lawe trewly sworn
    to telle þe trewth I have no fer
    in galelye · I know þat he was born   335
    I can telle in what place · and where
    Aȝens þis no man may Answere
    for he was born in bedlem jude
    and þis ȝe knowe now AH and haue don here
    þat it stant in þe lond of galelye.   340

Pylat

¶ Weyl serys syn þat I knowe · þat it is so
    þe trewth of þis I must nedys se
    I vndyrstand ryth now · what is to do
    þe Jugement of jhesu · lyth not to me
    herowde is kyng of þat countre   345
    to jewge þat regyon in lenth and in brede
    þe jurysdyccyon of jhesu · now han must he
    þer fore · jhesu in hast · to hym ȝe lede
    In haH þe hast þat ȝe may spede
    lede hym to þe herownde · A-non present   350
    and sey I comawnde me with worde and dede
    And jhesu to hym þat I haue sent.

i[us] doctor

¶ This erand in hast sped xal be
    in aH þe hast þat we can do
    We xal not tary in no degre   355
    tyl þe herowdys presens · we com̃ to

☞ here þei take jhesu and lede hym · in gret hast to þe herowde · And þe herowdys scafald xal vn-close shewyng herowdes[1] in astat aH þe jewys knelying · except Annas and cayphas þei xal stondyn etcetera

i[us] doctor

¶ heyl herowde most excyllent kyng
    we Arn comawndyd · to þin presens

---

[1] The contraction for (d)es is here probably a slip for the contraction for (d)e.

   pylat sendyth þe · be us gretyng
   And chargyth · us · be oure Obedyens.   360
             ij^{us} doctor

  Þat we xuld do oure dylygens
  to bryng jhesus of nazareth on-to þe
  And chargyth us to make no resystens
  be-cawse he was born in þis countre.
             Annas

Fo. 173 ¶ we knowe he hath wrowth · gret fole   365
  A-geyns þe lawe · shewyd present
  Ther-fore pylat sent hym on to þe
  þat þou xuldyst gyf hym jugement.
           Herowde Rex

 ¶ Now be Mahound my god of grace
  of pylat þis is A dede ful kende   370
  I for-gyf hym now his gret trespace
  And schal be his frend with-owtyn ende.

 ¶ Jhesus to me þat he wole sende
  I desyred ful sore hym for to se
  gret ese in þis pylat xal fynde   375
  and jhesus þou art welcome to me.
            i^{us} judeus

 ¶ My sovereyn lord þis is þe case
  þe gret falsnesse of jhesu · is opynly knawe
  þer was nevyr man dede so gret trespas
  ffor he hath · Al-most · dystroyd oure lawe.   380
            ij^{us} judeus

 ¶ ȝa be fals crafte of soserye
  wrowth opynly to þe pepyll Alle
  and be sotyl poyntys of nygramancye
  many thowsandys fro oure lawe be falle.
            Cayphas

 ¶ Most excellent [1] kyng · ȝe must take hede   385
  he wol dystroye all þis countre · both elde and ȝyng
  yf he ten monthis more procede
  be his meraclys and fals prechyng
  he bryngyth þe pepyl in gret fonnyng
  And seyth dayly A-mong hem Alle   390

   [1] *excelyng* first written, and *yng* crossed through.

The Trial before Herod

173ᵛ   That he is lord *and* of þe jewys kyng
       and þe sone of god he doth hym calle.

                                                         Rex Her*owd*e

     ¶ Serys Alle þese materys · I haue herd sayd
       and meche more þan ȝe me telle
       Alle to-gedyr þei xal be layde   395
       A*nd* I wyl take þ*er* on cow*n*celle.

     ¶ Jhe*s*us þ*ou* Art wel-come to me
       I kan pylat gret thank for his sendying
       I haue desyryd ful longe þe to se
       and of þi meracles · to haue knowyng.   400

     ¶ It is told me þ*ou* dost many A wondyr thyng
       Crokyd to gon *and* blynd men to sen
       and þei þ*at* ben dede gevyst hem levyng
       A*nd* makyst lepers · fayr*e and* hool to ben.

     ¶ These Arn wondyr werkys · wrougth of þe   405
       be what wey · I wolde knowe þe trew sentens
       Now jhe*s*u I p*r*ay the · lete me se
       O meracle wrougth in my p*r*esens.

     ¶ In hast now · do þi dylygens
       A*nd* p*er*-Aventure · I wyl shew favo*ur* to the   410
       for now þ*ou* art in my p*r*esens
       thyn lyf *and* deth · here lyth in me.

α   *And here jh*esus *xal not speke no word to þ*e *herowde.*

     ¶ Jhe*s*us why spekyst not to þi kyng
       what is þe cawse þ*ou* stondyst so stylle
       þ*ou* knowyst I may deme Aƚƚ thyng   415
       thyn lyf *and* deth · lyth At my wylle.

f. 174 ¶ What spek jhe*s*us · *and* telle me why
       þis pepyl do þe so · her*e* Acuse
       spare not but telle me now · on hey
       how þ*ou* canst þ*i*-self · excuse.   420
                                                           Cayphas

     ¶ loo serys þis is of hym · A false sotylte
       he wyl not speke · but whan he lyst
       þus he dysceyvyth þe pepyl in eche degre
       he is ful fals · ȝe veryly tryst.

                                                    Rex her*owde*
  What þou on-hangyd harlot · why wylt þou not speke 425
  hast þou skorne to speke on to þi kyng
  be-cawse þou dost oure lawys breke
  I trowe þou art A-ferd · of oure talkyng.

              Annas
¶ Nay he is not Aferde · but of A fals wyle
  be-cawse we xuld not hym A-cuse 430
  if þat he Answerd 3ow on-tylle
  he knowyth he can not hym¹-self excuse.

            Rex her*owde*
¶ what spek I say · þou foulyng · evyl mote þou fare
  loke up · þe devyl mote þe cheke
  Ser*ys* bete his body w*ith* scorg*ys* bare 435
  And A-say to make hy*m* for to speke.

             i*us* jude*us*
¶ it xal be do w*ith*-outyn teryeng
  come on þou treto*ur* evyl mot þou þe
  whylt þou not speke on-to oure kyng
  A new lesson we xal lere þe. 440

ℭ here þei pulle of jhesus clothis and betyn hym with whyppys.

            ij*us* jude*us*
Fo. 174ᵛ ¶ Jhesus þ*i* bonys we xal not breke
  but we xal make þe to skyppe
  þou hast lost þ*i* tonge · þou mayst not speke
  þou xalt a-say now of þis whippe.

            iij*us* jude*us*
¶ Serys take þese whyppys in 3o*ur* hande 445
  and spare not whyl þei last
  and bete þis tretoure þat here doth stonde
  I trowe þa*t* he wyl speke in hast

ℭ and qwan þei han betyn hym tyl he is alle blody þan þe herownde seyth

¶ Sees ser*ys* · I comawnde 3ou · be name · of þe devyl of helle
  Jhesus thynkyst þis · good game 450
  þou Art strong · to suffyr schame
  þou haddyst levyr be betyn lame
  þan þ*i* defawtys for to telle.

    ¹ This *y* is written over an *e*.

The Trial before Herod

¶ But I wyl not þi body · Aḧ spyl
  nor put it here · in-to more peyn̄
  Serys takyth jhesus · At ȝour owyn wyl   455
  and lede hym to pylat hom Ageyn̄
  Grete hym weyl *and* telle hym serteyn̄
  Aḧ my good frenchep xal he haue
  I gyf hym powere of jhesus · þus ȝe hym seyn   460
  Whether he wole hym dampne or save.

i^{us} doctor

¶ Sere at ȝour request it xal be do
  we xal lede jhesus at ȝour demawde¹
  and delyver hym Pylat · on-to
  And telle hym aḧ · as ȝe comawnde.   465

175 here enteryth Satan in to þe place in þe most orryble wyse · and qwyl
þat he pleyth þei xal don on jhesus clothis · and ouerest A whyte clothe
and ledyn hym A-bowth þe place and þan to pylat be þe tyme þat
hese wyf hath pleyd.

Sathan

¶ Thus I reyne As A rochand · with A rynggyng rowth
  As A devyl most dowty · dred is my dynt²
  Many A thowsand develys · to me do þei lowth
  brennyng in flamys · as fyre out of flynt
  ho so serve me sathan · to sorwe is he sent   470
  with dragonys in doungenys · *and* develys fu³ derke
  in bras *and* in bronston · þe brethellys be brent
  þat wone in þis werd · my wyl for to werke.

31

¶ with myschef on moolde · here membrys I merke
  þat japyn with jhesus þat judas solde   475
  be he nevyr so crafty · nor conyng clerke
  I harry þem to helle · as tretour bolde.

¶ But þer is o thyng · þat grevyth me sore⁴
  Of A prophete · þat jhesu men calle
  he peynyth me every day · more *and* more   480
  with his holy meraclis *and* werkys Alle.

¶ I had hym onys · in A temptacyon
  with glotenye with covetyse · *and* veyn glorye

---

¹ So in MS.   ² *dyth* first written.   ³ So in MS.
⁴ *myn hert* first written instead of *me sore*.

I ha-sayd hym be All weys þat I cownde¹ don
and vttyrly he refusyd hem · and gan me defye. 485

¶ Þat rebuke þat he gaf me · xal not be vn-qwyt
Som what I haue be-gonne · and more xal be do
ffor All his barfot goyng · fro me xal he not skyp
but my derk dongeon I xal bryngyn hym to.

¶ I haue do made redy his cros · þat he xal dye up-on 490
And thre nayles to takke hym with þat he xal not styrte
be he nevyr so holy he xal not fro me gon
but with a sharpe spere · he xal be smet to þe herte.

¶ And sythyn he xal come to helle · be he nevyr so stowte
And ȝet I am Aferd and he come · he wole do som wrake 495
þerfore I xal go warnyn helle · þat þei loke A-bowte
þat þei make redy chenys · to bynd hym with in lake.

¶ Helle helle · make redy · for here xal come A gest
Hedyr xal come jhesus · þat is clepyd goddys sone
And he xal ben here · be þe oure of none 500
And with þe here · he xal wone
And han ful shrewyd rest.

here xal A devyl spekyn in helle   demon
Out upon þe · we conjure þe
þat nevyr in helle we may hym se
ffor And he onys in helle be 505
he xal oure power brest.

     Sathan
¶ A · A · than haue I go to ferre
but som wyle help I haue a shrewde torne
My game is wers þan I wend here
I may seyn · my game is lorne. 510

¶ lo A wyle ȝet haue I kast
if I myth jhesus lyf save
helle gatys xal be sperd fast
And kepe stylle all þo I haue.

¶ to pylatys wyff I wele now go 515
And sche is A-slepe A-bed ful fast

¹ So in MS.

Pilate's Wife's Dream

  And byd here wit*h*-owtyn wordys mo
  to pylat *þ*at sche send in hast.

 ¶ I xal A-say *and* *þ*is wol be
  to bryng pylat in belef   520
  wit*h*-inne A whyle ȝe xal se
  how my craft I wole go pref.

æ *here xal þe devyl gon to pylatys wyf · þe corteyn drawyn as she lyth in bedde* and *he xal no dene make but she xal sone After þat he is come in · makyn a rewly noyse · comyng* and *rennyng of þe schaffald* and *here shert · and here kyrtyl in here hand ·* and *sche xal come beforn pylat leke A mad woman · seyng þus*

               Vxor pilaty

 ¶ Pylat I charge *þ*e · *þ*at *þou* take hede
  deme not jh*es*u · but be his frende
  ȝyf *þou* jewge hy*m* to be dede   525
  þou art da*m*pnyd wit*h*-owtyn ende.

 ¶ A fend · Aperyd me befor*n*
  As I lay in my bed slepyng fast
  Sethyn *þ*e tyme *þat* I was bor*n*
  was I nevyr so sore A-gast.   530

 ¶ As wylde fyre *and* thondyr blast
  he cam cryeng on to me
  he seyd *þ*ei *þ*at bete jh*es*u · or bownd hy*m* fast
  wit*h*-owtyn ende da*m*pnyd xal be.

 ¶ *þ*er-fore A wey here-in *þou* se   535
  and lete jh*es*u from *þ*e cler*e* pace
  *þ*e jewys *þ*ei wole be-gyle *þ*e
  and put on *þ*e Aḥ *þ*e trespace.

               Pylat

176v ¶ Gramercy my*n* wyf for evyr ȝe be trewe
  ȝo*ur* cowncel is good *and* evyr hath be   540
  now to ȝo*ur*ˊ chawmer ȝe do sewe
  and aḥ xal be weyl dame as ȝe xal se.

æ here þe jewys bryng jhesus Aȝen to pylat.

              i[us] doctor

 ¶ Sere pylat gode tydandys · *þou* here of me
  Of herowd *þ*e kyng *þou* hast good wyl

And jhesus he sendyth Aȝen to the 545
And byddyth þe chese · hym to save or spylle.

              ij[us] doctor

¶ ȝa ser all þe poer lyth now in þe
and þou knowyst oure feyth · he hath ner schent
þou knowyst what myschef · þer of may be
we charge þe · to gyf hym jwgement. 550

              Pylat

¶ Serys trewly ȝe be · to blame
Jhesus þus to bete · dyspoyle or bynde
or put hym to so gret schame
ffor no defawth · in hym I fynde.

¶ Ne herowdys nother · to whom I sent ȝow 555
defawte in hym cowde fynde ryth non
but sent hym Aȝen · to me be ȝow
As ȝe knowe wel everychon.

* Ther fore vndyrstande what I xal say
ȝe knowe þe custom is in þis londe 560
of ȝour pasche day · þat is ner honde
what þeff or tretore · be in bonde
   with-out Any price
for worchep of þat day[1] · xal go fre Away

Fo. 177   Now þan me thynkyth · it were ryth 565
to lete jhesus · now go qwyte
And do to hym no mo dyspyte
   serys þis is myn A-vyse

α[2]   I wolde wete what ȝe say.

α   here Alle þei xul cryen ‖ Nay nay nay. 570

              i[us] doctor

α   delyvere us · þe þeff barabas
þat for mansclawth presonde was.

              Pylat

α   What xal I þan with jhesu do
Whethyr xal he A-byde or go.

              ij[us] doctor

α   Jhesus xal on þe cros be don 575
Crucifigatur we crye echon.

---

* Paragraph mark omitted here.
[1] Some miswritten letters (ff ?) crossed out before xal.
[2] This and the following similar signs against the couplets are in red ink.

The Trial of Christ and the Thieves before Pilate

                                                    pylat

α   Ser*ys* what hath jhesus don A-mys
      Crucifigatur · we sey At onys [1].   *populus clamabit.*

                                                  Pylat

    ¶ Serys syn Al gatys · ȝe wolyn so
       puttyn jhesu to wo *and* peyn   580
       jhesus A wyle wi*th* me xal go
       I wole hy*m* examyne betwyx us tweyn.

α here pylat takyth jhesu and ledyth hy*m* in to *þ*e cowncel hous and seyth

    ¶ Jhesus what seyst now · lete se
       This matere now *þ*ou vndyrstonde
       In pes *þ*ou myth be for me   585
       but for *þi* pepyl of *þi* londe.

    ¶ Busshoppys *and* prestys · of *þ*e lawe
       *þ*ei love *þ*e not as *þ*ou mayst se
       and *þ*e comon [2] pepyl A-ȝens *þ*e drawe
       In pes *þ*ou myth A be for me   590
       *þ*is I telle *þ*e pleyn.

177ᵛ ¶ [3] What seyst jhesus · whi spekyst not [4] me to
       knowyst not I haue power on *þ*e cros *þ*e to do
       A*nd* also I haue power to lete *þ*e forth go
       what kanst *þ*ou here to seyn.   595

                                                 Jhesus

    ¶ On me poer *þ*ou hast ryth non
       but *þ*at my fadyr hath grawntyd be-forn
m̃   I cam my faderys wyl to fult-fylle
       *þ*at mankynd xuld not spylle
m̃   he *þ*at hath betrayd me · to *þ*e at *þ*is tyme   600
       his trespas is more *þ*an is *þ*ine.

                                            i[us] doctor

    ¶ ȝe pryncys *and* maysterys · takyth hed *and* se
       how pylat in *þ*is matere is favorabyl
       and *þ*us oure lawys dystroyd myth be
       And to vs Alle vn-recurabyl.   605

α here pylat letyth jhesus A-lone *and* goth in to *þ*e jewys *and* seyth

---

[1] *Omnes* written in the right-hand margin and crossed through in red ink.
[2] MS. comoñ.   [3] This paragraph sign seems inserted by mistake.
[4] *spekyst not* written twice and crossed through in red ink; *to me* written first, corrected, and finally *me to* written above the line.

292 *The Trial of Christ and the Thieves before Pilate*

                                                                                     Pylat

¶ Serys what wole ȝe now · with jhesu do
  I can fynde in hym but good
  it is my cownce¹ ȝe lete hym go
  it is rewthe to spylle his blood.

                                                       Cayphas

¶ Pylat me thynkyth · þou dost gret wrong   610
  Aȝens² oure lawe þus to fortefye
  and þe pepyl here is so strong
  bryngyng þe lawful · testymonye.

                                                  Annas

¶ ȝa and þou lete jhesu fro us pace
  þis we welyn up-holdyn Alle   615
  þou xalt Answere for his trespas
  and tretour to þe emperour we xal þe kalle.

                                                  Pylat

Fo. 178 Now þan syn ȝe wolne · non other weye
  but in Alwyse þat jhesus must deye
  Artyse bryng me watyr I pray þe   620
  and what I wole do ȝe xal se   *hic vnus afferet aquam.*
  As I wasche with watyr my handys clene
  so gyltles of hese deth I mut ben.

                                                  i[us] doctor

þe blod of hym mut ben on vs ⎫
and on oure chyldyr Aftyr vs ⎬ *et clamabunt* ȝa · ȝa · ȝa ·

þan pylat goth A-ȝen to jhesu and bryngith³ hym þus seyng

                                                Pylat

lo serys I bryng hym here to ȝour presens   626
þat ȝe may knowe I fynde in hym non offens.

                                                ij[us] doctor

dylyuere hym · delyuere hym · and lete us go
on þe crosse þat he were do.

                                                  pilat

Serys wolde ȝe ȝour kyng · I xulde on þe cros don.   630
                                                iij[us] doctor

Sere we seyn þat we haue no kyng but þe emperour a-lon.

                                                  pilat

Serys syn Al-gatys it must be so
We must syt and our offyce do

---

¹ So in MS.   ² The *A* is written over some other letter.
³ The contraction is not clear. The scribe apparently first wrote *bringys* and then crossed the top of the contraction without making any erasure.

The Trial of Christ and the Thieves before Pilate

α	brynge forth to þe barre þat Arn to be dempt	
	and þei xal haue here jugement.	635
α	here þei xal brynge barabas · to þe barre and jhesu and ij þewys[1] in here shertys bare leggyd and jhesus standyng at þe barre be-twyx them · and annas and cayphas xal gon in to þe cowncelle hous qwan pylat sytty[th.	

Pylat

α	Barabas hold up þi hond	
178ᵛ	for here at þi delyvere[2] dost þou stond — and he halt up his honde
α	Serys qwhat sey 3e of barabas · thef and tretour bold	
	xal he go fre or he xal be kept in holde.	

iᵘˢ doctor

α	Sere for þe solennyte[3] of oure pasche day	640
	be oure lawe he xal go fre A-way.	

Pylat

α	Barabas þan I dymysse[4] þe	
	and 3eve þe lycens to go fre	*et curret.*[5]
α	dysmas and jesmas · ther as 3e stondys	
	þe lawe comawndyth 3ou · to hald up 3our hondys	645
α	Sere what sey 3e of þese thevys tweyn.	

ijᵘˢ doctor

Sere þei ben boþ gylty we seyn.

Pylat

α	And what sey 3e · of jhesu of nazareth.

iᵘˢ doctor

Sere we sey he xal be put to deth.

Pylat

α	And kone 3e put A-3ens hym no trespas.	650

ijᵘˢ doctor

Sere we wyl All þat he xal be put upon þe crosse.[6]

α *et clamabunt omnes · voce magna dicentes* 3a · 3a · 3a ·

Pylat

α	jhesu þin owyn pepyl han dysprevyd
	Al þat I haue for þe seyd or mevyd.
¶	I charge 3ou All at þe be-gynnyng
	as 3e wole Answere me be-forn 655

---

[1] So in MS. for þerys. Halliwell prints Jewys, but the initial letter is clear.
[2] The word he (?) has been erased before dost.   [3] Or possibly solonnyte
[4] So in MS.   [5] Not underlined in red.
[6] This speech is underlined in red as if it were a stage direction.

þat þer be no man · xal towch ȝour kyng
but yf he be knyght · or jentylman born.

¶ fyrst his clothis ȝe xal of don
and makyn hym nakyd · for to be
bynde hym to A pelere · as sore as ȝe mon   660
þan skorge hym with qwyppys · þat al men may se

Fo. 179α   Whan he is betyn · crowne hym for ȝour kyng
and þan to þe cros ȝe xal hym bryng

α   And to þe crosse þou xalt be fest
And on thre naylys þi body xal rest   665

α   On xal thorwe þi ryth hand go
Anothyr thorwe þi lyfte hand Also

α   þe thred xal be smet thour bothe þi feet
Whech nayl þer-to be mad ful mete

α   And ȝet þou xalt not hange A-lone   670
but on eyther syde of þe xal be on

α   Dysmas now I deme þe
þat on hese ryth hand þou xalt be

α   And Jesmas on þe left hand hangyd xal ben
on þe mownth[1] of caluerye þat men may sen.   675

α   *here pylat xal rysyn and gon to his schaffalde, and þe busshoppys
with hym and þe jewys xul crye for joy with a gret voys · and Arryn
hym and pullyn of his clothis and byndyn hym to A pelere and skorg-
yn hym on seyng þus*

i^{us} judeus

Doth gladly oure kyng
for þis is ȝour fyrst begynnyng.

α   *and qwan he is skorgyd · þei put upon hym A cloth of sylk and settyn
hym on a stol and puttyn A kroune of þornys on hese hed with
forkys and þe jewys knelyng to cryst takyng hym A septer and
skornyng hym · and þan þei xal pullyn of þe purpyl cloth and don
on A-geyn his owyn clothis and leyn þe crosse in hese necke to berynt
and drawyn hym forth with ropys · And þan xal come to women
wepyng and with here handys wryngyn seyng þus*

i^{us} [2] mulier

Fo. 179ᵛ ¶ Allas jhesus · Allas jhesus · wo is me
þat þou art þus dyspoylyd Allas

---

[1] MS. *mowth* (with a stroke over the *th*).

[2] So undoubtedly in MS., though on Fo. 169, p. 277, the distinction in the contraction marks is observed.

and ȝet nevyr defawth · was fownd in the 680
but evyr þou hast be fole of grace.

ij^{us} mulier

¶ A here is a rewful syth · of jhesu so good
þat he xal þus dye Aȝens þe ryth
A wykkyd men ȝe be more þan wood
to do þat good lord so gret dyspyte. 685

☧ here jhesus turnyth a-ȝen to þe women with his crosse þus seyng

jhesus

¶ Dowterys of hierusalem · for me wepyth nowth
but for ȝour-self wepyth · and for ȝour chyldyr Also
for þe days xal come · þat þei han aftyr sowth
here synne and here blyndnesse xal turne hem to wo.

¶ þan xal be sayd · blyssyd be þe wombys þat bareyn be 690
and wo to þe tetys · tho days · þat do ȝevyn sokyng
and to here faderys þei xul seyn · wo to þe tyme þat þou be-gat me
and to here moderys · Allas · wher xal be oure dwellyng.

¶ þan to þe hyllys and mownteynes · they xal crye and calle
oppyn and hyde us from þe face · of hym syttyng in trone 695
or ellys ovyr-throwyth · and on us now come falle
þat we may be hyd · from oure sorweful mone.

☧ here jhesus turnyth fro þe women and goth forth and þer þei metyn
with symonem · in þe place þe jewys seyng to hym

i^{us} jude[us

¶ Sere to þe A word of good
a man is here þou mayst se
beryth hevy of A Rode 700
where-on he xal hangyd be.

Fo. 180 ¶ Therefore we prey all the   þer fore we prey [1]
(T quire)   þou take þe crosse of þe man
bere it with vs to kalvarye
and ryth gret thank þou xalt han. 705

Symon

¶ Serys I may not in no degre
I haue gret errandys for to do
þerfore I pray ȝow excuse me
and on my herand · lete me go.

---

[1] Written at the bottom of Fo. 179ᵛ as catchword.

                                                    ij^us judeus
¶ what harlot hast þou skorne   710
  to bere þe tre whan we þe preye
  þou xalt berynt haddyst þou sworn
  And yt were ten tyme þe weye.

                                                    Symon
¶ serys I prey ȝou dysplese ȝou nowth
  I wole help to bere þe¹ tre   715
  in to þe place it xal be browth
  where ȝe wole comawnde me.

℞ here symon takyth þe cros of jhesus and beryth it forth.

                                                    Veronica
¶ A ȝe synful pepyl why fare þus
  Ffor swet and blood he may not se
  Allas holy prophete cryst jhesus   720
                    and sche whypyth his face
  Careful is myn hert for the.   *with here kerchy.*

                                                    Jhesus²
¶ veronyca þi whipyng doth me ese
  my face is clene · þat was blak to se
  I xal þem kepe from all mys-ese
  þat lokyn on þi kerchy · and remembyr me.   725

Fo. 180ᵛ *þan xul þei pulle jhesu out of his clothis and leyn them togedyr and þer þei xul pullyn hym down and leyn hym Along on þe cros · and after þat naylyn hym þeron.*

                                                    iᵘˢ Judeus
¶ Come on now here we xal A-say
  Yf þe cros for þe be mete
  Cast hym down here in þe devyl way
  how long xal he standyn on his fete.

                                                    ijᵘˢ judeus
¶ pul hym down evyl mote he the   730
  And gyf me his arm in hast
  And A-non we xal se
  hese good days þei xul be past.

                                                    iijᵘˢ judeus
¶ Gef hese other Arm to me
  A-nother take hed to hese feet   735

---

¹ Some word may have been written above þe in darker ink and smudged out.
² Jh written as correction over ve.

And A-non we xal se
yf þe borys be for hym meet.

*iiij*ᵘˢ *judeus*

¶ þis is mete take good hede
pulle out þat Arm to þe sore.

*i*ᵘˢ *judeus* 740

þis is short þe deuyl hym sped
be a large fote *and* more.

*ij*ᵘˢ *judeus*

¶ Ffest on A Rop · *and* pulle hym long
and I xal drawe þe A-geyn
spare we not þese ropys strong
þow we brest both flesch *and* veyn.

745
*iij*ᵘˢ *judeus*

¶ dryve in þe nayl a-non · lete se
And loke *and* þe flesch *and* senues well last.

*iiij*ᵘˢ *judeus*

þat I graunt so mote I the
lo þis nayl is dreve ryth wel *and* fast.*

*i*ᵘˢ *judeus* 750

f. 181 ¶ Ffest A rop þan to his feet
And drawe him don long Anow.

*ij*ᵘˢ *judeus*

here is a nayl for both good *and* greet¹
I xal dryve² it thorwe I make A vow.

☧ *here xule þei leve of and dawncyn a-bowte þe cros shortly.*

*iij*ᵘˢ *judeus*

¶ Lo fela here A lythe · takkyd on A tre.

*iiij*ᵘˢ *judeus*³
755

ȝa *and* I trowe þou art A worthy kyng.

*i*ᵘˢ *judeus*

A good sere telle me now what helpyth⁴ þi prophecy þe.

*ij*ᵘˢ *judeus*

ȝa or Any of þi ffals prechyng.

*iij*ᵘˢ *judeus*

¶ Serys set up þe cros on þe hende
þat we may loke hym in þe face.

---

* The first six words of this line are scribbled again in the bottom margin in a slightly later hand.
¹ The *g* written over an *r*.   ² The *r* written over a *y*.
³ A piece of the original edge has been torn away here.
⁴ The *h* (initial) is written over some other letter.

iiij⁽ᵘˢ⁾ judeus

ȝa *and* we xal knelyn on to oure kyng so kend' 760
and preyn hym of his gret grace.

℺ here qwan *þei han set hym up þei xuln gon be-fore hyne¹ seyng eche affter other þus*

i⁽ᵘˢ⁾ judeus

¶ heyl kyng² of jewys · yf þou be.

ij⁽ᵘˢ⁾ judeus

ȝa · ȝa · sere as þou hangyst þere flesche *and* bonys.

iij⁽ᵘˢ⁾ judeus

Com now down of þat tre.

iiij⁽ᵘˢ⁾ judeus

And we wole worchepe þe aft Atonys. 765

℺ *here xul poer comonys stand and loke upon þe jewys iiij or v and þe jewys xul come to theme and do theme hange þe þevys.*

i⁽ᵘˢ⁾ judeus

¶ come on ȝe knavys · *and* set up þise³ ij crosses ryth
and hange up þese to thevys Anon.

ij⁽ᵘˢ⁾ jud[eus

Fo. 181ᵛ ȝa *and* in þe worchep of þis worthy knyth
on eche syde of hym xal hangyn on.

℺ *here þe sympyl men xul settyn up þese ij crossys and hangyn up þe thevys be þe Armys and þer whylys xal þe jewys cast dyce for his clothis and fytyn and stryvyn and in þe mene tyme xal oure lady come with iij maryes with here and sen Johan with heme settyng hem down A-syde A-fore þe cros · oure lady swuonyng⁴ and mornyng and leysere seyng*

Maria

¶ A my good lord my sone so swete 770
what hast þou don why hangyst now þus here
is þer non other deth to þe now mete
but þe most shamful deth · Among þese thevys fere.

¶ A out on my hert whi brest þou nowth 774
And þou art maydyn *and* modyr *and* seyst þus þi childe spylle

---

¹ So in MS. for *hyme*.
² *kyng* is written as correction over some erased word.
³ The medial vowel is not clear and might be *e*, but is perhaps to be read as an *i* with an unusually marked dot.
⁴ There is a blotted stroke between the *w* and the *u*.

The Crucifixion

how mayst þou a-byde þis sorwe *and* þis woful þowth
A deth · deth · deth · why wylt þou not me kylle.

α here oure lady xal swonge A-ȝen and *ore lord xal seyn þus*

Jhesus

¶ O Ffadyr Al-mythy · make*re* of ma*n*
Ffor-gyff þese jewys þat don me wo
Ffor-geve hem fadyr for-geve hem þan   780
Ffor thei wete notwh[1] what þei do.

i*us* Jude*us*

¶ ȝa vath[2] · vath · now · here is he
þ*at* bad us dystroye oure tempyl on A day
and wit*h*-inne days thre
he xulde reysynt A-ȝen in good A-ray.   785

ij*us* Jude*us*

¶ Now *and* þou kan do swech A dede
help now þi-self yf þ*at* þou kan
and we xal be-levyn on þe wit*h*-outyn drede
and seyn þou art a mythty man.

iij*us* Judeus

ȝa yf þu be goddys sone · as þou dedyst teche   790
ffrom þe cros come now dow*n*
þan of mercy we xal þe beseche
and seyn þou art a lord of gret renown.

jestes

¶ Yf þou be goddys sone as þou dedyst seye
helpe here now both þe *and* vs   795
but I fynde it not al in my feye
þat þou xuldyst be cryst goddys sone jh*e*sus.

Dysmas

¶ Do wey fool why seyst þou so
he is þe sone of god I be-leve it wel
and synne dede he nevyr !o   800
þat he xuld be put þis deth tyl.

¶ but we ful mech wrong han wrowth
he[3] dede nevyr þing A-mys
now mercy good lord mercy · *and* for-gete me nowth
whan þou comyst to þi kyngham *and* to þi blysse.   805

---

[1] So in MS.   [2] *w* written before *vath* and crossed out.
[3] Some word before *dede* erased and *he* written beyond the line in the margin.

The Crucifixion

Jhesus

¶ Amen Amen · þou art ful wyse
þat þou hast Askyd I grawnt þe
þis same day in paradyse
with me þi god þou xalt þer be.

Maria

Fo. 182ᵛ ¶ O my sone my sone · my derlyng dere   810
[marked 180—some older mark ?100 or 180 smudged out.]
what haue I defendyd þe
þou hast spoke to alle þo · þat ben here
and not o word þou spekyst to me.

¶ To þe jewys þou art ful kende
þou hast for-gove al here mysdede   815
and þe thef · þou hast in mende
for onys haskyng mercy · hefne is his mede.

¶ A my sovereyn lord why whylt þou not speke
to me þat am þi modyr · in peyn for þi wrong
A hert hert why whylt þou not breke   820
þat I were out of þis sorwe so stronge.

Jhesus

¶ A woman woman · be-hold þer þi sone
And þou jon take¹ her for þi modyr
I charge þe to kepe here as besyly as þou kone
þou A clene mayde xal kepe A-nother.   825

¶ And woman þou knowyst þat my fadyr of hefne me sent
to take þis manhod of þe · Adam ys rawnsom to pay
Ffor þis is þe wyl · and my faderys intent
þat I xal þus deye · to delyuere man · fro þe develys pray.

¶ Now syn it is þe wyl of my fadyr · it xuld þus be   830
Why xuld it dysplese þe modyr · now my deth so sore
And for to suffre Al þis for man · I was born of the
to þe blys þat man had lost · man A-ȝen to restore.

ɔ her oure lady xal ryse and renne and halse þe crosse.

Maria Magdalena

Fo. 183 ¶ A good lady why do ȝe þus
ȝour dolfol cher · now cheuith² us sore   835
And for þe peyne of my swete lord jhesus
þat he seyth in ȝou it peyneth hym more.

¹ This *a* is corrected from another letter.   ² MS. *cheuᵗ*.

The Crucifixion

Maria virgo

¶ I pray ȝow Alle lete me ben here
and hang me up here on þis tre
be my frend and sone þat me is so dere   840
ffor þer he is þer wuld I be.

Johannes

¶ Jentyl lady now leve ȝour mornyng
and go with us now we ȝou pray
And comfort oure lord at hese departyng
ffor he is Al-most redy to go his way.   845

℞ here þei xal take oure lady from þe crosse · and here xal pylat come down from his shaffald with Cayphas and Annas and all here mene and xul come and lokyn on Cryst and annas and cayphas xul skornfully sey[n

Cayphas

¶ lo serys lo · be-heldyth and se
here hangyth he þat halpe many A man
And now yf he goddys sone be
helpe now hym-self yf þat he kan.

Annas

¶ ȝa and yf þou kyng of israel be   850
come down of þe cros Among us alle
And lete þi god now delyuere the
and þan oure kyng we wole þe calle.

℞ here xal pylat Askyn penne and inke and A tabyl xal be take hym wretyn A-fore · hic est jhesus nazarenus rex judeorum. and he xal make hym to wryte and þan gon up on A leddere and settyn þe tabyl abovyn crystys hed and þan cayphas xal makyn hym to redyn and seyn

Cayphas

℞ Sere pylat we merveylyth of þis
þat ȝe wryte hym to be kyng of jewys   855
℞ þerfore we wolde þat ȝe xuld wryte þus
þat he namyd hymself kyng of jewus.

Pylat

℞ þat I haue wretyn · wretyn it is
and so it xal be for me i-wys.

℞ and so forth all þei xal gon Aȝen to þe skaffalde and Jhesus xal cryen[1]

---

[1] *Jhesus* is enclosed in a red loop as the name of a speaker, and the red underlining of the stage direction is continued under *xal cryen*.

¶ heloy · heloy · lama3abathany ‖ 860
my fadyr in hevyn on hy
why dost þou me for-sake¹
The frelte of my mankende
With stronge peyn yt gynnyth to peynde
ha dere fadyr haue me in mende 865
    and lete deth my sorwe slake.

                           ij^us judeus

¶ me thynkyth he this doth calle hely
lete us go nere and a-spy
and loke yf he come preuely
    from cros² hym down to reve. 870

                           Jhesus

So grett a thrust dede nevyr man take
as I haue man now for þi sake
for thrust a-sundyr my lyppys gyn crake
    for drynes þei do cleve.

                           iij^us judeus

¶ 3our thrust sere hoberd for to slake 875
ey3il and galle here I þe take
what me thynkyth a mowe 3e make
    is not þis good drynk
to crye for drynke 3e had gret hast
and now it semyth it is but wast 880
is not þis drynk of good tast
    now telle me how 3e thynk.

                           4us judeus

¶ on lofte sere hoberd now 3e be sett³
we wyl no lenger with 3ou lett
we grete 3ou wel on þe newe gett 885
and make on 3ou a mowe.

                           i^us judeus

we grete 3ou wel with a scorn
and pray 3ou bothe evyn and morn

---

¹ The ink changes and the writing becomes rather smaller with this line, which is written as one with the preceding line. Note also that the next speaker is 'ii^us judeus' where 'i^us' might be expected.

² This *r* is written over an *o*.

³ From line 883 to line 890 two lines are written in one.

  take good eyd to oure corn
  and chare awey þe crowe.   890
             Jhesus
¶ In manus tuas domine
  holy fadyr in hefly se
  I comende my spyryte to þe
    for here now hendyth my fest
  I xal go sle þe fende þat freke   895
  ffor now myn herte be-gynnyth to breke
  wurdys mo xal I non speke
    Nunc consummatum est.
             Maria
¶ Alas Alas I leve to longe
  to se my swete sone with peynes stronge   900
  As a theff on cros doth honge
  And nevyr ȝet dede he synne
  Alas my dere chyld to deth is dressyd
  now is my care wel more in-cressyd
  A myn herte with peyn is pressyd   905
  Ffor sorwe myn hert doth twynne.
             Johannes
¶ A blyssyd mayde chaunge ȝour thought
  Ffor þow ȝour sone with sorwe be sought
  ȝitt by his owyn wyl þis werk is wrought
  And wylfully his deth to take   910
  ȝow to kepe he chargyd me here
  I am ȝour servaunt my lady dere
  wherfore I pray ȝow be of good chere
  And merthis þat ȝe make.
             Maria
¶ Thow he had nevyr of me be born   915
  And I sey his flesch þus al to torn
  on bak be-hyndyn on brest be-forn
  Rent with woundys wyde
  Nedys I must wonyn in woo
  to se my ffrende with many a ffo   920
  all to rent from top to too
    his flesch with-owtyn hyde.

[1] This and the following folio (185) are interpolated in this quire, being of different paper. The writing on them is firmer and more regular.

Johannes

¶ A blyssyd lady as I ȝow telle
  had he not deyd we xuld to helle
  Amonges ffendys þer evyr to dwelle   925
  In peynes þat ben smert
  he sufferyth deth for oure trespace
  and thorwe his deth we xal haue grace
  to dwelle with hym in hevyn place
  þerfore beth mery in hert.   930

Maria

¶ A dere ffrende weel woot I this
  þat he doth bye us to his blys
  but ȝitt of myrth evyr mor I mys
  whan I se þis syght.

Johannes

  Now dere lady þerfore I ȝow pray   935
  Ffro þis dolful dolour wende we oure way
  Ffor whan þis syght ȝe se nought may
  ȝoure care may waxe more lyght.

Maria

¶ Now sythe I must parte hym fro
  ȝit lete me kysse or þat I go   940
  his blyssyd ffeyt þat sufferyn wo
  naylid on þis tre
  So cruelly with grett dyspyte
  þus · shamfully was nevyr man dyghte
  þerfore in peyn myn hert is pyghte   945
  al joye departyth fro me.

hic quasi semi nortua[1] *cadat prona in terram et dicit johannes*

Johannes

¶ Now blyssyd mayd come forth with me
  no lengere þis syght þat ȝe se
  I xal ȝow gyde in þis countre
  where þat it plesyth ȝow best.   950

Maria

  Now jentyl Johan my sonys derlyng
  to goddys temple þou me brynge
  þat I may prey god with sore wepynge
  And mornynge þat is prest.

---

[1] So in MS. (seminor tua).

The Crucifixion 305

            Johannes

¶ Aƚƚ ȝour desyre xal be wrought   955
 wit*h* herty wyƚƚ I werke ȝour thought
 now blyssyd mayde taryeth nowth
 In þe temple þat ȝe ware
 Ffor holy prayere may chaunge ȝour mood
 and cawse ȝour chere to be more good   960
 whan ȝe se notȝ ȝour childys blood
 þe lasse may be ȝour care.

*Tunc transiet maria ad templu*m *cu*m *Io*ha*nne et cetera.*
            Maria

¶ Here in þis temple my lyff I lede
 And serue my lord god wit*h* hertyly drede
 now xal wepynge me fode *and* fede   965
 Som̃ comforte tyƚƚ god sende
 A my lord god I þe pray
 Whan my childe ryseth þe iij[de] day
 Comforte tha*n*ne thy*n* hand-may
 my care for to Amende.   970

        33[1] *ani*m*a* Chr*i*sti

¶ Now aƚƚ mankende in herte be glad
 wit*h* aƚƚ merthis þat may be had
 ffor ma*n*nys sowle þat was be-stad
   in þe logge of helle ;
185ᵛ now xal I ryse to lyve agayn   975
 from peyn to pleys of p*a*radyse pleyn
 þer*fo*re man in hert be fayn
   in m*er*the now xalt þou dwelle.

¶ I am þe sowle of cryst jh*es*u
 þe whicħ is kynge of aƚƚ vertu   980
 my body is ded þe jewys it slew
   þat hangyth ȝitt on þe rode
 rent *and* torñ all blody red
 ffor ma*n*nys sake my body is deed
 ffor ma*n*nys helpe my body is bred   985
   And sowle drynk my bodyes blode.

¶ þow my body be now sclayn

---

[1] This number is written partly against the preceding speech.

þe thrydde day þis is certayn
I xal reyse my body a-gayn
   to lyve as I ȝow say
now wole I go streyth to helle   990
and feche from þe fendys felle
all my frendys þat þer-in dwelle
   to blysse þat lestyth Ay.

Nota anima latronis

The sowle goth to helle gatys and *seyth*
℣ ² Attollite portas principes vestras et eleuamini porte eternales et introibit rex glorie.

¶ Ondothe ȝoure ȝatys of sorwatorie
On mannys sowle I haue memorie   995
here comyth now þe kynge of glorye
   these gatys for to breke
ȝe develys þat arn here with-inne
helle gatys ȝe xal vn-pynne
I xal delyvere mannys kynne   1000
   ffrom wo I wole hem wreke.

            Belyall

Alas Alas out and harrow
Onto þi byddynge must we bow
þat þou art god now do we know
Of þe had we grett dowte   1005
Aȝens þe may no thynge stonde
All thynge obeyth to thyn honde
bothe hevyn and helle watyr and londe
All thynge must to þe lowte.

          Anima Christi

Fo. 186 ¶ Aȝens me it wore but wast   1010
   to holdyn or to stondyn fast
helle logge may not last
   Aȝens · þe kynge of glorye
þi derke dore down I throwe
My fayr ffrendys now wele I knowe   1015

---

* These words are written in the margin apparently in the hand of the scribe of Ff. 95, 96.

¹ This contraction for *Versus* is otherwise only used to mark the Latin versicles in the Assumption play.

I xal hem brynge reknyd be rowe
  Out of here purcatorye.
           Centurio

¶ In trewth now I knowe with ful opyn syght
that goddys dere sone is naylid on tre
these wundyrful tokenys Aprevyn ful ryght   1020
quod vere filius dei erat iste.
        Alius miles 2

The very childe of god I suppose þat he be
and so it semyth wele be his wundyrful werk
þe erth sore qwakyth and þat agresyth me
With myst and grett wedyr it is woundyr dyrk.   1025
        Alius miles 3

¶ Soch merveylis shewe may non erthely man
þe eyr is ryght derke þat fyrst was ryght clere
The Erth-qwave is grett þe clowdys waxe whan
those tokenys preve hym a lorde with-out Any¹ pere.
          Centurio
his fadyr is pereles kyng of most empere   1030
bothe lorde of þis world and kynge of hevyn hyӡe
ӡit out of all synne to brynge us owt of daungere
he soferyth his dere sone for us all to dye.
          Nichodemus

¶ Alas Alas what syght is this   1035
to se þe lorde and kynge of blys
þat nevyr synnyd ne dede Amys
þus naylid vpon a rode
Alas ӡewys what haue ӡe wrought
A ӡe wyckyd wytys what was ӡour thought   1040
Why haue ӡe bobbyd and þus betyn owth
all his blyssyd blood.
          Senturyo²

¶ A now trewly telle weyl I kan
þat þis was goddys owyn sone
I knowe he is both god and man   1045
be þis wark þat here is done.

¶ þer was nevyr man but god · þat cowde make þis werk
þat evyr was of woman born

---

¹ A large *A* seems to have been written over a small one.
² This name stands at the top of the page. No name of next speaker is given at the bottom of Fo. 186.

were he nevyr so gret A clerk
it passeth hem all þow þei had sworn. 1050

¶ hese lawe was trewe I dare wel saye
þat he tawth us here A-monge
þerfore I rede ʒe turne ʒour faye
and amende þat ʒe han do wronge.

Joseph of Ara .. [¹

¶ O good lord jhesu · þat deyst now here on rode 1055
haue mercy on me · and for-gyf me my mys
I wold þe worchep here with my good
þat I may come to þi blysse.

¶ to pylat now wole I goon ²
and aske þe body of my lord jhesu 1060
to bery þat now wold I soon
in my grave þat is so new.

¶ heyl sere pylat þat syttyth in sete
heyl justyce of jewys men do þe calle
heyl with helth · I do þe grete 1065
I pray þe of A bone what so befalle.

¶ to bery jhesu is body I wole þe pray
þat he were out of mennys syth
for to-morwyn xal be oure holyday
þan wole no man hym bery I þe plyth. 1070

Fo. 187 ¶ And yf we lete hym hange þer stylle
Some wolde seyn þer of A-now
þe pepyl þer-of wold seyn ful ylle
þat nother xuld be ʒour-worchep nor prow.

Pylat

¶ Sere joseph of baramathie · I graunt þe 1075
With jhesu is body do þin intent
but fyrst I wole wete þat he ded be
as it was his jugement.

¶ Sere knytys I comawnd ʒow þat ʒe go
in hast with josepht of baramathie 1080

¹ Strokes of m lost in hole in paper—rest of the word or mark of contraction cut off with margin.
² god first written and crossed through.

and loke ȝe take good hede þer-to
þat jhesu suerly ded be.

¶ se þat þis comawndement ȝe fulfylle
with-out wordys ony mo
and þan lete joseph do his wylle 1085
what þat he wyl *with* jhesu do.

α *here come to knytys be-forn pylat At-onys þus seyng*

i^us miles

¶ Sere we xal do oure dylygens
with joseph goyng to Caluerye
be we out of þi *p*resens
sone þe trewth we xal aspye. 1090

Joseph Ab[. . . .

¶ Gramercy pylat of ȝo*ur* jentylnesse
þ*a*t ȝe han grawntyd me my lyst
Any thyng in my *p*rovince
ȝe xal haue at ȝo*ur* resquest [1]

Pylat

187v ¶ Sere all ȝo*ur* lest ȝe xal haue 1095
with jhesu is body do ȝo*ur* intent
whethyr ȝe bery hy*m* in pyt or grave
þe powere I grawnt ȝow here *p*resent.

α *The ij knygtys go with joseph to jhesus and stande and heldy*n *hy*m *in þe face.*

ij^us miles

¶ Me thynkyth jhesu is sewre anow
it is no ned his bonys to breke 1100
he is ded how þinkyth ȝow
he xal nevyr go nor speke.

i^us miles

¶ We wyl be sure or þan we go
Of A thyng I am be-thowth [2]
ȝondyr is a blynd knyth I xal go to 1105
And sone A whyle here xal be wrowth.

α *here þe knyth goth to blynde longeys and seyth*

¶ heyl sere longeys þ*ou* gentyl knyth
þe I prey now ryth hertyly

---

[1] *rest* first written as the first syllable and the *t* adapted to a *q*.
[2] The final *h* is blotted and seems to have been altered with darker ink.

þat þou wylt wend with me ful wyth
it xal be for þi prow veryly. 1110

longeus

¶ Sere at ȝour comauwndement with ȝow wyl I wende
in what place ȝe wyl me haue
for I trost ȝe be my frend
lede me forth sere · oure sabath ȝou save.

i^ms miles
1115

¶ lo sere longeys here is a spere
bothe long and brood and sharp a-now
heve it up fast þat it wore þere
for here is game show man show.

a here longeys showyth þe spere warly · and þe blood comyth rennyng to his hand and he Auantorysly xal wype his eyne.

longeys

¶ O good lord · how may þis be
þat I may se so bryth now 1120
þis thretty wyntyr I myth not se
and now I may se I wote nevyr how
but ho is þis þat hangyth here now
I trowe it be þe mayndonys[1] sone
and þat he is now · I knowe wel how 1125
þe jewys to hym þis velany han don.

a here he ffallyth down on his knes.

¶ Now good lord fforgyf me that
þat I to þe now don have
for I dede I wyst not what
þe jewys of myn ignorans dede me rave 1130
Mercy mercy mercy I crye.

a þan joseph doth set up þe lederys and nychodemus comyth to help hym.

Nicodemus

¶ Joseph ab Aramathy · blyssyd þou be
ffor þou dost a fol good dede
I prey the · lete me help þe
þat I may be partenere of þi mede. 1135

Joseph

¶ Nychodemus welcome · indede
I pray ȝow ȝe wole help þer-to

[1] So in MS.

The Burial

he wole Aqwyte us ryth weyl oure mede
and I haue lysens for to do.

℀ here joseph and nychodemus takyn cryst of þe cros on on o ledyr and þe tother on An-other leddyr and qwan is had down joseph leyth hym in oure ladys lappe · seyng þe knyts turnyng heme · and joseph seyth

                Joseph
¶ Lo mary modyr good and trewe   1140
here is þi son blody and bloo
ffor hym myn hert ful sore doth rewe
kysse hym now onys · eer he go.

               Maria virgo
¶ A Mercy Mercy myn owyn son so dere
þi blody face now I must kysse   1145
þi face is pale with-owtyn chere
of meche joy now xal I mysse
þer was nevyr modyr þat sey this
so here sone dyspoyled · with so gret wo
and my dere chylde nevyr dede A-mys   1150
A mercy fadyr of hefne · it xulde be so.

               Joseph
¶ Mary ȝour sone ȝe take to me
in-to his grave it xal be browth.

               Maria
joseph blyssyd evyr mot þou be
for þe good dede þat ȝe han wrowth.   1155

℀ here þei xal leyn cryst in his grave.
               Joseph
¶ I gyf þe þis syndony þat I haue bowth
to wynde þe in whyl it is new.

               Nichodemus
here is An onyment þat I haue browth
to Anoynt with all myn lord jhesu.

               Joseph
¶ Now jhesu is with-inne his grave   1160
wheche I ordeyn som tyme for me
on þe lord I vowche it save
I knowe my mede ful gret xal be.

               Nichodem[us
¶ now lete us leyn on þis ston Ageyn
And jhesu in þis tombe stylle xal be   1165

   And we wyl walke hom ful pleyn
   þe day passyth fast I se
   Fare wel joseph and wel ȝe be
   no lengere teryeng here we make.

             Joseph

   Sere almythy god be with þe   1170
   in-to his blysse he mote ȝou take.

             Maria

Fo. 189 ¶ Ffare wel ȝe jentyl princys kende
   in joye evyr mote ȝe be
   þe blysse of hefne with-owtyn ende
   I knowe veryly þat ȝe xal se.   1175

  α here þe princys xal do reuerens to oure lady and *gon here way* and
* nota leve þe maryes at be sepulcre. Cayphas goth to Pylat seyng þus*[1]
Incipit hic
            Cayphas[2]

   ¶ herk sere pylat lyst to me
   I xal þe telle tydyngys new
   of o thyng · we must[3] ware be
   er ellys here after · we myth it rewe.

   ¶ þou wotyst weyl þat jhesu   1180
   he seyd to us with wordys pleyn
   he seyd we xuld fynd it trew
   þe thryd day he wold ryse agey[4]
   yf þat hese dyscyplys come serteyn
   and out of his graue stele hym away[5]   1185
   þei wyl go preche and pleyn seyn
   þat he is reson þe thryd day.

   ¶ þis is þe cowncel þat I gyf here
   take men and gyf hem charge þerto
   to weche þe grave with gret power   1190
   tyl þe thryd day be go.

---

[*] These words *nota* and *Incipit hic* are written in the margin in another hand —apparently that of the scribe of Ff. 95, 96—and blacker ink. There are also some smudged indecipherable words (? scribblings) on the right-hand margin.

[1] These stage directions are written partly at the side of the text, and the *α* stands before the last line only.

[2] *Cayphas* is in paler ink and may have been added.

[3] The *st* is blotted as if corrected from another letter.   [4] So in MS.

[5] From line 1185 to line 1215 two lines are written in one.

The Guarding of the Sepulchre

                Pylat

¶ Sere Cayphas · it xal be do
for as ȝe say þer is peryl in
And it happend þat it were so
it myth make our lawys for to blyn   1195
ȝe xal se ser er þat ȝe go
how I xal þis mater saue
And what I xal sey þer-to
and what charge þei xal haue.

hic ¶ Come forth ȝe ser Amorawnt   1200
... hic and ser Arphaxat com ner also
Ser Cosdram and ser Affraunt
and here þe charge þat ȝe must do
Serys to Jhesu is grave ȝe xal go
tyl þat þe thryd day be gon   1205
and lete nother frend nor fo
in no wey to towche þe ston.

¶ yf ony of hese dyscipelys come þer
to fech þe body fro ȝou a-way
bete hym down have ȝe no fere   1210
with shamful deth do hym day
in payn of ȝour godys and ȝour lyvys
þat ȝe lete hem nowth shape ȝou fro
and of ȝour chyldere and ȝour wyfys
for al ȝe lese and ȝe do so.   1215
           i^{us} miles

189ᵛ ¶ Sere pylat we xal not ses
we xal kepe it strong A-now.
           ij^{us} miles
ȝa and An hunderyd put hem in pres
þei xal dey I make A vow.
           iij^{us} miles
¶ And han honderyd fy on An C. and an C. þer-to   1220
þer is non of hem xal us with-stonde.
           iiij^{us} miles
ȝa and þer com̃ An hunderyd thowsand and mo
I xal hem kylle with myn honde.

---

* This note in the margin, the first word of which is doubtful—probably Nota—is apparently in the same hand as that on p. 312.

                Pylat
¶ wel ser*ys* þan 3o*ur* part 3e do.
And to 3o*ur* charge loke 3e take hede   1225
Wi*th*-owtyn wordys ony mo *here þe knytys gon out of þe place*
Wysly now þat 3e procede.

¶ lo s*er* cayphas how thynky*th* 3ow
is not þis wel browth Abowth.
               Cayphas
in feyth s*er* it is sure A-now   1230
hardely haue 3e no dowth.
            i*us* Arfaxat [1] ij
¶ let se ser amarau*n*t where wele 3e be
wole 3e kepe þe feet or þe hed.
           ij*us* Amerau*n*t
At þe hed so mote I the [2]
and ho so come here he is but ded.   1235
            i*us* Arfaxat ii
¶ *And* I wole kepe þe feet þis tyde
þow þ*er* come both jakke *and* gylle.
           iij*us* cosdram iii
And I xal kepe þe ryth syde
and ho so come I xal hym kylle.
          iiij*us* Affrau*n*t 4
And I wole on þe lefte hand ben   1240
and ho so come here he xal nevyr then
fful sekyrly his bane xal I ben
 * *nota* wi*th* dyntys of dowte⸱/   *syr pylat* [3]
Fo. 190 Syr pylat haue good day
(V quire) We xal kepyn þe body in clay   1245
And we xal wakyn wele þe way
    and wayte*n* all abowte⸱/
               Pylatus
Now jentyl ser*ys* wole 3e vouch-saffe
to go wi*th* me *and* sele þe [4] graffe
þ*at* he ne aryse out of þe grave   1250
   Þ*at* is now ded

---

[1] These names were originally numbered to the right Amerau*n*t being 1, Arfaxat 2 — the order in which Pilate calls upon them. These numbers have been partly cut away with the margin, and the names have been renumbered to the left, Arfaxat the first speaker being 1, Amerau*n*t 2.

* Marginal note apparently in the same hand as those above.

[2] *So mote I the I wole be at þe h.* first written and crossed through.

[3] Catchword.   [4] Some miswritten letter crossed through after þe.

The Guarding of the Sepulchre

Cayphas

We graunte wel lete us now go
Whan it is selyd and kepte Also
Than be we sekyr with-owtyn wo
    And have of hym no dred.   1255

Tunc ibunt ad sepulcrum pilatus Cayphas Annas et omnes milites et dicit

Annas

¶ Loo here is wax ful redy dyght
Sett on ȝour sele anon ful ryght
þan be ȝe sekyr I ȝow plyght
    he xal not rysyn agayn.

Pilatus

On þis corner my seal xal sytt   1260
And with þis wax I sele þis pytt
now dare I ley he xal nevyr flytt
    out of þis grave serteayn.

Annas

¶ Here is more wax fful redy loo
all þe cornerys ȝe sele Also   1265
And with a lokke loke it too
Than lete us gon oure way
And lete þese knytys abydyn þer by
And yf hese dysciplys com preuyly
to stele awey þis ded body   1270
to vs they hem brynge with-out delay.

Pilatas[1]

¶ On every corner now is sett my seale
now is myn herte in welthe and wele
This may no brybour a-wey now stele
þis body from vndyr ston   1275
Now syr buschoppe I pray to the
And Annas also com on with me
Evyn to-gedyr all we thre
    hom-ward þe wey we gon.

¶ As wynde wrothe[2]   1280
knyghtys now goht

---

[1] A stroke in the *a* may indicate a correction of *as* to *us*.
[2] From line 1280 to line 1320 two lines are written in one. From line 1321 to line 1343 three lines are written in one.

clappyd in cloth
and kepyth hym well
loke ȝe be bolde
wi*th* me for to holde   1285
ȝe xul haue gold
And helme of stele.

Pylat Annas and *cayphas go to* þ*er skaffaldys* · and þ*e knyghtys sey*[1] . . [

                                     Affraunt 4

¶ Now in þis grownnde[2]
he lyeth bounde
þ*at* tholyd wounde   1290
ffor he was ffals
þis lefft[3] corner*e*
I wyl kepe her*e*
Armyd cler*e*
bothe hed *and* hals.   1295

                                       Cosdram 3

¶ I wyl haue þis syde
what so betyde
If any man ryde
to stele þe cors
I xal hym chyde   1300
wi*th* woundys wyde
Amonge hem glyde
wi*th* fyne fors.

                                     Ameraunt ij[us][4]

¶ The hed I take
her*e* by to wake ⸳   1305
A stele stake
I holde in honde
Maystryes to make
crownys I crake ⸳

---

[1] Fourth letter blotted, the rest cut off with margin.
[2] *tyde* first written and crossed through.
[3] *left* or *leftt* first written—corrected to *lefft*.
[4] The first of the two strokes is a later addition making i^{us} into ij^{us}.

Schafftys to shake 1310
And Schapyn schonde¹.

Arfaxat i^{us} ²

¶ I xal not lete
to kepe þe fete
they ar ful w . . .³
walterid in blood 1315
He þat wyll stalke
be brook or balke
hedyr to walke
þo wrecchis be wood.

i^{us} miles
Myn heed dullyth 1320
myn herte ffullyth
of sslepp
Seynt Mahownd
þis bereynge grownd
þou kepp.

1325
ij^{us} miles
I sey þe same
ffor Any blame
I falle
Mahownde whelpe
Aftyr þin helpe 1330
I calle.

3us miles
I am hevy as leed
ffor Any dred
I slepe
Mahownd of myght 1335
þis ston to nyght
þou kepe.

4us miles
I haue no foot
to stonde on root
by brynke 1340

---

¹ *sle fre* and *bonde* written above the last two words of this line in another ink and in the Ff. 95, 96 hand.

² An original 2 scratched out and i^{us} written in in darker ink.

³ There has been some erasure and rewriting here by the later corrector; perhaps *wete* altered to *white*.

here I Aske
to go to taske
A wynke.

Tunc dormyent milites et ueniet Anima Christi *de inferno cum* Adam *et* Eua[1] · *Abraham johan baptista et Alijs.*

                                              Anima Christi ⁖

¶ Come forthe Adam *and* Eue wi*th* the
And all my fryndys þat here-in be   1345
to paradys come forthe wi*th* me
In blysse for to dwelle
þe fende of helle þat is ȝo*ur* ffoo
he xal be wrappyd *and* woundy*n* in woo
Ffro wo to welthe now xul ȝe go   1350
Wi*th* myrthe evyr more to melle.

                                                Adam

¶ I thanke þe lord of þ*i* grett grace
that now is for-ȝovyn my grett trespace
now xal we dwellyn in blysful place
In joye *and* endeles myrthe   1355
Thorwe my synne man was fforlorn
and man to saue þou wore all torn
and of a mayd in bedlem born
þat evyr blyssyd be þ*i* byrthe.

                                                Eua

Fo. 191ᵛ ¶ Blyssyd be þou lord of lyff   1360
I am Eue Adam is wyff
þou hast soferyd strok *and* stryff
Ffor werkys þat we wrought
þ*i* mylde mercy haht[2] All[3] for-ȝovyn
Dethis dentys on þe were drevyn   1365
now wi*th* þe lord we xul levyn
þ*i* bryght blood hath us bowth.

                                                Johannes bap*t*ista

¶ I am þ*i* cosyn my name is Johan
þ*i* woundys hath betyn þe to þe bon
I baptyzid þe in flomjordon   1370
And ȝaff þ*i* body baptyze

---

[1] MS. Euā.   [2] An *h* between *haht* and *Alle* has been crossed through.
[3] This A seems to have been changed from a small to a larger form.

With þi grace now xul we gon
Ffrom oure enmyes every-choñ
And fyndyn myrthis many on
In play of paradyse.   1375

Abraham

¶ I am Abraham fadyr trowe
þat reyned after noes flowe
A sory synne Adam gan sowe
þat clad us all in care
A sone þat maydenys mylk hath sokyn   1380
and with his blood oure bonde hath brokyn
helle logge lyth vnlokyn
Ffro fylth with freude we fare.*

anima christi

¶ Ffayre ffrendys now be ȝe wunne
on ȝow shyneth þe sothfast sunne   1385
þe gost þat all grevaunce hath gunne
Fful harde I xal hym bynde
As wyckyd werme þou gunne Apere
to tray my chylderyn þat were so dere
þerfore traytour heuer-more here   1390
newe peynes þou xalt evyr ffynde.

¶ Thorwe blood I took of mannys kynde
Ffals devyl I here þe bynde
In endles sorwe I þe wynde
þer-in evyr-more to dwelle   1395
now þou art bownde þou mayst not fle
Ffor þin envyous cruelte
In endeles dampnacion xalt þou be
And nevyr comyn out of helle.†

Beliall

¶ Alas herrow now am I bownde   1400
In helle gonge to ly on¹ grownde
In hendles sorwe now am I wounde

---

* The words *anima caym* are here written in the margin in the hand of Ff. 95, 96, and in the left-hand margin *Nota anima caym* followed by some erased words ending: *as folow fayere frendys*.

† A word or two, of which the first is *thowght*, is written here in the margin in the Fo. 95, &c., hand and smudged out.

¹ *on* is written (by the scribe) over the line above an erasure.

In care evyr more to dwelle
In helle logge I lyȝ a-lone
now is my joye a-wey al gone 1405
Ffor all fendys xul be my fone
I xal nevyr com from helle.*

Anima Christi

¶ Now is ȝour ffoo boundyn in helle
þat evyr was besy ȝow for to qwelle
now wele I rysyn flesch and felle 1410
    þat rent was for ȝour sake
myn owyn body þat hynge on rode
And be þe jewys nevyr so wode
It xal a-ryse both flesch and blode
    my body now wyl I take. 1415

Tunc transiet anima christi ad resuscitandum corpus quo resuscitato dicat Jhesus

Fo. 192ᵛ (marked 190)

Jhesus

¶ harde gatys haue I gon
And peynes sofryd many on
Stomblyd at stake and at ston
    nyȝ thre and thretty ȝere
I lyght out of my faderys trone 1420
ffor to Amende mannys mone
my flesch was betyn to þe bon
    my blood I bledde clere.

¶ Ffor mannys loue I tholyd dede
and for mannys loue I am rysyn up rede [1] 1425
ffor man I haue mad my body in brede
    his sowle for to fede
Man and þou lete me þus gone
and wylt not folwyn me a-none

---

* The hand of Fo. 95, &c., has interlined here and written down the left-hand margin:
'nota þe devyll   thowght many &c.
    Thowght many begon I am glad &c.   hens I wyll þe bere. þan Crist.
    and þan cayme xall sey his spech. And þan crist xall sey now ys your foo. etc.'

[1] The r is written above the line over an erased letter.

suchͪ a frende fyndyst þou nevyr none   1430
to help þe at þi nede.

¶ Salue sancta parens ⸴ my modyr dere
Aͫ heyl modyr wi*th* glad che*re*
Ffor now is A-resyn wi*th* body cle*re*
þi sone þ*at* was dolve depe   1435
þis is þe thrydde day þ*at* I ȝow tolde
I xuld a-rysyn out of þe cley so colde
now am I he*re* wi*th* brest ful bolde
þerfo*re* no mo*re* ȝe wepe.
                                Maria

¶ Welcom my lord welcom̄ my grace   1440
welcome my sone *and* my solace
I xal þe wurchep in every place
Welcom̄ lord god of myght
mekel sorwe in hert I leed
whan þou we*re* leyd in dethͪ is beed   1445
but now my blysse is newly breed
Aͫ men may joye þis syght.
                                Jhesus

¶ Aͫ þis werlde þ*at* was forlorn
Shal wurchepe ȝou¹ bothe evy*n* *and* morn
Ffor had I not of ȝow be born   1450
man had be lost in helle
I was deed *and* lyff I haue
A*nd*² thorwe my dethͪ man do I saue
Ffor now I a*m*³ resyn out of my graue
In hevyn man xal now dwelle.   1455
                                Maria

¶ A de*re* sone þese wurdys ben⁴ goode
þou hast wel comfortyd my morny*n*g moode
blyssyd be þi precyous bloode
þ*at* mankende þus doth saue.
                                Jhesus

Now de*re* modyr my leve I take   1460
joye in hert *and* myrthͪ ȝe make

---

¹ Omitted and written above the line.   ² MS. A͂d.
³ MS. a͂resyn, with some mark under *a*, possibly indicating error.
⁴ Some miswritten letter obliterated before *ben*.

Ffor deth is deed *and* lyff doth wake
now I am resyn fro my graue.

Maria

¶ Ffare wel my sone fare wel my childe
Ffare wel my lorde my god so mylde 1465
my*n* hert is wele þ*at* ffyrst was whylde
Ffare wel my*n* owyn der*e* love
Now all ma*n*kynde beth glad w*ith* gle
Ffor deth is deed as ȝe may se
and lyff is reysed endles to be 1470
In hevyn dwellynge Above.

¶ whan my sone was naylyd on tre
All women myght rewe w*ith* me
Ffor grett*ere* sorwe myght nevyr non be
than I dede suffyr i-wys ? 1475
but þis joy now passyth all sorwe
þ*at* my childe suffryd in þ*at* hard morwe
Ffor now he is our*e* alderers borwe
to brynge us all to blys.

Tunc evigilabunt milites sepulcri et dicit primus miles

i*us* miles

¶ Awake Awake¹ 1480
hillis gyn qwake
And tres ben shake
ful nere a-too
Stonys clevyd
wyttys ben revid² 1485
Erys ben devid
I am scrvid soo.

2*us* miles

¶ he is a-resyn þis is no nay
þ*at* was deed *and* colde in clay
now is resyn be-lyve þis day 1490
grett woundyr it is to me
He is resyn by his owyn myght
And fforth he goth his wey ful ryght

---

¹ From line 1480 to line 1487 two lines are written in one.
² *rewi* first written and crossed through.

The Resurrection

  how xul we now us qwytte
  Whan Pylat doth us se.   1495
            3us miles

¶ lete us now go [1]
 pilat on-too
 *An*d ryght evyn so
 as we han sayn
 þe trewtħ we sey   1500
 þ*at* out of clay
 he is resyn þis day
 þ*at* jewys han slayn.
            4us miles

¶ I holde it best
 lete us nevyr rest   1505
 but go we prest
 þ*at* it wer*e* done
 Aħ heyl pilatt
 in þi*n* A-stat [2]
 he is resyn up latt   1510
 þ*at* þ*ou* gast dome.
            Pilat

¶ What What What What
 Out upon the why seyst þ*ou* þat
 Ffy vpon the harlat
 how darst þ*ou* so say   1515
 þ*ou* dost my*n* herte ryght grett greff
 þ*ou* lyest vpon hym fals theff
 howe xulde he rysyn ageyn to lyff
 þ*at* lay deed in clay ⸵
            1us miles [3]

¶ ȝa þow þ*ou* be nevyr so wrotħ   1520
 And of these tydandys nevyr so lotħ
 ȝitt goodly on ground on lyve he gotħ
 qwycke *and* levynge man
 Iff þ*ou* haddyst a ben þ*er* we war*e* [4]
 in hert þ*ou* xuldyst han had gret car*e* 1525

---

[1] From line 1496 to line 1511 two lines are written in one.
[2] Three red dots against this line.
[3] Some miswritten number or letter (? 2 or 3) crossed through between 1us and *miles*.
[4] The *a* is written as a correction over another (? *e*) letter.

And of blysse a ben ryght bare
Of colore bothe pale and whan.

Pilatus

¶ Or ӡe com there [1]
ӡe dede all swere
to fyght in fere
and bete and bynde
All þis was trayn
ӡour wurdys wore vayn
þis is sertayn
ӡow fals I fynde.

1530

1535
2ᵘˢ miles

Be þe deth þe devyl deyd
we were of hym so sore Atreyd
þat ffor ffer we us down leyd
Ryght evyn vpon oure syde
whan we were leyd vpon þe grounde
stylle we lay as we had be bounde
we durst not ryse for a thowsand pounde
ne not for all þis worlde so wyde.

1540

Pilatus

¶ Now ffy upon ӡour grett bost
all ӡour wurchep is now lost
in felde in town and in every cost
    men may ӡou dyspravyn ?
now all ӡour wurchep it is lorn
And euery man may ӡow we [2] scorn
And bydde ӡow go syttyn in þe corn
    And chare a-wey þe ravyn ?

1545

Fo. 194ᵛ

1550

3ᵘˢ miles

¶ ӡa it was hyӡ tyme to leyn oure bost
Ffor whan þe body toke aӡen þe gost
he wold a frayd many An ost
    kynge knyght and knave
ӡa whan he dede ryse out of his lake [3]
þan was þer suche An erthe-quake
þat all þe worlde it gan to shake
    þat made us ffor to rave.

1555

---

[1] From line 1528 to line 1535 two lines are written in one.
[2] So in MS. for *wel*.   [3] Some letter (? *k*) crossed through before *lake*.

¶ ȝa ȝa · herke ffelawys what I xal say   1560
late us not ses be nyght nor day
but telle þe trewth ryght as it lay
In countre where we goo
And than I dare ley myn heed
þat þei þat crystys lawys leed   1565
they wyl nevyr ses tyl they be deed
his deth þat brought hym too.

                                i^{us} miles

¶ Be belyall þis was now wele ment
to þis cownceîl lete us consent
lett us go tellyn with on Assent   1570
he is resyn up þis day.

                               2us miles

I grawnt þer-to and þat forth ryght
þat he is resyn by his owyn myght
Ffor þer cam non be day nor nyght
to helpe hym owte of clay.   1575
                               Pilatus

fo. 195 ¶ Now jentyl serys I pray ȝow All
A-byde stylle a lytyl thrall
whyll þat I myn cowncel call
And here of þer counceîl.

                               i^{us} miles
Syr att ȝour prayour we wyl abyde   1580
here in þis place a lytel tyde
but tary not to longe ffor we must ryde
we may not longe dwelle.

                               Pilatus

¶ Now jentyl serys I pray ȝow here
Sum good cowncel me to lere   1585
Ffor sertys serys without dwere
We stounde[1] in right grett dowte.

                               Cayphas

Now trewly sere I ȝow telle
þis matere is both ffers and felle
combros it is þerwith to melle   1590
And evyl to be browth a-bowte.

[1] Or *stonnde*.

(Top right: 4us miles)

Annas

¶ Syr pylat þou grett justyse
þow þou be of wittys wyse
ȝit herke fful sadly with good devyse
what þat þou xalt do   1595
I counsel þe be my reed
þis wundyrful tale pray hem to hede
and upon þis ȝeve hem good mede
bothe golde and sylver also.

¶ And sere I xaII telle ȝow why   1600
in ȝoure erys prevyly
be-tweyn us thre serteynly
  now herk serys in ȝour erys.

hic faciant pilatus cayphas et Annas priuatim inter se consilium quo finito dicat

Annas

Fo. 195ᵛ   Ffor mede doth most in every qwest
and mede is mayster bothe est and west   1605
now trewly serys I hold þis best
With mede men may bynde berys.

Cayphas

¶ Sekyr sere þis counseII is good
pray þese knyhtys to chaunge þer mood
ȝeve them golde ffeste and ffood   1610
And þat may chaunge þer wytt.

Pylatt.

Serys ȝoure good councel I xaII fulfylle
now jentyl knyhtys come hedyr me tylle
I pray ȝow serys of ȝour good wylle
no ferther þat ȝe fflytt.   1615

¶ Jentyl knyhtys I ȝow pray
A bettyr sawe þat ȝe say
Sey þer he was cawth away
with his dyscyplis be nyght
Sey he was with his dyscyplis ffett   1620
I wolde ȝe worn in ȝoure sadelys ssett
And haue here gold in a purs knett
and to rome rydyth ryght.

⁋ Now syr pylatt¹
we gon oure gatt
We wyll not prate
no lengere now
now we haue golde
no talys xul be tolde
to whithtys on wolde
we make þe A vow.

Now ȝe men of² myth
as ȝe han hyght
Euyn so forth ryght
ȝoure wurdys not falle
And ȝe xul gon
with me A-non
All every-chon
in to myn halle.

Now hens we go
As lyth as ro
And ryght evyn so
As we han seyd
We xul kepe counsel
where so evyr we dwell
We xul no talys tell
be not dysmayd.

hic uenient ad sepulcrum maria magdalene maria jacobi et maria Salome · et dicit maria magdalene

Magdalen

⁋ Swete systeryn I ȝow besech
heryght now my specyal speche
Go we with salvys ffor to leche
Cryst þat tholyd wounde
he hath us wonnyn owt of wreche
the ryght wey god wyl us teche

4ᵘˢ miles
1625

1630

Pilatus

1635

iᵘˢ miles
1640

1645

36

5

---

¹ From line 1624 to line 1647 two lines are written in one, divided as before by double strokes to 1639, then by points.
² *men of* written twice and crossed through.
* This marginal direction is apparently in the Ff. 95, 96 hand. It seems to refer to the end of one day's performance.

ffor to seke my lorde my leche
his blood hath me vnbownde.

¶ vij develys in me were pyght
my loue my lord my god Almyght 10
A-wey he weryd þo ffyndys wight
with his wyse wurde
he droff fro me þe fendes lees
in¹ my swete sowle his chawmere I ches
In me be-levyth þe lord of pes 15
I go to his burryenge boorde.

        Maria jacobi

¶ My systerys sone I woot he was
he lyth in here as sunne in glas
þe childe was born by oxe and Asse
Vp in a bestys stall 20
thow his body be gravyd vndyr gres

Fo. 196ᵛ þe grete godhede is nevyr-þe-lasse²
þe lord xal rysyn and gon his pas
and comfortyn his ffrendys all.

        Maria Salome

¶ My name is mary Salome³ 25
his modyr and I systerys we be
Annys dowterys we be all thre
Jhesu we be þin Awntys
The naylis gun his lemys feyn
and þe spere gan punche and peyn 30
on þo woundys we wold haue eyn
þat grace now god graunt vs.

        Maria Magdalene

¶ Now go we stylle⁴
with good wyll
þer he is leyd 35
he deyd on crowch
we wolde hym towch
as we han seyd.

Tunc respicit maria magdalene in sepulcro dicens

 ¹ Omitted and added against the line in the margin.
 ² *a* written as a correction over an *e*.
 ³ Next line first written *systerys dowterys bothe* and crossed through.
 ⁴ Lines 33 to 38 are written in two lines divided by points.

¶ Where is my lord þat was here
þat for me bledde bowndyn in brere 40
his body was beryed rygh¹ by þis mere
þat Ffor me gan deye
þe jewys ffekyll and ffals ffownde
where haue þei do þe body with wounde
he lyth not upon þis grownde 45
þe body is don A-weye.
       Maria jacobi
¶ To my lorde my love my ffrende
Ffayn wolde I salve A spende
and I myght aught A-mende
  his woundys depe and wyde 50
to my lord I owe lowlyte
both homage and fewte
I wolde with my dewte
a softyd hand and syde.
       Maria Salome
p. 197 ¶ To myghtfful god omnypotent 55
I bere a boyst of oynement
I wold han softyd his sore dent
his sydys al a-bowte
Lombe of love with-owt loth
I ffynde þe not myn hert is wroth 60
in þe sepulcre þer lyth a cloth
And jentyl jhesu is owte.
       Angelus
¶ wendyth fforth ȝe women thre
In-to þe strete of Galyle
ȝour savyour þer xul ȝe se 65
walkynge in þe waye
ȝour Ffleschly lorde now hath lyff
þat deyd on tre with strook and stryff
wende fforth þou wepynge wyff
and seke hym I þe saye. 70

¶ Now goth fforth ffast all thre
to his dyscyplys ffayr and fre
and to petyr þe trewth telle ȝe

    ¹ So in MS.

330 The Announcement to the Three Maries

  þer of haue ȝe no dreed
  Spare ȝe not þe soth to say   75
  he þat was deed and closyd in clay
  he is resyn þis same day
  And levyth with woundys reed.

             Maria Magdalen
 ¶ A myrthe and joye in herte we haue
  Ffor now is resyn out of his graue   80
  he levyth now oure lyf to saue
  þat dede lay in þe clay ·/

             Maria jacoby
Fo. 197ᵛ   In hert I was ryght sore dysmayd
  the Aungel to us whan þat he sayd
  þat cryst is resyn I was affrayd   85
  þe Aungel whan I say ·/

             Maria Salome
 ¶ Now lete us all thre fulfylle
  þe Angelys wurde and goddys wylle
  lett us sey with voys wul shrylle
  Cryst þat jewys dede sle   90
  oure lord þat naylyd was on þe rode
  And betyn out was his bodyes blode
  he is a-resyn þough they ben wode
  A lorde ȝitt wele þou be.

Maria magdalene ‖ *dicit petro et ceteris apostolis.*

 ¶ Bretheryn all in herte be glad   95
  bothe blythe And joyful in herte ful fayṅ
  Ffor ryght good tydandys haue we had
  þat oure lord is resyn Agayn [1]
  An Aungel [2] us bad [2] ryght þus sertayn
  to þe petyr þat we xulde telle   100
  how cryst [2] is resyn þe which was slayn
  A levynge man evyr more to dwelle.

[1] The scribe of Ff. 95, 96, 112 has written against this line : *and aperyd to us sertayne*, which he afterwards crossed out, writing longitudinally in the left-hand margin for insertion here:

  † lyk as he dyede nakyd as he was borne
   and commande us to go to peter and john and hys dyscipulys all
   and tell to yow he wolde apere in lyknes as he was befo[rn.

[2] The same hand has marked *aungel, bad,* and *cryst* with dots for deletion, and has written *bade* over *aungel* and *he* over *cryst.*

                            Maria jacobi

¶ To lyve is resyn a-geyn þat lorde
  the qwych judas to jewys solde
  of þis I bere ryght trewe¹ recorde   105
  by wurdys þat þe Aungel tolde *
  now myrth and joye to man on molde
  Euery man now myrth may haue
  he þat was closyd in cley ful colde
  This day is resyn owt of his grave.   110
                            Petrus

¶ Sey me systeryn with wurdys blythe
  may I troste to þat 3e say
  Is cryst resyn ageyn to lyve
  þat was ded and colde in clay.

                            Maria Salome

  3a trostyth us truly it is no nay   115
  he is a-resyn it is no les
  and so An Aungel us tolde þis day²
  with opyn voys and speche expres.

                            Johannes

¶ 3a þese³ be tydyngys of ryght gret blys
  þat oure mayster resyn xulde be   120
  I wyl go renne in hast i-wys
  and loke my lord yf I may se.

---

¹ Some miswritten letter crossed through before *trewe*.
* The scribe of Ff. 95, 96 has written *for* in the margin just above *by* at the beginning of this line, and has marked for insertion here a passage written in the bottom margin:

for .. aperyd to us with handys fytte and hert borde
and ... he schowyd us his woundys fyve
both handys and fytte and þe wound in his syde
and þerfor be-leve us þat he is man a-lyve.
                                    [?]

In a darker ink but probably by the same hand, *for* in the first line has been crossed out and *All so* written before it, and *he* over the illegible second word; the last five words of the line have been crossed out, and *with body bolde* (or *belde*) written over them; the third line has been crossed out.

² An alternative line is written by the scribe of Fo. 95 against this longitudinally in the margin:

   And so he badd us teñ yow þis daye.

³ *be 3e* first written and crossed through.

Petrus

Ffor joye also I renne with the
my brothyr johan as I þe say
In hast anon evyn forth go we 125
to his grave we renne oure way.

hic currunt Johannes et Petrus simul ad sepulcrum et Johannes prius venit Ad monumentum sed non intrat.

Johannes

The same shete here I se
þat crystys body was in wounde
but he is gon where so ever he be
he lyth not here up on þis grownde. 130

Petrus intrat monumentum et dicit Petrus

Petrus

in þis cornere þe shete is fownde
and here we fynde þe sudary
Fo. 198ᵛ   In þe whiche his hed was wounde
whan he was take from calvary[1].

hic intrat Johannes monumentum dicens

Johannes

¶ The same sudary and þe same shete 135
here with my syth I se both tweyn
now may I wele knowe and wete
þat he is rysyn to lyve ageyn
On to oure bretheryn lete us go seyn
þe trewth ryght hevyn as it is 140
Oure mayster lyvyth þe whech was slayn
All myghty lorde and kynge of blys.

Petrus

¶ No lengere here wyll we dwelle
to oure bretheryn þe wey we take
the trewth to them whan þat we telle 145
grett joye in hert þan wul þei make.

hic petrus loquitur omnibus apostolis simul collectis

Beth mery bretheryn for Crystys sake
þat man þat is oure mayster so good

---

[1] A *u* has been altered to *v* in this word, apparently by the scribe.

Ffrom deth to lyve he is A-wake
þat sore was rent up on þe rood. 150

Johannes

¶ As women seyd so haue we fownde
remevyd Awey we saw þe ston
he lyth no lengere vndyr þe grownde
out of his graue oure mayster is gon.

Omnes congregati[1] **Thomas**

We haue grett woundyr everychon 155
of þese wurdys þat ʒe do speke
A ston ful hevy lay hym up on
Ffrom vndyr þat ston how xulde he breke.

Petrus

199 ¶ The trewth to tellyn it passyth oure witt
Wethyr he be resyn thorwe his owyn myght 160
Or ellys stolyn out of his pitt
be sum man prevely be nyght
That he is gon we saw with syght
Ffor in his graue he is nowth
we can not tellyn in what plyght 165
Out of his graue þat he is browth.

Maria magdalene goth to þe graue and wepyth and seyth

Maria Magdalen

¶ Ffor hertyly Sorwe myn herte doth breke
with wepynge terys I wasch my face
Alas ffor sorwe I may not speke **37**
my lorde is gon þat here-inne wase
Myn owyn dere lorde *and* kyng of gras 5
þat vij develys ffro me dyd take
I kan nat se hym Alas Alas
he is stolyn awey owt of þis lake.

Aungelus

¶ Woman þat stondyst here Alone
Why dost þou wepe *and* morne *and* wepe so sore 10
What cawse hast þou to make such mone
Why makyst þou such sorwe *and* where-fore.

---

[1] MS. congregat̃—? so for *omnibus congregatis*.

Maria Magdalene

I haue gret cawse to wepe evyr more
my lord is take out of his graue
Stolyn Awey and fro me lore 15
I kan not wete where hym to haue.

hic parum deambulet A sepulcro dicens

¶ Alas Alas what xal I do
my lord awey is fro me take
Fo. 199ᵛ A woful wrecche whedyr xal I go
My joye is gon owth of þis lake. 20

Jhesus

Woman suche mornynge why dost þou make
Why is þi chere so hevy and badde
Why dost þou sythe so sore and qwake
Why dost þou wepe so sore and sadde.

Maria Magdalene

¶ A grettyr cawse had nevyr woman 25
Ffor to wepe both nyth and day
than I myself haue in serteyn
for¹ to sorwyn evyr and Ay
Alas ffor sorwe myn hert doth blede
my lorde is take fro me A-way 30
I muste nedys sore wepe and grede
where he is put I kan not say.

¶ but jentyl gardener I pray to the
If þou hym took out of his graue
telle me qwere I may hym se 35
þat I may go my lorde to haue.

Jhesus
M. A. R. I. A. *spectans*
Maria Magdalene

A mayster and lorde to þe I crave
As þou art lord and kynge of blys
Graunt me lord and þou vowchesave 40
thyn holy ffete þat I may kys.

---

¹ *And* first written before *for* and crossed through.

             Jhesus
¶ Towche me not As ȝett ⁄ Mary
 Ffor to my fadyr I haue not Ascende
 but to my bretheryn in hast þe hyȝ
 with these gode wurdys here care Amende 45
 Sey to my bretheryn þat I intende
 to stey to my fadyr and to ȝowre
 to oure lord both god and frende
 I wyl Ascende to hevyn towre.

¶ In hevyn to ordeyn ȝow A place 50
 to my ffadyr now wyl I go
 to merth and joye and grett solace
 And endeles blys to brynge ȝow to
 Ffor man I sufferyd both schame and wo
 more spyteful deth nevyr man dyd take 55
 ȝit wyl I ordeyn ffor all this lo
 In hevyn an halle for mannys sake.
         Maria Magdaly[n

¶ Gracyous lord at ȝour byddyng[1]
 to all my bretheryn I xal go telle
 how þat ȝe be man levynge 60
 quyk and qwethynge of flesch and ffelle
 Now all hevynes I may expelle
 And myrth and joy now take to me
 my lord þat I haue louyd so wele
 with opyn syght I dede hym se. 65

¶ Whan I sowght my lord in grave
 I was fful sory and ryght sad
 Ffor syght of hym I myght non haue
 Ffor mornynge sore I was nere mad
 Grettere sorwe ȝit nevyr whith had 70
 Whan my lord A-wey was gon
 But now in herte I am so glad
 so grett a joy nevyr wyff had non.

¶ how myght I more gretter joye haue
 than se þat lorde with opyn syght 75

---
[1] The colour of the ink changes with this line from the darker colour which has prevailed since Fo. 193.

the whiche my sowle from synne to saue
From develys sefne he made me qwyght.

¶ There kan no tounge my joye expres
now I haue seyn my lorde on lyve
to my bretheryn I wyl me dresse 80
and telle to hem A-non ryght be-lyve
With opyn speche I xal me shryve
and telle to hem with wurdys pleyn
how þat cryst ffrom deth to lyve
to endles blys is resyn Ageyn. 85

¶ Bretheryn all bllyth ȝe be
ffor joyful tydyngys tellyn I kan
I saw oure lord cryst · lyste wel to me
of flesch and bon · quyk levynge man
beth glad and joyful as for than 90
Ffor trost me trewly it is ryght thus
Mowth to mowth þis sertayn [1]
I spak ryght now with cryst jhesus.

Petrus

¶ A woundyrful tale for-sothe is this
ever onowryd oure lorde mote be 95
we pray þe lord and kynge of blys
Onys þi presence þat we may se
Fo. 201   Ere thu Ascende to thi mageste
Gracyous god if þat ȝe plese
late us haue sum syght of the 100
oure careful hertys to sett in ease   Amen.

Explicit apparicio Marie magdale[ne.

[The remainder of Fo. 201 is filled up with what appears to be a roughly scribbled copy of part of Magdalene's speech on Fo. 200ᵛ.* Fo. 201ᵛ is blank.]

*[But now in herte I am so glad
So grete a jooy [2] nevyr wyff had non
how myght I more gretter haue than se
þat lorde with opyn syght the wyche my soule

---

[1] So, with *is* omitted, in MS.   [2] The first *o* blotted.

from synne to saue from develys sefne
he mad' me qwyght there kan no tounge
my joy expres now I haue seyn my lorde
on lyve to my brethryn I wyll me dresse
and thell to hem with wurdys pleyn hwow
þat cryst from deth to lyve to endles blys
ys resyn agayen [1] bretheryn all bllyth
ye be for joyfull tydyngys tellyn I kan I saw
Oure lord cryst lyste wel to me of flesch and bon
quyk levyng man beth glad and Joyfull as for than
ffor trost me trewly it ys ryght thus
Mowth to mowth þis ys sartayne I [2] spak rght now
with cryst Jhesus.]

. 202 hic incipit aparicio cleophe et luce.

¶ Cleophas

M Y brothir lucas I ȝow pray
plesynge to ȝow if þat it be
To þe castel of Emawus A lytyl way
þat ȝe vowche-saf to go with me.

38

lucas

All redy brother I walke with the
to ȝone castell with ryght good chere
Euyn to-gedyr Anon go we
brother cleophas we to infere.

5

Cleophas

¶ A brother lucas I am sore mevyd
Whan cryst oure mayster comyth in my mynde
whan that I thynke how he was grevyd
Joye in myn herte kan I non fynde
he was so lowlye so good so kynde
holy of lyf and meke of mocd
Alas þe jewys þei were to blynde
hym for to kylle þat was so good.

10

15

lucas

¶ Brothyr cleophas ȝe sey [3] ful soth
they were to cursyd and to cruell

---

[1] Some blotted letter after *agayen*. [2] Some crossed out letters after *I*.
[3] The *e* is written over some erasure.

And judas þat traytour he was to loth
Ffor gold and sylvyr his Mayster to selle 20
the jewys were redy hym for to qwelle
With skorgys bete out all his blood
Alas þei were to fers and ffelle
Shamfully þei henge hym on a rood.

Cleophas

¶ ȝa be-twen to Thevys Alas for shame 25
they henge hym up with body rent
Alas alas they were to blame
to cursyd and cruel was þer intent
Whan for thurste he was nere shent
Eyȝil and galle þei ȝovyn hym to drynke 30
Alas for ruthe his deth thei bent
in a ffowle place of horryble stynke.

lucas

¶ ȝa And cawse in hym cowde they non fynde
Alas for sorwe what was here thought
And he dede helpe bothe lame and blynde 35
And all seke men þat were hym browght
A-ȝens vice Alwey he wrought
synfull dede wold he nevyr do
ȝit hym to kylle þei sparyd nought
Alas Alas why dede they so. 40

Jhesus

¶ Well ovyr-take ȝe serys in same
to walke in felachep with ȝow I pray.

lucas

welcom serys in goddys name
of good felachep we sey not nay.

Jhesus

Qwat is ȝour langage to me ȝe say 45
that ȝe haue to-gedyr ȝe to
sory and Evysum ȝe ben Alway
ȝour myrthe is gon why is it so.

Cleophas

¶ Sere me thynkyth þou art a pore pylgrym
here walkynge be þi selfe A-lone 50
and in þe cete of jerusalem
þou knowyst ryght lytyl what þer is done

Ffor pylgrymys comyn and gon ryth sone
Ryght lytyl whyle pylgrymes do dwelle
In all jerusalem as þou hast gone 55
I trowe no tydyngys þat þou canst telle.

Jhesus

Why in Jerusalem what thynge is wrought
What tydyngys fro thens brynge ȝe.

lucas

A ther haue they slayn a man for nought
gyltles he was as we telle the 60
An holy prophete with god was he
myghtyly in wurde and eke in dede
of god he had ryght grett pooste
Amonge þe pepyl his name gan[1] sprede.

he hyght jhesu of nazareth 65
A man he was of ryght grett fame
the jewys hym kylde with cruel deth.
with-out trespas or Any blame
hym to scorne they had grett game
And naylid hym streyte on tyll a tre 70
Alas Alas me thynkyth grett shame
With-out cawse þat this xulde be.

Cleophas

¶ ȝa sere and ryght grett trost in hym we had
All Israel countre þat he xulde saue
the thrydde day is this þat he was clad 75
in coold cley and leyd in grave
ȝitt woundyrful tydyngys of hym we haue
Of women þat sought hym be-forn day lyth
wethyr they sey truthe or ellys do raue
We can not telle þe trew verdyth. 80

¶ Whan cryst in grave þei cowde not se
they comyn to us And Evyn thus tolde
How þat An Aungell seyd to them thre
that he xuld leve with brest fful bolde
ȝitt petyr and johan preve this wolde 85
to Crystys graue they ran thei tweyne

---

[1] The *a* is written over some other letter (? *r*).

and whan they come to þe graue so coolde
They fownde þe women fful trewe sertayne.

Jhesus

¶ A ȝe Ffonnys And Slought of herte
Ffor to be-leve in holy scripture   90
haue not prophetys with wurdys smerte
Spoke be tokenys in signifure
That Cryste xuld deye ffor ȝour valure
And syth entre his joye and blys
why be ȝe of herte so dure   95
And trust not in god þat myghtful is.

¶ Bothe Moyses and Aaron and othyr mo
in holy scripture ȝe may rede it
of Crystis deth thei spak Also
And how he xuld ryse out of his pitt   100
Owt of ffeyth than why do ȝe fflitte
Whan holy prophetys ȝow teche so pleyne
turne ȝoure thought and chaunge ȝour witte
And truste wele þat cryst doth leve a-geyne.

lucas

¶ Leve Ageyn : man be in pes   105
how xulde A ded man evyr A-ryse
I cowncell þe such wurdys to ses
Ffor dowte of pylat þat hyȝ justyce
Fo. 204   he was slayn At þe gre ¹ A-syse
be cowncell of lordys many on   110
Of suche langage · take bettyr A-vise
In every company þer þou dost gon.

Christus

¶ Trewth dyd nevyr his maystyr shame
Why xulde I ses than trewth to say
Be Jonas þe prophete I preve þe same   115
Þat was in a Whallys body iij nyghtis and iij day
So longe Cryst in his grave lay
As Jonas was with-inne þe se
his grave is brokyn þat was of clay
to lyff Resyn Aȝen now is he.   120

¹ So in MS.

Cleophas

¶ Sey nott so man it may not be
 thow thyn exaunple be sumdele good
 Ffor jonas on lyve evyr more was he
 And Cryst was slayn · vpon A rood
 The Jewys on hym they were so wood   125
 Þat to his herte A spere they pyght
 he bled owt All his herte blood
 how xulde he thanne ryse with myght.

Christus

¶ Take hede at Aaron and his dede styk
 Which was ded of his nature   130
 And ȝit he floryschyd with flowrys ful thyk
 and bare Almaundys of grett valure
 The dede styk was signifure
 how cryst þat shamfully was deed and slayn
 As þat dede styk bare frute ful pure   135
 So cryst xuld ryse to lyve a-geyn.

lucas

¶ That A deed styk ffrute xulde bere
 I merveyle sore þer of i-wys
 but ȝitt hym sylf ffro deth to rere
 And leve A-geyn more woundyr it is   140
 That he doth leve I trost not[1] this
 Ffor he hath bled his blood so[2] red
 but ȝitt of myrthe evyr moor I mys
 Whan I haue mende þat he is ded.

Christus

¶ Why be ȝe so hard of truste   145
 dede not Cryst reyse thorwe his owyn myght
 lazare þat deed lay vndyr þe duste
 And stynkyd ryght foule as I ȝow plyght
 To lyff Cryst reysid hym a-ȝen ful ryght
 out of his graue þis is serteyn   150
 why may nat Cryste hym self þus qwyght
 and ryse from deth to lyve Ageyn.

Cleophas

¶ Now trewly sere[3] ȝour wurdys ben good
 I haue in ȝow ryght grett delyght

---

[1] An *l* after *not* crossed through.   [2] A *d* after *so* crossed through.
[3] The vowel is indistinct but should be *e*.

I pray ȝow sere w*ith* mylde mood 155
to dwelle w*ith* vs aff þis nyght.

           Chri*s*tus

I must gon hens a-non ful ryght
Ffor grett massagys I haue to do
I wolde abyde yf þ*at* I myght
but at þis tyme I must hens go. 160

           lucas

¶ ȝe xal not gon fro us þis nyght
It waxit aff derke gon is þe day
þ*e* sonne is downe lorn is þe lyght
ȝe xal not gon from vs A-way.

           Chri*s*tus

Fo. 205 I may not dwelle As I ȝow say 165
I must þis nyght go to my ffrende
þ*er*fore good bretheryn I ȝow pray
lett me not my wey to wende.

           Cleophas

¶ Trewly from vs ȝe xal not go
ȝe xal abyde w*ith* us here stylle 170
ȝo*ur* goodly dalyaunce plesyth us so
We may nevyr haue of ȝow oure fylle
We pray ȝow sere w*ith* herty wylle
Aff nyght w*ith* us abyde And dwelle
more goodly langage[1] to talkyn vs tylle 175
and of ȝo*ur* good dalyaunce more ffor to telle.

           lucas

¶ ȝa brothy*r* cleophas be my*n* Assent
lete us hym kepe w*ith* strenth *and* myght
Sett on ȝoure hand w*ith* good entent
And pulle hym w*ith* us þe wey weff[2] ryght 180
The day is done sere *and* now it is nyght
Why wole ȝe hens now from us go
ȝe xal abyde as I ȝow plyght
ȝe xal not walke þis nyght vs ffro.

           Cleophas

¶ This nyght fro us ȝe go not Away 185
we xal ȝow kepe be-twen us tweyne
to vs þ*er*fore ȝe sey not nay
but walke w*ith* us þe wey is pleyne.

---

[1] The second *g* is written over another letter.
[2] Or woff—the vowel is blotted.

The Appearance to Cleophas and Luke 343

             Chr*is*tus
 Sythyn ȝe kepe me wi*t*h myght *and* mayn̄
 Wi*th* hérty wyll I xal abyde.   190
             lucas
 of ȝo*ur* abydyng we be fulfayn̄
 no man more wel-kom · in þis werd wyde.
             Cleophas

fo. 205ᵛ ¶ Off oure maystyr Cryst Jhesu
 Ffor ȝe do speke so mech good
 I loue ȝow hertyly trust me trew   195
 he was bothe meke *and* mylde of mood
 Of hym to speke is to me food
 If ȝe had knowe hym I dare wel say
 And in what plyght wi*t*h hym it stood
 ȝe wold haue thought on hy*m* many A day. 200
             lucas

¶ Many A day ȝa · ȝa · i-wys
 he was a man of holy levynge
 Thow he had be þe childe of god in blys
 bothe wyse *and* woundyrfull was his werkynge
 But Aftere ȝo*ur* labour and feire walkynge 205
 takyth þis loff *and* etyth su*m* bred
 And than wyl we haue more talkynge
 Of Cryst oure maystyr þat is now ded.
             Chr*is*tus

¶ Beth mery *and* glad wi*t*h hert fful fre
 ffor of cryst jhesu þat was ȝo*ur* ffrende 210
 ȝe xal haue tydyngys of game *and* gle
 wi*th*-inne A whyle or ȝe hens wende
 wi*th* myn hand þis bred I blys
 And breke it here as ȝe do se
 I ȝeve ȝow parte Also of þis   215
 this bred to ete *and* blythe to be.
hic subito discedat christus ab oculis eorum.
             [Cleophas]

¶ A mercy god what was oure happ*e*
fo. 206 was not oure hert wi*t*h loue brennynge
 Whan cryst oure mayster so nere oure lapp*e*
 Dede sitte *and* speke such suete talkynge 220

    ¹ No name of speaker in the MS.

He is now quyk and man lyvenge
þat fyrst was slayn and put in grave
now may we chaunge all oure mornynge
Ffor oure lord is resyn his seruauntys to saue.

                                                        Lucas

¶ Alas for sorwe what hap was this   225
   Whan he dyd walke with vs in way
   he prevyd by scripture ryght wel i-wys
   þat he was resyn from vndyr clay
   We trustyd hym not but evyr seyd nay
   Alas for shame why seyd we so   230
   he is Resyn to lyve þis day
   out of his grave oure lord is go.

                                                Cleophas

¶ latt us here no lengere dwelle
   but to oure bretheryn þe wey we wende
   With talys trewe to them we telle   235
   that cryst doth leve oure mayster and frende.

                                              lucas

   I graunt þer-to with hert ful hende
   lete us go walke forthe in oure way
   I am ful joyfull in hert and mende
   þat oure lord levyth þat fyrst ded lay.   240
                                           Cleophas

¶ Now was it not goodly don
   of cryst jhesu oure mayster dere
   he hath with us a large wey gon
   And of his vprysyng he dede us lere
Fo. 206ᵛ   Whan he walkyd with us in fere   245
   And we supposyd hym both dede and colde
   þat he was A-resyn ffrom vndyr here
   be holy scripture þe trewth he tolde.

                                              lucas

¶ Ryght lovyngely don for-sothe this was
   What myght oure mayster tyl us do more   250
   Than us to chere þat fforth dede pas
   and ffor his deth we murnyd ful sore
   Ffor loue of hym oure myrthe was lore
   We were ffor hym ryght hevy in herte
   but now oure myrth he doth restore   255
   Ffor he is resyn bothe heyl and qwert.

The Appearance to Cleophas and Luke

Cleophas

¶ That he is þus resyn I haue grett woundyr
An hevy ston ovyr hym þer lay
how shulde he breke þe ston A-soundyr
þat was deed and colde in clay   260
Euery man þis mervayle may
And drede þat lord of mekyl myght
but ȝit of þis no man sey nay
Ffor we haue seyn hym[1] with opyn syght.

lucas

¶ That he doth leve I woot wel this   265
he is A-resyn with flesch and blood
A levynge man for-sothe he is
þat rewly was rent upon a rood
All heyl dere brothyr and chaunge ȝour mood
Ffor cryst doth levyn and hath his hele   270
We walkyd in wey with cryst so good
and spak with hym wurdys fele.

Cleophas[2]

o. 207 ¶ Evyn tyll Emawus þe grett castell
ffrom jerusalem with hym we went
Syxti ffurlonge as we ȝow telle   275
we went with hym evyn passent
he spak with us with good entent
þat Cryst xuld leve he tolde tyll us   278
And previd it be scripture verament   *vade worlych*[3]
trust me trewe it is ryght thus.   *not[a] worlych*

lucas

¶ ȝa and whan he had longe spokyn vs tylle
he wold ffrom vs agon his way
with strenght and myght we keptyn hym stylle
And bred we tokyn hym to etyn in fay
he brak þe loff As Evyn on tway   285
As ony sharpe knyff xuld kytt[4] breed
þer-by we knew þe trewth þat day
þat cryst dede leve and was not deed.

---

[1] The *y* is written over some other letter—? a half-written *e*.
[2] Under *Cleophas* in the bottom margin is written and smudged out *Vade Worlych*.
[3] These words seem to be written by the scribe of Ff. 95, 96.
[4] *be* crossed through after *kytt*.

                                                    Petrus[1]

¶ Now trewly serys I haue grett woundyr
　of these grete merveylis þat ȝe vs telle   290
　In brekynge of bred fful evyn A-soundyr
　oure mayster ȝe knew · and lord ryght well
　ȝe sey Cryst levith þat jewys dyd qwelle
　tyll us glad tydyngys þis is serteyn
　and þat oure mayster with ȝow so longe dede dwelle   295
　it doth well preve þat he levith a-geyn.

¶ A brother Thomas we may be ryght glad
　of these gode Novell þat we now haue
　þe grace of oure lorde god is ouer vs all sprad
　Oure lord is resyn his seuauntys[2] to saue.   300
                                                    Thomas
Fo. 207ᵛ Be in pes petyr þou gynnyst to rave
　thy wurdys be wantowne and ryght vnwyse
　how xulde A deed man þat deed lay in grave
　with qwyk fflesche and blood to lyve ageyn ryse.
                                                    Petrus
¶ ȝis Thomas dowte þe not oure maystyr is on lyve   305
　Record of Mawdelyn and of here systerys too
　Cleophas and lucas þe trewthe ffor to contryve
　Ffro jerusalem to Emaws with hym dede they go.
                                                    Thomas
　I may nevyr in hert trust þat it is so
　he was ded on cros and colde put in pitt   310
　kept with knyhtys iiij. his grave sealyd Also
　how xulde he levyn Ageyn þat so streyte was shitt.
                                                    Petrus
¶ Whan Mawdelyn dede tell us þat cryst was a-resyn
　I ran to his graue and Johan ran with me
　In trewth þer we ffownde he lay not in presyn[3]   315
　gon out of his graue and on lyve þan was he
　Therfore dere brother thomas I wole rede the
　Stedfastly þou trust þat cryst is not deed

---

[1] *Nota Worlych* is again written under this name of speaker, and crossed through.
[2] So in MS.
[3] This *y* is written over an *o*.

Ffeythfully be-leve a qwyk man þat he be
A-resyn from his deth by myght of his godhed. 320

Thomas

¶ I may nevyr be-leve these woundyr merveles
tyl þat I haue syght of Euery grett wounde
and putt in my ffyngyr in place of þe nayles
I xal nevyr be-leve it ellys ffor no man on growunde
And tyll þat myn hand þe sperys pytt hath fowunde 325
Which dede cleve his hert and made hym sprede his blood
I xal nevyr be-leve þat he is qwyk and sownde
In trewth whyl I knowe þat he was dede on rood.

Petrus

fo. 208 ¶ Cryst be þi comforte and chawnge þi bad witt
Ffor ffeyth but þou haue þi sowle is but lorn 330
With stedfast beleve god enforme þe ȝitt
of A meke mayde As he was ffor us born.

Christus

Pees be Amonge ȝow · be-holde how I am torn
take hede of myn handys my dere brothyr thomas.

Thomas

My god and my lorde nyght and every morn 335
I[1] Aske mercy lorde ffor my grett trespas.

Christus

¶ Be-holde wele thomas my woundys so wyde
which I haue sufferyd ffor All mankynde
Put þin hool hand in to my ryght syde
And in myn hert blood þin hand þat þou wynde 340
So ffeythffull a ffrend were mayst þou fynde
be stedfast in feyth be-leve wel in me
be þou not dowtefful of me in þi mynde
but trust þat I leve þat deed was on A tre.

Thomas

¶ my lord and my god with syght do I se 345
þat þou art now quyk which henge deed on rode
more feythful þan I · ther may no man be
Ffor myn hand haue I wasch in þi precyous blode.

Christus

Ffor þou hast me seyn þerfore þi ffeyth is good
but blyssyd be tho of þis þat haue no syght 350

---

[1] *Ah* first written and crossed out after *I*.

    And be-leve in me they ffor here meke mood
    Shall com in to hefne my blysse þat is so bryght.

Thomas

¶ As A ravaschyd man whos witt is all gon
    grett mornynge I make ffor my dredfful dowte
    Alas I was dowteful þat cryst from vndyr ston   355
    be his owyn grett myght no wyse myght gon owte
    Alas what mevyd me thus in my thought
    my dowtefful be-leve ryght sore me Avexit
    the trewthe do I knowe þat god so hath wrought
    Quod mortuus et sepultus· nunc resurrexit [1].   360

¶ He þat was bothe deed and colde put in grave
    to lyve is A-resyn [2] by his owyn myght
    In his dere herte blood myn hand wasch I haue
    Where þat þe spere poynt was peyn-fully pyght
    I take me to feyth ffor-sakynge All vn-ryght   365
    þe dowte þat I had fful sore me Avexit
    Ffor now I haue seyn with ful opyn syght
    Quod mortuus et sepultus nunc resurrexit.

¶ I trustyd no talys þat were me tolde
    tyll þat myn hand dede in his hert blood wade   370
    My dowte doth Aprevyn cryst levynge fful bolde
    And is a grett Argument in feyth us to glade
    þou man þat seyst þis ffrom feyth nevyr þou ffade
    my dowte xal evyr chere the ‖ þat sore me Avexit
    trust wele in cryst þat such meracle hath made   375
    Quod mortuus et sepultus nunc resurrexit.

¶ The prechynge of petir myght not conuerte me
    tyll I felyd þe wounde þat þe spere dyde cleve
    I trustyd nevyr he levyd þat deed was on A tre
    tyll þat his herte blood dede renne in my sleve   380
    Thus be my grett dowte · oure feyth may we preve
    be-hold my blody hand · to feyth þat me Avexit

---

[1] *Surreait* apparently first written, and the initial *r* with a blotted over-written letter added. This line, in this stanza only, is written in slightly larger form.

[2] A small fragment of printed paper adhering to the MS. hides the last stroke of the *n*.

be syght of þis myrroure · ffrom feyth not remeve
Quod mortuus et sepultus nunc resurrexit.

¶ Thow þat Mary Magdalyn in cryst dede sone be-leve 385
And I was longe dowteful ȝitt putt me in no blame
Ffor be my grett dowte oure ffeyth we may preve
Aȝens all þe Eretykys¹ þat speke of cryst shame.
Truste wel Jhesu cryst · þe jewys kyllyd the same
The ffende hath he fferyd · oure feyth² þat evyr a-vexit 390
To hevyn ȝow brynge ; and saue ȝow all in same
That mortuus et sepultus Iterum resurrexit.   Amen.

[Remainder of Fo. 209—4½ inches—and Fo. 209ᵛ left blank.]

Scribble on Fo. 209. That mortuus et se.
Scribble on Fo. 209 ᵛ. hic Incipit Ascencio.

ɔ. 210 hic incipit Ascencio domini nostri · cum maria et vndecim
quire] discipulis et duobus angelis sedentibus in albis et Jhesus dicit
discipulis suis &c.³

¶ Jhesus
39

PAx vobis⁴ ; Amonge ȝow pes
bothe love and reste and charyte
Amonge All vertues lete it not ses
Ffor Amonge All vertues prynspal his he
ȝe be to blame I may wel preve 5
Ffor I wyl vse to ȝow wordys pleyn
þat ȝe be so hard of herte to be-leve
þat from deth to lyve I am resyn Ageyn
Not-with-stondynge As ȝe knowe serteyn
to ȝow viij sythys · Aperyd haue I 10
be soundry tymes the trewth to seyn
and þis is þe ix · tyme sothly
 Evyn and no mo
but Now sum mete
A-non doth gete 15
Ffor I wyl Ete
 with ȝow And goo

¹ The y of the final syllable has been written over an i.
² Written feyyth with first y marked for deletion.
³ The title and stage direction are written as one in the MS.
⁴ Pax vobis is written in larger form.

¶ My dyscyplis here what I sey
And to my wourdys ȝevyth Attencion
From jerusalem loke ȝe go nott Awey   20
but mekely A-bydyth my fadyres promiscion
Off whiche be my mowth ȝe haue had informacion
whyll bodyly with ȝow I was dwellynge
Ffor johan Sothly ffor mannys Saluacion
Onlye in watyr was me baptysynge   25
but I ȝow be-hete
With-inne ffewe days þat ȝe
In þe holy goost xul baptyzid be
therfore rysyth up and ffolwyht me
On-to þe mownte of Olyvete.   30

          Jacobus maior

¶ O lord vowche-saff vs for to telle
Iff þou wylt now with-owte more delay
restoryn þe kyngdam of israell
And ȝeve vs þe joye lord þat lestyth Ay.

          Jhesus

Serys þe tymes and þe monthis knowe ȝe ne may   35
whiche my fadyr hath put in his owyn power
but ȝe xul take with-in short[1] day
of þe holy goost þe vertu cler
thorwe whiche xul ȝe
In jerusalem And in jury   40
And more ovyr Also in samary
And to þe worldys ende vttyrly
My wyttnes only be.

¶ lovyth no wrath nor no wronge
but levyth in charyte with mylde stevyn   45
With myrthe and melody and Aungell songe
now I stey streyte ffrom ȝow to hevyn.

hic Ascendit ab oculis eorum et in celo cantent etcetera.

         Angelus

¶ Returnyth ageyn to ȝour loggynge
to jerusalem ffor he wyl thus
his promys mekely þer Abydynge   50

---

[1] The *t* is blotted and written again—small—above the line.

Ffor dowteles þis forseyd jhesus
Whiche from ȝow is take ꝫ¹
In a clowde As ȝe hym seyñ
Steyng vp so xal comyn A-geyñ
Of Al mankynde þis is serteyn 55
jugement xal he make.²

¶³O ȝe bretheryn Attendyth to me
And takyth good hede what I xal seyñ
it be-hovyth þe scripture ffulfyllyd to be
þat of dauyd was seyd with wourdys pleyn 60
Of judas whiche was þe gyde serteyn
Of hem þat cryst slow cruelly
Which Aftyr ffrom deth ros vp ageyñ
and hath abedyn in erthe · fful days fourty
and Aftyr All this 65
Before oure eye ꝫ⁴
in A bryght skye
he dede up stye ꝫ
to hevyn blys.

¶ This seyd judas was Amonges us 70
noumbryd Apostyll and had lych dygnyte
but whan he be-trayd oure lord jhesus
he hynge hym-self vpon A tre
in whos sted muste nedys ordeyned be
A-nother oure noumbre ffor to restore 75
On of þo whiche As weel knowe we
han be conuersaunt here longe before
in oure company
Whiche xal wyttnes ꝫ⁵
berun expresse ꝫ 80
to more and lesse ꝫ
of crystys resurrexioñ stedfastly.

---

¹ *Hic ascendit ab oculys* is scribbled at the bottom of the page.
² There is a larger space than usual between this and the next speech.
³ This paragraph sign is slightly more elaborate.
⁴ Lines 66 to 69 are written on two lines divided by the ꝫ stop.
⁵ Lines 79 to 82 are similarly written.

hic statuent duos joseph justum[1] *et mathiam etcetera.*

Fo. 211ᵛ ¶ O sovereyn lorde whiche of Every man
The hertys dost knowe most inwardly
With all þe lowlyness we may or kan 85
to þe we prey fful benygnely
That þou vowche-saff thorwe þi mercy
Vs hym to shewe · whiche in þis cas
þou lykyst to chesyn effectuously
To ocapye[2] þe lott of judas plas. 90

hic dabunt sortes et cadet super Mathiam etcetera.

Now gramercy lord
And to fulfylle[3]
þin holy wylle
As it is skylle
We All Accorde.[4] 95

[Remainder of 211ᵛ—4⅝ inches—left blank.]

Fo. 212 ¶ *Modo de die pentecostes. Apostoli dicat*[5] *genuflectentes*
Spiritus sanctus descendat super eos etcetera.

Petrus[6]	Andreas	jacobus major	40
¶ Honowre ?	Wurchippe	and reverens	
johannes[7]	Philippus	jacobus minor	
Glorye ?	Grace ?	And goodnes ?	
Thomas	Bartholomeus	Symon	
Dygnyte ?	Vertu ?	and excellence ?	
matheus	Judas	Matheas	
Bewte ?	Blyssynge	and bryghtnes	

            Petrus
Be to that lord heye wurthynes. 5
            Andreas
Whiche hath performyd þat he vs hyght.
            Jacobus major
And vs enbawmyd with suche swetnes.

---

[1] MS. just̃.   [2] Written in larger form.
[3] Lines 96 to 100 are written in two lines divided by parallel strokes.
[4] Scribbled under last line *modo de die* and *hic dabunt so*.   [5] So in MS.
[6] These names are written in red, in larger but not more elaborate letters.
[7] The name *Johannes* was first written to the right of *Jacobus major*, but was erased; and the number *40* is written over it.

The Day of Pentecost

                                                                         Johannes

Whiche to dyscrye ffer passyth oure myght.

                                                                         Philippus

This we All wel kenne.

                                                                      Jacobus minor   10

Now gracyous lord jhesu.

                                                                         Thomas

Conferme us in þi vertu.

                                                                      Bartholomeus

And graunt us grace evyr it to sew.

                                                                        Symon

Sey we All togedyr Amen · Amen.

Et omnes osculant terram.

                                                                      primus judeus

¶ Now ffelawys take hede ffor be my trewthe
ʒondyr syttyth A dronkyn ffelacheppe.

                                                                      ij^{us} judeus   15

To don hem good it were grett ruthe.

                                                                      iij^{us} judeus

ʒa I prey god ʒeve hem all shenscheppe.

                                                                      i^{us} judeus

Muste in here brayn so sclyly doth creppe
þat þei chateryn And chateryn As they jays were.

                                                                      2us judeus   20

ʒa were they ony wel browth A-sclepe
it wore Almes to þe Revere hem to bere
there hem to baptyze.

                                                                      i^{us} judeus

þat were as thynkyth me [1]
A jentyl sport to se
A bettyr game to be   25
Cowde no man devyse.

                                                                         Petrus

¶ Serys Alas what do ʒe mene
why scorne ʒe now þus goddys grace
It is no thynge as ʒe do wene
þer is no drunke man in þis place   30
Where-fore ryght grett is ʒowre trespace
but syrys lyst what it doth sygnyfye
Ffulfyllyd is now to mannys solace
Of Johel þe pregnaunt prophecye

---

[1] Lines 23 to 26 are written in two lines divided by parallel strokes.

In whiche þat he 35
That ȝe han seyn
In wourdys pleyn
Declaryth serteyn
Now blyssyd god be
    Amen.

The remainder of 212ᵛ—2¾ inches—left blank. An interpolated quire—unmarked—begins with Fo. 213 (Ff. 213 to 222ᵛ). On Fo. 213 the words: ' *The Lord be thanked for his q.*' are written in a sixteenth-century hand. And there are some scribbled single letters.

On Fo. 213ᵛ the words *Ad mea facta* are copied in a contemporary hand from the beginning of the next play.

Fo. 214[1] *Ad mea facta pater assit deus et sua mater.*

                                                                          Doctor

¶[2] Ryht worchepful souereynes · liketh yow to here   **41**
of the assumpcioñ of the gloryous moder mary
that seynt Jhoñ the euangelist · wrot and tauht as I lére
in a book clepid Apocriphum · wyth-outyn dyswary
At fourten yer sche conseyved cryste in hire matere clere   5
and in the fiftene yer sche childyd · this avowe dar I
here lyvyng wyth that swete sone · thre and thretty yere
And after his deth in erthe · xii yer dede sche tary
Now acounte me thise[3] yeris wysely
and I sey the age was of this maide Marye   10
when sche assumpte above the ierarchye
thre score yer · as scripture dothe specyfye
legenda sanctorum autorysyth this trewely.

¶ She was inhabith in Jure by the mounte of syoñ
after the assencioñ of hir sone · conseyved in spoused   15
alle the holy placys in erthe · that criste duellyd[4] on
devouthly sche went hem · honouryng the godhed.
Fferste to the place there criste cristenyd was clepid fflum Jordoñ
there he fastyd and takyn was · by malicious falshed
there he beryed was and roos · vittoryously alon   20
there he assendid alle hevenys · god in his manhed

[1] This play is written in a different handwriting from the body of the MS.
[2] This paragraph sign is small and appears an afterthought.
[3] This *s* is written over some other letter.
[4] The *e* omitted and written above the line.

Thus was sche ocupyed I rede
and meche sche was in the temple preyand
now blissid mot sche be · we owe to be seyand
how sche was assum̄pte · here men schul be pleyand 25
preyng you of audience · now ses and tak hede.

                                            mi[. . .

. 214ᵛ ¶ Pes now youre blaberyng in the develis name
what lousy begchis mow ye not se
owre worthy prynsis lo are gaderid in same
that are statis of this lond · hye men of degre
by there hye wisdam they schal now attayne 5
how alle Jure beste gouernyd may be
and of this pillid prechouris · that oure¹ lawis defame
they schul ben slayn as they say · or fayn for to fle
Wherfore in pes be ye
and herkenyth on to hem moste stille I² 10
ffor what boy bragge outh³ · hym spilly I
as knave wyth this craggyd knad · hym kylle I
now herkenyth oure pryncis alle kneland on kne.

                                        Episcopus legis

¶ Now ye prysis⁴ · I prest of the lave
of this demaunde responcyon̄ · I aske here anon 15
ys there ony renogat among vs · fer as ye knawe
or any that peruertyth the pepil · wyth gay eloquens alon
yif there be we muste on to hem set awe
for they feyne falsly oure feyth · hem preve I houre fon
Sweche schul ben bounden vp be the beltys · til flyes hem
    blawe 20
and gnaggyd vp by the gomys · tyl the devyl doth hem grone
We may not won
to sweche harlotis settyn̄ reddure
that geynseyn oure lawe and oure scripture

---

¹ The *u* written over an erased *r*.
² *stilly* originally written, the *e* being written over the erased *y*. The rhyme line linking this with *spilly I* and *kylle I* passes through the following *I* which may therefore have been added.
³ The *u* in this word appears a correction, and there are marks—possibly connected with the correction—under *o* and *t*.
⁴ So in MS.

The Assumption of the Virgin

now let sere pryncis in purpure 25
In savynge of oure lawys now telle on.

primus princeps

¶ Sere syn we slew hym · that clepid hym oure kyng
and seyde he was goddis sone · lord ouer all
Syn his deth I herd of no maner rysyng
and lo yif he hadde levyd · he had mad vs his thrall. 30

Episcopus

Fo. 215 Therfore oure wysdam was to schortyn his endyng
Whoso clyme ouer hie · he hath a foule fall.

ij*us* princeps

ya yit of on thing I warne yow at the gynnyng
his dame is levyng mary that men call
Myche pepil halt hire wythall 35
wherfore in peyne of reprefe
yif we Suffre hyre thus to relefe
oure lawys sche schal make to myschefe
and meche schame don vs sche schall.

Episcopus

¶ A sere ye ben bolde i-now · art thou ferd of a wenche 40
What trowyste that sche myht don vs agayn.

iij prince[ps

Sere there are other in the contre that clenche
and prechyn he is levyng that we slewe they seyn
and yif they ben sufferyd thus this will bredyn a stench
for thorow here fayre speche oure lawys they steyn 45
and therfore devyse we now · vp on this pleyn bench
what is beste for to do · hem for to atteyn
we are but loste yif they reyn.

Episcopus

why let se than · sey me youre ententis.

Prim[us . . .

let vs preson hem til here myht schent is. 50

Secund[us . . .

bettyr is to slen hem wyth dentis.

---

[1] The *m* is obliterated by a smear of later ink.
[2] In this and the following three lines the rhyme words are joined by red lines instead of black. This occurs only on Fo. 163 elsewhere in this MS.
[3] The word has been corrected and is not clear—possibly *that* corrected to *than* or *thou*.
[4] The *is* is written over an erasure.
[5] *will* omitted and written over the line.

The Assumption of the Virgin

              Terc[ius . . .
  nay best is to hang hem wyth peyn.
               Ep[iscopus

¶ Nay seris[1] nowth so · youre better a-vyse
 haue in syth before · what after may tide
 yif we slewe[2] hem · it wolde cause the comownys to ryse   55
 and rathere the devyl sle hym · than we schulde that abid[e
. 215ᵛ But be that senstere ded mary that fise[3]
 We shal brenne here body · and the aschis hide
 and don here aH the dispith · we can here devise
 and than sle tho disciplis · that walkyn so wyde   60
 and here bodyes devyde
 halde ye not this beste as is sayde.
               Primus
Wyth youre wysdam sere we are wel payed.
               Episcopus
than ye knyhtis I charge yow beth arayed[4]
and ye turmentouris · redy that tyde   65
α   When mary is ded
and but she deye the sunere · the devyl smyte of here hed.
hic est maria in templo orans et dicens
               Maria
¶ O hye wysdam in youre dygne deyte
 youre infynyth lovnesse mad oure saluacyoñ
 that it lyst you of me sympilest · to take here humanite   70
 Wyth dew obeschyauns · I make you gratulacyoñ
 and gloryous lord and sone · yif it like youre benygnyte
 nouth to ben displesid wyth my desideracyoñ
 me longith to youre presense now conjunct[5] to the vnyte
 Wyth aH myn herte and my sowle · be natures excitacyoñ   75
 To youre domynacyoñ
 Ffor aH creaturis in you don affye
 and myche more owe I · youre moder be alye
 syn ye wern born god · and man of my bodye
 to desyre yowre presens · that were oure ferste[6] formacyoñ.   80

  [1] A smear of later ink over *se*.
  [2] The *s* is incomplete and the *e* is written over an erasure.
  [3] Some word has been erased after *fise*.
  [4] The *yed* is written over an erasure.
  [5] An extra stroke of the *u* before *c* is perhaps cancelled.
  [6] The *s* is written over some correction.

                                                      Sapi*entia*
¶ My suete mod*er*is preyere on to me doth assende
  here holy herte and here love · is only on me
  Wherfore aungyl to here thou schalt now dyssende
  seyinge here sche shal com*y*n to myn eternyte.

Fo. 216 ¶ myn habundau*n*t mercy · on here I extende   85
  resseyuy*n*ge here to joye · from worldly p*er*plexite
  and in tokyn ther of this palme now pretende
  Seyinge here sche fere no man*er* [1] of diu*er*cyte.
                                                      Angelus i[
  By youre myth I dissende to youre mod*er* in v*ir*ginite.
                                                      Angel*us* ij[
  Ffor qwyche message injoyeth the hefnely consorcyte.   90
*hic discendet angelus ludenti*bus [2] *citha*ris [2] *et dicet marie*
                                                    pri*mus* A[ngelus
¶ Heyl excellent prynces · mary most pure
  Heyl radyant sterre · the su*n*ne is [3] not so bryth
  Heyl moder of m*er*cy and mayde most mure
  the blessyng that god yaf Jacob vp on you now is [3] lyth.
                                                         Maria
  Now wolcom bryth berde · goddis au*n*gel I seu*er* [4]   95
  ye ben messager of all myhty · wolcom wyth my myhtis
  I beseke you now say me · vp-on youre hie nortur
  What is the very name · that to youre p*er*sone dith is.
                                                         Angel*us*
 x What nedith you lady my name ben desyrand.
                                                         Maria
  A yis gr*a*cyows au*n*gyl I beseke you requyrand.   100
                                                         Angelus
¶ My name is gret and m*er*veylous · treuly you telland
  the hye god youre sone abidyth you in blis
  the thrydde day hens · ye schul ben expirand
  and assende to the presence · there my god youre sone is.
                                                         Ma[ria
  Mercy and grom*er*cy god now may I be seyand   105
  thankyng you suete au*n*gyl for this message i-wys.

[1] MS. *man̄*.   [2] MS. *luden͡t͡ citha͡r*.
[3] *is* written over the line; originally written after *bryth* and *lyth* respectively and erased.
[4] MS. *sen̄* or *seu͡er*; cf. *Seuere*, Fo. 221, l. 411.

The Assumption of the Virgin

An[gelus

    In tokenyng where-of lady I am here presentand
    A braunce of a palme · owth of paradis com this
a before youre bere god biddith it be bore.

Mari[a

    now thanke be to that lord · of his mercy euermore.   110

Angel[us

¶ yowre meknesse youre lovnesse and youre hie lore
   is most acceptable in the trynite syth
   youre sete ryall in hefne apparaled is thore
   now dispose yow to deye · youre sone wyl thus rith.

Mar[ia

f. 216v I obbeye the commaundement · of my god here before   115
    but on thyng I beseke · that lord of his myth
    that my brether the appostelis · myht me be before
    to se me and I hem · or I passe to that lyth
a But they ben so deseverid · me thynkyth it nyl be.

Angelus

    A yis lady inpossible to god nothyng trowe ye.   120

¶ Ffor he that sent Abbacuc with mete to babylonye from Jure
   In to the lake of lyonys to danyel the prophete
b Se the same myht god make may the appostolis here mete
a be an her of his hed lo · so myhty was he [1]

¶ And therfore abasche you not lady · in yowre holy mende.   125

Maria

    no more I do glorious aungyl in kynde.

¶ also I beseke my sone · I se not the fende
   What tyme outh of this word · I schal passe hens
   his horible lok wold fere me so hende
   ther is no thyng I dowte · but his dredfull presens.   130

Angelus

    What nedith [2] it to fere you empres so hende
    syn be the fruth of youre body · was convycte his vyolens
    that horible serpent · dar not nyhyn youre kende
    and yowre blosme · schal make hym recistens
    that he schal not pretende   135
    Desyre ye outh ellys now rythis.

---

[1] So written in reverse order and so corrected in MS.
[2] *yow lady* first written after *nedith* and crossed through.

                Maria
  nouth but blessyd be my god in his myhtys.
               Angelus
  to yow I recomaunde me than · moste excellent in sithis
  and wyth this ageyn [1] to god I assende.
 hic ascendit [2] angelus.
                Maria
 ¶ Now lord thy swete holy name · wyth lovnesse I blysse   140
  of qwyche hefne and erthe · eche tyme pshalmodyeth
  that it lykyth youre mercy [3] · me to you to wysse
  my sympil sowle in serteyn · youre name magnefyeth
  Now holy maydenys the seruauntis of god as I gysse
  I schal passe from this world · as the aungyl sertefyeth   145
  therfore to my sympil habitacyon · I telle you now this
  I purpose me to go · besekyng yow replyeth
Fo. 217 *a* And assedually wachith me be dayes and nythis.
             Prima virgo
 ¶ We schal gracyous lady · wyth alle oure mythis
  schal ye from vs passe swete sonne of socoure   150
  that are oure sengler solas · radyant in youre lythis
  youre peynful absence · schal make me doloure.
             virgo secunda
  Moste excellent princes in all vertu that is [4] dith
  alle hefne and erthe · lady you doth honure
  We schal wachyn and wake as oure dewe and [5] ryth   155
  In-to the tyme ye passe to that hye toure
  with [6] [
               Ma[ria
  God thanke you and so do I
  now I wyl dispose me to this Jurne redy
  so wolde god my brether were here me by   160
  To bere my body · that bare jhesu oure savyoure.
 hic subito apparet sanctus Iohannes euangelista ante portam marie.
               Johannes
 ¶ A myrable god meche is thy myth [7]
  many wonderis thou werkyst evyn as thi wyll is

---

[1] This *e* is corrected from an *a*.
[2] The *i* may have been written over an *e*.
[3] A miswritten letter crossed through before *mercy*.
[4] *is* written above the line, having been first written after *dith* and erased.
[5] *and* written above the line, an original *is* having been erased after *ryth*.
[6] Some word (? *honure*) rhyming with *savyoure* cut away here.
[7] Some word (? *now*) is erased after *myth*.

The Assumption of the Virgin

In pheso I was prechyng · a fer contre ryth [1]
and by a whyte clowde · I was rapt to these hyllys 165
here dwellyth cristis moder · I se wel in syth.[1]
Sum merveylous message is [2] comyn̄ that mayde tyll
I wyl go saluse that berde that in vertu is [2] moste brith
and of my sodeyn comyng · wete what is [2] the skele.

hic pulsabit super portam intrante domum marie sibi [3] dicente

α heyl moder mary mayden perpetuall. 170

             Maria

¶ A wolcome mayde Johan · wyth all myn herte in specyall
Ffor Joye of youre presence myn herte gynnyth sweme
thynke ye not Johan how my child eternall
When he hynge on cros · sayd vs this teme
lo here thy sone woman · so bad he me you call
and you me moder · eche othir to queme 175
He betok you the gouernayl there of my body terestyall
on mayde to another as convenyens wold seme

217v α and now that gracyows lord · hath sent me yow sone.

            Johannes

¶ Now good fayr lady · what is ther to done
tellyth the cause why I am heder sent. 180

            Maria

swete sone Johan so wyll I a-none
oure lord god sent to me an aungyl [4] that glent
and sayde I schulde pass hens · where thre were in one
tho I askyd the aungyl to haue you present.

            Johannes

A [5] holy moder schul ye from [6] vs gone 185
My brether of this tydyngis sore wyl repent þ^t ȝe schuld ben
euer trybulacyon̄ lord meche þu vs sendyst Absent [7]

---

[1] Some word (? *now*) is erased after *ryth* and *syth*.
[2] The *is* on each of these lines is written as a correction above the lines, having been originally written at the end of the lines and erased.
[3] MS. *s* with overwritten *i*.
[4] An original *e* changed to *y*. [5] A small *a* changed to a large.
[6] *fron* corrected to *from*.
[7] This, the missing ninth line of the stanza, is written in the right-hand margin in a different hand—apparently that of the scribe of the rest of the MS. The same words were written in the opposite margin by the scribe of this play, and crossed through in red ink—*ye schulde . . . bsen . . .* is legible.

b	and now oure Ioye thy moder to take thou pretendist [1]	
a	thou oure mayster and oure comfort from [2] us ascendist	
	thanne all oure comfort is from us detent	190
α	but what seyde then [3] aungyl moder on-to you more.	

                                                               Maria

¶ he brouth me this palme from my sone thore
qwyche I beseke as the aungyl me bad
that a-forn my bere · by you it be bore
saynge my dirige · devouthly and sad   195
α   Ffor Johan I haue herde the Jewys · meche of me spelle.

                                                              Johannes

A good lady what likyth it you to telle.

                                                              Maria

¶ Secretly they ordeyne in here conseytis felle
When my sowle is paste where god is liste [4] is
to brenne my body · and schamly it quelle   200
For Jhesu was of me born · that they slew with here fistis
And therfore I beseke you Johan · both flech and felle
Helpe I be beryed · for yn yow my tryst is.

                                                            Johannes

Ffere yow not lady · for I schal wyth you duelle
wolde god my brether were here now and wyst this.   205

hic subito omnes apostoli congregentur ante portum [5] *mirates.*

                                                             Petrus [6]

¶ A holy brether wyth grace · be ye met here now
lord god what menyth · this sodeyne congregacyon
now swete brother powle wyl ye take this vp-on yow
preye to god for vs all · we may have relacyon.

                                                             Paulus

Fo. 218   Good brother [7] peter [8] how schuld I here pray now   210
that am lest and most vnworthy of this congregacyon [9]
I am not worthy to ben clepyd apostle sothly I say yow

---

[1] So written in reverse order and so corrected in MS.
[2] So in MS. for *from*.   [3] So in MS.
[4] *Wytt* first written and crossed through in red ink; *liste* written above the line by the scribe.
[5] So in MS.
[6] *Petrus* written as name of speaker against preceding stage direction and crossed through.
[7] The *e* is written in between *h* and *r* as if a correction.
[8] *Powle* first written and crossed through.   [9] MS. *congregacyn*.

The Assumption of the Virgin 363

Ffor as a woodman ageyn holy cherche I mad persecucyoñ
α   but neuertheles I am the grace of god in that þat [1] I am lo.
                                       Petrus
A gret is youre lownesse powle brother euer-mo.   215
                                       Paulus
¶ the keyes of heuene peter · god hath you betake [2]
and also ye ben peler of lith · and prynce of vs all
it is most sittyng to you this preyere to make [2]
and I vnworthy wyth yow · preyen here schall.
                                       Petrus
I take this vp-on me poule for youre sake [2]   220
now almythty god that sittiste · aboue cherubyn halle
In sygne of thyn holy cros · oure handis we make [2]
besekyng thy mercy · may vp-on vs falle
α   And why we ben thus met yif it lyke vs lare.
                                       Johannes
A holy brether alle welcom ye are   225
¶ Why ye be [3] met here I schal you declare
Ffor mary goddys moder by message is sent
that from this wrechid world · to blysse sche schal fare
and at here deying sche desyryth to haue vs present.
                                       Petru[s
A brother Johan we may syhyn and care   230
yif it displese not god · for these tydyngis ment.
                                       Paulus
Fforsothe so we may hevyin euermare
that oure moder and oure comfort schuld ben vs absent
α   but neuertheles the wyl of god fulfyllid mot be.
                                       Johannes
¶ that is wel seyd poule but her-of be-war ye   235
that non of you for here deth schewe hevy speche
Ffor a-non to the Jewys it schuld than notyd be
that we were ferd of deth and that is [4] a-geyn that we teche
Ffor we seyn all tho belevyn in the hol Trynyte
they schul ever leve and nouth deye this truly we preche   240
And yif we make hevynesse for here than wyl it seyd be

---

[1] The thorn letter þ is used here for the only time in the MS. except in two notes on Ff. 144ᵛ and 145; the form y is otherwise used.

[2] Some erasure (? *now*) has been made at the end of each of these lines before the rhyme lines were drawn; cf. ll. 360, 2.

[3] *sent* first written after *be* and crossed through.

[4] Omitted and written over the line.

lo yone prechouris to deye · they fere hem ful meche
Fo. 218ᵛ α And therfore in god now beth glad euerychoñ.

                Petrus

¶ We schal don as ye sey vs holy brother Johan
 now we beseke you let vs se · oure¹ moder marie. 245
               Johannes
 now in goddys name to here² than all let vs gon
 Sche wyl ben ful glad to se · this holy companye.
               Petrus
 Heyl moder and maydyn · so was neuer non
 but only ye most blissid treulye.
               Paulus
 heyl incomparabil quen · goddis holy tron 250
 of you spreng salvacyoñ · and all oure glorye
 heyl mene for mankynde · and mendere of mys³.
               Maria
¶ A wyth all myn hol herte brether ye are wolcom i-wys
 I beseke you now to telle me · of youre sodeyne metyng.
               Petrus
 In dyueris contreys we prechid · of youre sone *and* his blis 255
 Diueris clowdys eche of vs was · sodeynely curyng
 w⁴ in on were brouth before youre yate here i-wys
 the cause why no man cowde telle of oure comyng.
               Maria
 now I thanke god of his mercy · an hy merakle is this
 now I wyl telle yow the cause · of my sonys werkyng 260
α I desyrid his bodily presence to se.
               Johannes
 no wonder lady · thow so dedo ye.
               Maria
¶ Tho my sone jhesu · of his hye pete
 sent to me an aungyl · and thus he sayd
 that the thredde nyth I schuld assende to my sone in deite 265
 thanne to haue youre presence brether · hertly I prayed
 And thus at my request · god hath you sent me.
               Petrus
 Wys gracyous lady · we are ryth wel payed.
               Maria
 blissid brethere I beseke you than tent me

   ¹ A capital *M* crossed through after *oure*.
   ² *let* first written and crossed through after *here*.
   ³ A paragraph sign has been erased at the beginning of the line.
   ⁴ So in MS. for *we*.

The Assumption of the Virgin

      now wyl I rest me in this bed · that for me is rayed   270
c   Wachith me besily wyth youre laumpys and lithtis.

                                                             Paulus
      We schal lady redy all thyng for you dith is.
                                                             Maria
      ¶ now sone schul ye se what god is myth is
      my flech gynnyth feble be nature.
   hic erit decenter ornatus in lecto.[1]
                                                             Petrus
219   Brether eche of you a candele takyth now rithis   275
         And lith hem in haste whil oure moder doth dure
         and bisyli let vs wachyn in this virgyne sythis
         That when oure lord comyth in his spoused pure
         he may fynde vs wakyng · and redy wyth oure lithtis
         for we knowe not the hour of his comyng now sure   280
c   and yn clennesse alle · loke ye be redy.
                                                             Maria
      ¶ A swete sone Jhesu now mercy I cry
      ouer alle synful thy mercy let sprede.
   hic dissendet dominus cum omni celesti curia et dicet
                                                           Dominus
      the voys of my moder me nyhith fulny
      I am dyssend on to here of whom I dede sede.   285
   hic cantabunt organa.
                                                            Maria
      ¶ A wolcom gracyous lord · Jhesu sone and god of mercy
      an aungyl wold a ssuffysed me hye kyng at this nede.
                                                           Dominus
      In propure persone moder I wyl ben here redy
      Wyth the hefnely quer yowre dirige to rede
   ℣ Veni tu electa mea et ponam in te thronum meum   290
      quia concupiuit rex speciem tuam.
                                                            Maria
   ℣ Paratum cor meum deus paratum cor meum
      cantabo et psalmum dicam domino.
                                                           Apostoli
   ℣ hec est que nesciuit thorum in delictis
      habebit requiem in respectu animarum sanctarum.   295
                                                            Maria
   ℣ Beatam me dicent omnes generaciones
      quia fecit michi magna qui potens est et sanctum nomen eius.

---

[1] This stage direction is not underlined in red.

Dominus

℣ Veni de libano sponsa mea veni Coronaberis.

 Ecce venio quia in capite libri scriptum est de me
 vt facerem voluntatem tuam deus meus   300
 quia exultauit spiritus meus in deo salutari meo [1].

hic exiet anima marie de corpore in sinu dei.

Dominus.

Fo. 219ᵛ ¶ Now com my swete soule in clennesse most pure
 and reste in my bosom [2] brithtest of ble
 alle ye myn apostelis · of this body taketh cure
 In the valle of Josephat · there fynde schul ye   305
 A grave new mad for maryes sepulture
 there beryeth the body wyth all youre solempnite
 and bydyth me there stylle thre dayes severe [3]
 and I schal pere ageyn to yow · to comfort your aduercyte
 Wyth this swete soule now from you I assende.   310

Petrus

 In oure tribulacyones [4] lord thou vs defende
 We haue no comfort on [5] erthe · but of the alon
 O swete soule of mary prey thy sone vs defende
 haue mynde of thy pore brether · when thou comyst to þi tron.

Chorus martyrum

℣ Que est ista que assendit de deserto   315
 deliciis affluens innixa super dilectum suum.

Ordo angelor[um

℣ Ista est speciosa inter filias Jerusalem sicut vidistis eam
 plenam caritate et dilectione sic que in celum gaudens suscipitur
 et a dextris filii in trono glorie collocatur.

hic cantabit omnis celestis curia.

Prima virgo *

¶ Now suster I beseke you · let vs do oure attendaunce   320
 and wasche this gloryous body · that here in oure sith is
 as is the vse among vs wyth-outyn ony varyaunce
 now blessid be this persone that bar god of mythtis.

---

[1] MS. apparently *mes*.
[2] *brist* miswritten after *bosom* and crossed through.
[3] So in MS. for *seuere*? = *sure*, rhyming with *pure, cure, sepulture*.
[4] MS. *tribulacyoūs*.   [5] Or *in*.
* *prima virgo* is written as name of speaker against the preceding stage direction and crossed through in red.

Secunda virgo

I am redy sus*ter* wyth all myn hol affyaunce
to wesche and worschepe · this body so brith is 325
alle creaturys therto owyn dew obeschaunce
ffor this body resseyved · thē holy gostis slithtis.

et osculabunt corpus marie.

Johannes

¶ Now holy brother peter · I hertely you pray
to bere this holy palme · before this gloryous body
Ffor ye ben *p*rince of apostelis · and hed of oure fay 330
therfore it semyth you best to do this offis treuly¹.

Petrus

Sere and ye slept on cristis brest · seyng all celestly
ye are goddis clene² mayde · wyth-outyn ony nay
this observaunce is most like · you to do dewly
Wherfore tak it vpon you · brother we pray. 335
and I schal helpe for to bere the bere.

*Paulus*³

¶ and I peter wyth oure brether in fere
this blessid body schal helpe to the ground
this holy cors now take we vp here
Seyng oure observaunce · wyth devouth sound. 340

hic portabunt corpus versus sepulturam cum eorum luminibus.

Petru[s

℣ Exiit israel de egipto · domus iacob de populo barbaro. Alleluia.

Apostoli

℣ Ffacta est iudea sanctificacio eius israel potestas eius. Alleluia.

hic angeli dulciter cantabunt in celo · Alleluia.

Episcopus

¶ herke sere *p*rincys · what noyse is all this
the erthe and the eyer · is ful of melodye
I herde neu*er* er · swyche a noyse now i-wys 345
con ye outh say · what they signefye.

primus princeps

I not be my god that of myht meche is
What-sum*euer* they be · hougely they crye
I am aferd⁴ there wyll be sum thyng a-mys
It is good prevely among vs we spye 350
      wyth-owte.

---

¹ *tru* written before *treuly* and crossed out.   ² The *l* corrected from an *h*.
³ Name not looped in red; perhaps added in a different ink.
⁴ The *a* is squeezed in between the *m* of *am* and the *f* of *ferd*.

Secundus princeps

Now I haue levyd this thre skore yer
but sweche another [1] noyse · herd I neuer er
myn herte gynnyth ogyl · and quake for fer
there is sum newe sorwe · sprongyn I dowte. 355

Tercius princ[eps

¶ ya that there is sothly I say yow
the prophetis moder [2] mary is ded
the disciplis here beryn in gret aray now
and makyn alle this merthe in spyth of oure hed.

Episcopus

ffy on yon [3] lousy doggys · they were better nay [4] 360
outh harrow · the devyl is in myn hed
ye dodemvsyd prynces · faste yow aray [4]
or I make avow · to mahound youre bodyes schul blede

Fo. 220v Now that quene is ded
ye coward knytys in plate 365
and ye tormentours · thryfe schul ye late
Ffaste harlotys · go youre gate
and brynge me that bychyd body I red.

Primus princeps

¶ dowte you not sere byschop in peyne of repref
We schal don schame to that body · *and* to tho prechours. 370

Secundus

Sere I schal geyne tho glabereris or gramly hem gref
tho teynt tretouris schul tene · yif my loke on hem louris.

Tercius princeps

To hurle wyth tho harlotys · me is ful lef
I schal snarle [5] tho sneveleris · wyth rith scharp schowris.

Episcopus

hens than a devylis name · and take me that thef 375
and brnge [6] me that bygyd body · evyn to-fore these touris
and here disciplis ye slo
hye you hens harlotys at-onys
the devyl boyes mot breke youre bonys
Go stent me yone body wyth youre stonys 380
Outh harrow · al wod now I go.

---

[1] *another* first written and crossed through.
[2] *is* first written after *moder* and crossed through.
[3] Or *you*. [4] Some erasure at the end of these two lines; cf. ll. 216, 218, 220, 222.
[5] *snrle* first written and crossed through.   [6] So in MS. ? bringe.

Secundus princeps

hic discendunt principes cum suis ministris vt feroci percutientes petras cum eorum capitibus.

¶ What devyl where is this mene
  I here · here noyse · but I se ryth nouth
  allas I haue clene lost my poste
  I am ful wo · mad is my þowth. 385

Tercius princeps

  I am so ferd I wold feyn fle
  the devyl hym spede · hedyr me brouth
  I renne I rappe · so wo is me
  Wyndand wod wo hath me wrouth
  To deye I ne routh. 390

Primus princeps

  A cowardis vpon you now fy
  are ye ferd of a ded body
  I schal sterte ther-to manly
  alle that company fere I ryth nouth.

o. 221 *hic saltat insanus af¹ feretrum Marie et pendet per manus.*

¶ Allas my body is ful of peyne 395
  I am fastened sore to this bere
  myn ² handys are ser · bothe tweyne
  o peter now prey thy god for me here
  In cayfas halle · when thou were seyne
  And of the peter a mayde · acusid there 400
  I halpe the tho · now helpe me ageyne
  that I were hol · outh of this fere
  sum medycyn me lere.

Petrus

  I may not tend to the sere at this hour
  Ffor ocupacyon of this body of honour 405
  but neuertheles · beleue in Jhesu criste oure saveyour
  and that this was his moder that we bere on bere.

¶ I beleue in Jhesu mannys saluacyon.³

Petrus

  In goddis name go doun than and this body honure.

---

[1] So in MS.   [2] A miswritten word (*hŏdys?*) crossed through after *myn*.
[3] No name of speaker for the speech in MS.

Primus princeps

now mercy god and gromercy of this savacyon 410
In Jhesu and his moder to beleve euer I seuere.

Petrus

Than take yone holy palme · and go to þi nacyon
and bid hem beleve in god yif they wyl be pure
and towche hem ther wyth both hed · hand and facyon
and of her sekenesse · they schal have cure 415
and ellis in here peynys indure.

Primus princeps

Gromercy holy fader peter
I schal do as ye me teche her
thankyng god euer in my speche her
Wyth hye repentaunce · and herte most mure. 420

hic portabunt feretrum ad locum sepulture.

Petrus

¶ Now holy brether this body let vs take
and wyth alle the worschepe we may ley it in the graue
kyssyng it alle atonys · for here sonys sake
now insence ye · and we schal put here in this cave.

hic ponent corpus in sepulcrum insensantes et cantantes.

Johannes

¶ De terra plasmasti me *et* carne induisti me 425
redemptor meus domine resuscita me in novissimo die.
Now god blysse this body · and we oure synge make

hic vnanimiter benedicent corpus · In nomine patris et filii et spiritus sancti.

the fruth that it bar oure soules schal sauc
Now reste we vs brether [1] vp-on this pleyn lake
tyl from oure god and oure lord · tydyngis we haue 430
Here must we belave.

Paulus

so muste we Johan as ye say
thanne byde we here and pray
besekyng hym of comfort taht best may
restyng here abowtyn this graue. 435

---

[1] Or *brother*.

hic vadit princeps ad Iudeos cum palma.   Primus princeps

¶ ye Jewys that langour in this gret Infyrmyte
belevyth in crist Jhesu and ye schal haue helthe
throw vertu of this holy palme · that com fro the trinyte
yowur sekenesse schal aswage · and restore yow to welthe.
            Secundus princeps
I beleve in crist Jhesu · goddis sone in vnyte   440
and forsake my mavmentryes · fals in here felthe

hic tangat credentes cum palma et sanati sunt.

A I thanke the gracyous lord · and thy moder of pete
now are we hol of oure seknesse · and of oure foul belthe.
            Tercius princeps
What harlotys forsake oure lawe.
            Secundus princeps
So hald I best the do.   445
            Tercius princeps
hens fro me in the develis name ye go
I deye outh outh harro
the wylde develys · mot me to drawe.
            Primus demon
¶ herke belsabub and belyal sere sathan in the herne
vs fettyn oure servauntis to this preson   450
blow flamys of fer to make hem to brenne
mak redy ageyn we com to this demon.
            Secundus demon
Ffaste for tho harlotis now let vs renne
to caste hem in this pet here that depe is adon
they schul brenne and boyle and chille in oure denne   455
go we now a dewelys name as faste as we mon
Harrow harrow · we com to town.
            Primus demon
Drag we these harlotis in hye
In to the pet of helle for to lye.
            ij^{us} demon
Gowe now helle houndis[1] ye crye   460
Sere sathan may heryn oure son.

---

[1] MS. *hondis*. The similar flourish on the rhyme words *preson, adon, mon, son* in this stanza might also mark a contracted *u*, but it occurs also in *town* and not in *demon* the other rhyme words.

Dominus

¶ Now auⁿgyl and alle this court celestyall
In to herthe now discendith wyth me
to reyse the body of my moder terestyall
and bryng we it to the blysse of my deyte 465
α   assent ye here to now the vnyte.

Angeli

ya for yowre hye mercy lord · al hefne makyth melode.

hic discendit et venit ad apostolos dicens   *Dominus*

¶ Pes be to yow alle · my postelis so dere
lo me here yowre lord · and youre god now rythtis.

Petrus

A wolcom criste oure comfort · in thy manhed clere 470
gret merveylous god · mekyl now thy myth is.

Dominus

What worschepe and grace · semyth you now here
that I do to this body · mary that hythtis.

Johannes

lord as thou rese from deth · and regnyst in thyn empere
so reyse thou this body · to thy blysse that lyth is 475
vs semyth this ryth is.

Mychael

ya gloryous god · lo the sowle here prest now
to this blissid body · likyth it you to fest now
hefne and erthe wold thynke this the best now
In as myche as sche bare you god in youre mythtis. 480

hic vadit anima in corpus marie.

Dominus

¶ Go thanne blyssid soule to that body ageyn[1]
arys now my dowe · my nehebour and my swete frende
tabernacle of Joye · vessel of lyf · hefnely temple to reyn
ye schal haue the blysse wyth me moder · that hath non ende
Ffor as ye were clene in erthe · of alle synnys greyn 485
so schul ye reyne in hefne clennest in mend.

Maria

A endles worchepe be to you Jhesu · relesere of peyn
I and alle erthe may blisse ye com of owre kend
Lo me redy wyth you for to wend.

Dominus

A-bouen hefnys moder assende than we 490
In endles blysse for to be.

[1] This *e* has been corrected from an *a*.

The Assumption of the Virgin

Michael

Hefne and erthe now injoye may ye
Ffor god throw mary is mad mannys frend
Et hic assendent in celum cantantibus organis.
¶ Assumpta es maria in celum.

Dominus

Yow to worchepe moder · it likyth the hol trinyte 495
Wherfore I crowne you here · in this kyndam of glory
of alle my chosyn · thus schul ye clepyd be
qwen of hefne · and moder of mercy.

Michael

Now blysid[1] be youre namys we cry
ffor this holy assumpcyon · alle hefne makyth melody. 500
¶ Deo gracias.

The remainder of Fo. 222ᵛ—4¾ inches—left blank and Fo. 223. On Fo. 222ᵛ *Deo gratias* enclosed in flourished lines is scribbled in another hand. On Fo. 223 are the scribbled words: 'lo me redy with wᵗ yow to wend'. Cf. line 489.

¶ hic incipit dies iudicii et *Ihesu descendente cum Michaele et Gabriele Archangelis* et *Michael dicet* etcetera[3]

o. 223ᵛ [2]
[marked 220]

¶ MICHAEL

42

Svrgite ·' All men Aryse
  venite Ad judicium
Ffor now is sett þe hy3 justyce
And hath Assygnyd þe day of dom
Rape 3ow redyly to þis grett Assyse 5
bothe grett *and* small All An su*m*
And of 3*our* Answere 3ow now Avyse
What 3e xal sey · whan þat 3e cu*m*
3owre Ansuere ffor to telle
Ffor whan þat god xal 3ow appose 10
ther is non helpe of no glose
the trewth fful trewlye he wyl tose
And send 3ow to hevyn or helle.

Gabryell

¶ Bothe pope[4] prynce and prysste[5] with crowne
Kynge *and* caysere *and* knyhty*s* kene 15

---

[1] The third letter is a *y* written over an *s* or vice versa.
[2] Original handwriting resumed.
[3] Title and stage direction written as one in the MS.
[4] This word is written in larger form.
[5] *prynce* first written and corrected.

Rapely ȝe renne ȝour resonys to rownc
Ffor this xal be þe day of tene
Nowther pore ne ryche of grett renowne
ne aīī þe develys in helle þat bene
Ffrom þis day ȝow hyde not mowne   20
Ffor aīī ȝour dedys here ‖ xal be sene
Opynly in Syght
Who þat is fowndyn in deedly gylte
he were bettyr to ben hylte
in hendeles helle he xal be spylte   25
his dedys his deth xal dyght.

Fo. 224 *Omnes resurgentes subtus terram clamauit*[1] ha aa · ha aa · ha aa. ‖
Deinde surgentes dicat[1] ha aa etcetera.

¶ Ha aa cleue a-sundyr ȝe clowdys of clay
A-sundyr ȝe breke *and* lete vs pas
now may oure songe be wele-Away
þat evyr we synnyd in dedly trespas.   30
     Omnes demones clamant[2]
Harrow *and* owt what xal we say
harrow · we crye owt And Alas
Alas harrow is þis þat day
to endles peyne þat vs must pas
Alas harrow *and* owt we crye   35
    Omnes anime resurgentes dicant etcetera
A mercy lorde ffor oure mysdede
And lett þi mercy sprynge *and* sprede
but alas we byden in drede
It is to late to Aske mercye.
        Deus
¶ Venite benedicti[3] ‖ my bretheryn aīī   40
Patris mei ‖ ȝe childeryn dere
Come hedyr to me to myn hyȝ haīī
Aīī þo myn suterys *and* servauntys be[4]
Aīī þo ffowle wyrmys ffrom ȝow ffalle
With my ryght hand I blysse ȝow here   45
my blyssynge burnyschith ȝow as bryght as beraīī

  [1] So in MS.   [2] MS. *claman?*.   [3] Or *beneditti*.
  [4] So in MS. for *were*.

Doomsday

As crystall clene it clensyth ȝow clere
All ffylth ffrom ȝow ffade
Petyr to hevyn ȝatys þou wende and goo
þe lokkys þou losyn and hem vndo   50
my blyssyd childeryn þou brynge me to
here hertys for to glade.

Petrus

¶ The ȝatys of hevyn I opyn þis tyde
Now welcome dere bretheryn to hevyn i-wys
Com on and sytt on goddys ryght syde   55
Where myrthe and melody nevyr may mys.

Omnes saluati

On kne we crepe we gon we glyde
to wurchepp oure lorde þat mercyfful is
Ffor thorwe his woundys þat be so wyde
he hath brought us to his blys   60
holy lorde we wurcheppe þe.

Deus

Welcome ȝe be in hevyn to sitt
wel cum fro me xul ȝe nevyr flitt
so sekyr of blys ȝe xul be ȝitt
to myrth and joye welcum ȝe be.   65

Anime dampnandum[1]

¶ Ha · Ha · mercy mercy we crye and crave
A mercy lorde for[2] oure mysdede
A mercy mercy we rubbe we rave
A help us good lord in þis nede.

Deus

How wolde ȝe wrecchis Any mercy haue   70
Why Aske ȝe mercy now in þis nede
What haue ȝe wrought ȝour sowle to saue
to whom haue ȝe don Any mercyful dede
Mercy for to wynne.

Primus diabolus

Mercy nay nay they xul haue wrake   75
And þat on here fforehed[3] wyttnes I take
Ffor þer is wretyn with letteris blake
Opynly all here synne.

---

[1] MS. *dāpnādu*, the final *u* with circumflex and dot.
[2] *ore* miswritten and crossed through after *for*.   [3] The *r* is blotted.

	Deus

Fo. 225 [The number is written on a patch in the paper.]

¶ To hungry *and* thrusty þat Askyd in my name
mete *and* drynke wolde ȝe ȝeve noñ 80
Of nakyd men had ȝe no shame
ȝe wold nott vesyte men in no *p*reson
ȝe had no pete on seke nor lame
Dede of mercy wold ȝe nevyr don
Vn-herborwed men ȝe servyd þe same 85
to bery the deed pore man ; wold ȝe not gon
These dedys doth ȝow spylle
ffor ȝoure love was I rent on rode
and for ȝoure sake I shed my blode
Whan I was so mercyfull *and* so gode 90
Why haue ȝe wrought aȝens my wylle.

2*us* diabolus

¶ I ffynde here wretyn in þi*n* fforheed
þ*ou* wore so stowte *and* sett in pryde
þou woldyst not ȝeve a pore man breed
but ffrom þi dore þou woldyst hy*m* chyde. 95

3*us* diabolus

And in þi face here do I rede
þat if a thrysty man com any tyde
Ffor thrust þow he xulde be deed
drynk ffrom hy*m* þou woldyst evyr hyde
On covetyse was all thy thought. 100

1*us* diabol*us*

In wratth þi neybore to bakbyte
them for to hangere was þi delyte
þou were evyr redy them to endyte
On þe seke man rewyst þou nought.

2*us* diabolus

Fo. 225ᵛ ¶ Evyr-mor on Envye was All þi mende 105
þ[ou]¹ woldyst nevyr vesyte no *p*resoner
to All þi neybórys þou were vnkende
þou woldyst nevyr helpe man in daunger.

3*us* diabol*us*

The synne of slauth þi sowle xal shende
masse nor*e* mateynes woldyst þ*ou* non here 110
to bery þe deed man þou woldyst not wende.

¹ The over-written letter is blurred by a stain, of which there are many on this page.

þerfore þou xalt to endles ffere
to slowth þou were ful prest.
                                    i^{us} diabolus

Thou haddyst rejoyse in glotonye
In dronkesheppe and in rebawdye   115
Vnherborwyd with velonye
þou puttyst from here rest.
                                    2us diabolus

¶ Sybile sclutte¹ þou ssalte sewe
All 3our lyff was leccherous lay
to all 3our neyborys 3e wore a shrewe   120
all 3our plesauns was leccherous play
Goddys men 3e lovyd but fewe
nakyd men and ffebyl of array ²
3e wolde nott socowre with a lytel drewe
nott with A thred þe soth to say   125
Whan they Askyd in godys name.
                              Omnes dampnandi

A mercy lord mekyl of myght
we Aske þi mercy and not þi ryght
not after oure dede so us quyth ³
we haue synnyd we be to blame.   130
                                         Deus

The rest is missing, this page forming the last of the W quire.

---

¹ The *l* in this word is faint, only just legible.
² *ray* first written and *ar* added above the line.   ³ The final *h* is blotted.

Scribbled on right-hand margin in (?) contemporary hand—1. at*que*. 2. I must go to þe most.

LIST OF PROPER NAMES OCCURRING IN THE *LUDUS COVENTRIÆ*

Aaron, 340.
Abacuth, 60. Habakkuk.
Abbacuc, 359.
Abdias, 60. Obadiah.
Abel (Abelle, Abeel, Abeeit), 2.
Abias, 59. Abia (Matt. i. 7).
Abraham, 3.
Abyacar, 4. Abiathar or Abiacar pontifex (*De Nativitate S. M.*).
Abysakar, 83. Isachar reading of some MSS. for Abiathar (*De Nat. S. M.*).
Achas, 61. Achaz (Matt. i. 9).
Ada, 29 (note). (Gen. iv. 19.)
Adam, 1.
Affraunt, 313. Pilate's 4th knight.
Aggeus, 61. Haggai.
Almonye, 198. Germany (Almayne in *C. of P.* list).
Alpheus, 62 (note). (Matt. x. 3.)
Amon, 61. (Matt. i. 10.)
Amorawnt (Ameraunt), 313. Pilate's 1st knight (cf. Poem on the Resurrection, MS. Ashmole 61).
Amos, 147.
Andreas, 250.
Andrewe, 269.
Anna, prophetissa, 164.
Annas, 230.
Anne, 63 (note).
Apocriphum, 354.
Aragon, 198.
Aran, 37 (note). Haran (Gen. xi).
Archage, 153, 198. Realm of Jasper and country seen from the Mount (cf. Artage apparently for Arcadia in *Parl. of Three Ages*, l. 347. Achaia and Arcadia occur together in geographical lists. 'Ibi est Achaia, ibi est Arcadia.' Honorius Augustodunensis *De imagine mundi*; cf. *Rel. Ant.* i, p. 272, and MS. Arundel 123).
Archas, 198. Country seen from the Mount (see last note).
Arfaxat, 37 (note). Arphaxad (Gen. xi).
Arfaxat (Arphaxat), 313. Pilate's 2nd knight (MS. Ashmole 61, Arfax).

Arfexe, 231. Messenger of Annas.
Artyse, 292. Pilate's servant.
Asa, 59. (Matt. i.)
Asmaria, 62 (note). Mother of Joachim—no authority has been found for this name. Esmeria is the name of Joachim's sister-in-law in *Legenda aurea*.

Babolony, 198.
Babylony, 74.
Bakbytere, 125. Also a character in *Castle of Perseverance*.
Balaam, 147.
Baltazare, 152. 1st king (of Saba).
Barpanter, 62 (note). Father of Joachim (see *Legenda aurea*, cap. 131).
Barrabas, 12.
Bartholomeus, 350.
Baruk, 61. Baruch.
Bedleem, 136. Bethlehem.
Bedlem jude, 283.
Belsabub, 193. Beelzebub.
Belyall (Beliall), 194. Belial.
Belyard, 193. Name of a devil (mistake for Belyall ?).
Bertylmew, 270. Bartholomew.
Bertylmewe þe bochere, 123.
Betany, 261.
Bette þe bakere, 123.
Bettrys Belle, 123.
Boosdras, 148, l. 62 (? Is. lxiii: 'Who is this that cometh with dyed garments from Bozrah').
Boosras, 146. 1st shepherd.
Boutyng þe brewstere, 123.

Calsydon, 244. (?)
Caluerye (Kalvarye), 294.
Caton, 178. Catonis Disticha.
Cavdas, 269. Candace.
Caym, 2.
Caynan, 31 (note). (Gen. v.)
Cayphas (Cayfas), 10.
Cenacle, 15. Upper Chamber (Luke xxii. 12, Vulgate 'Coenaculum').
Cephas, 269.
Cham, 37. Ham.

Cleophas, 62 (note). Second husband of Anne (*Legenda aurea*).
Cok crane, 123.
Colett Crane, 123.
Coleyn, 7. Cologne.
Colle Crakecrust, 123.
Contemplaciọ, 62.
Cosdram (Cosdran). Pilate's 3rd knight (cf. Poem on the Resurrection, MS. Ashmole 61).

Danyel, 59.
Dauyd, 5.
Davy Drydust, 123.
Diabolus, 176.
Doctrynal, 178. Name of part of the *Speculum* of V. de Beauvais.
Dysmas, 293. The penitent thief.

Ebrew, 10.
Egypth, 7.
Egythp (Egypt), 254.
Elyud, 62 (note). Son of Emeria and brother of Elizabeth, mother of John (*Legenda aurea*).
Elyzabeth, 105.
Emanuel, 58.
Emawus (Emaws), 337.
Emeria, 62 (note). Sister of Anne, mother of Elizabeth and Elyud; generally found as Esmeria or Ismaria (*Legenda aurea*).
Eminem, 62 (note). Daughter of Elyud, mother of St. Servasius.
Enoch, 29 (note). (Gen. iv.)
Enok, 31 (note). (Gen. v.)
Enos, 31 (note). (Gen. v.)
Erlonde, 199. Ireland.
Eve (Eua), 2.
Ezecheel, 59.
Ezechias, 61.

Ffraunce, 198.
Flomjordon, 188.

Gabryell, 5.
Galelye, 169.
Galys, 199. (Cf. *Castle of Perseverance* and *Croxton* play.) ? Galicia.
Gamalyel, 246.
Geffrey Gile, 123.
Geruasius, 62 (note). For Seruasius.
Grek, 15. Greek.
Grw, 169. Greek.
Gryscysme, 178. (?)
Gylle Fetyse, 123.

Heber, 37 (note). Eber (Gen. xi).
Herownde (Herodes, Herowde, Herowdys, Herowndys), 7.
Hierusalem, 185.

Jabel, 29 (note). (Gen. iv.)
Jacob, 147.
Jacobus major, 62 (note).
Jacobus minor, 62 (note).
Jak at the Style, 123.
Jamys, 269.
Jamys the lesser, 269.
Jane, 123.
Januense, 298. Genoa? (Jenyse and Genewaye occur in the *Croxton* list).
Japhet, 37.
Jared, 31 (note). (Gen. v.)
Jasper, 153. 3rd king (of Ypotan and Archage).
Jeremye (Jeremias), 58.
Jesmas 293; Jestes, 299. The impenitent thief. *Legenda aurea*, cap. liii, Gesmas; *Gesta Pilati*, Gestas.
Jesse, 4.
Jewry, 234.
Jewys (Juwys), 10.
Jherosolyman, 269.
Jherusalem, 8.
Joachym, 4. Father of the Virgin.
Joathan, 60. (Matt. i.)
Joel, 60.
Jhon, 8.
Jon, 300.
Johan Jurdon, 123.
Johannes apostolus, 239.
Johannes baptista, 62 (note).
Johannes Evangelista, 62 (note).
Jonas, 59.
Joras, 60. Joram (Matt. i).
Jordon, 8.
Joseph, 5.
Joseph ab Aramathy, 310.
Joseph of Baramathie, 308.
Josephat (Valley of), 366.
Josophat, 60. (Matt. i.)
Iradh, 29 (note). (Gen. iv.)
Isaye, Ysaias, 57.
Israel, 169.
Itayl, 198. Italy.
Juda, 62 (note). Jude.
Judas, 10.
Jude, 283. Judea.
Jurye (Jure, Jury, Jewry), 216.

Kate Kelle, 123.
Kytt Cakelere, 123.

Lamech, 29 (note). (Gen. iv.)
Lameth, 39.

Latyn, 15.
Lazarus (Lazare), 10.
Letyce lytyl trust, 123.
Leyon (Leon, Lyon) 231. Jewish judge.
Longeus (Longeys), 12. Longinus.
Luce lyere, 123.
Lucyfer (Locyfere), 1.
Luke, 14. (Lucas, 337.)
Lumbardye, 198.

Mabyle, 123.
Mahound (Mahownde), 154.
Malachel, 31 (note). Mahalaleel (Gen. v).
Malcheus (Malcus), 266.
Malkyn mylkedoke, 123.
Manasses, 61. (Matt. i.)
Martha, 210.
Mary (Maria), 6.
Maria jacobi, 327.
Maria magdalene, 327.
Mary Mawdelyn, 14.
Maria Salome, 327.
Matheus, 250.
Mathew, 270.
Mathias, 352.
Matussahel, 29 (note). Methusael (Gen. iv).
Matussalem, 31 (note). Methuselah (Gen. v).
Maunfras, 146. 2nd shepherd. Mauferas occurs as name of a devil in two French mysteries: *Le Martyre de S. Pierre et de S. Paul* and *Les Miracles de Ste Geneviére* (ed. Jubinal).
Mawdelyn, Magdalyn, 210.
Megge Mery weder, 123.
Melchizar, 152. 2nd king (of Tarys).
Melophat, 63 (note). Alternative name for the mother of Anne. (No source has been found for this name.)
Messy, 59. Messiah.
Miles the miller, 123.
Montana, 116. The · hill country (Luke i. 39, Vulgate 'in montana ').
Mors, 174.
Moyse, 147. 3rd shepherd.
Moyses, 3. Moses.
Mychael (Michael), 372.
Mycheas, 59. Micah.

Nachor, 37 (note). Nahor (Gen. xi. 22).
Nacor, 37 (note). Nahor (Gen. xi. 26).
Nasaphat, 63 (note). Alternative name for mother of Anne. ('Est tuus Anna pater Jozafath, Nazafath tua mater', *Rel. Ant.* i, p. 287. Cf. also Pedigree of our Saviour, Harl. 6148, fo. 114ᵛ: 'Ex vetusto libro manuscripto de vaticiniis et genealogis', which gives Isakar and Nazaphat as parents of Anne.)
Naverne, 198. Navarre (cf. *Croxton Play of the Sacrament*).
Nazareth, 96.
Neptalym, 198. Naphtali—seen from the Mount.
Noe, 3.
Normandye, 198.
Nycodemus (Nichodemus), 13.

Olyvet, 262.
Osyas (Ozyas), 61. Hosea.
Ozias (Ozyas), 60. Ozias (Matt. i).

Parys, 199.
Pernal Prane, 123.
Petyr (Petir, Peter), 11.
Petrus, 224.
Peyrs Pottere, 123.
Phaleg, 37 (note). Peleg (Gen. xi).
Pharao, 255.
Pharasy, 232. Land of Pharisees.
Phariseus, 202.
Pharysewys, 9.
Phelypp, 269. (Philippus, 250.)
Phelypp þe good flecchere, 123.
Pheso, 361. Ephesus.
Poperynge, 199. (Cf. *Sir Thopas*.)
Portyngale, 199. Portugal.
Poul, 270. Paul.
Powle, Paulus, 363.
Powle Pewterere, 123.
Pownteys, 199. ? Pontoise.
Pycardye, 199.
Pylat (Pilatus), 11.

Rachel, 68.
Raguel, 63 (note). (Tobit vi.)
Rebecca, 93. One of the Virgin's handmaidens. (Cf. Pseudo-Matthew.)
Reu, 37 (note). (Gen. xi.)
Rewfyn (Rufyne), 231. Jewish judge.
Reyse sclaundyr, 125.
Roboas, 59. Roboam (Matt. i).
Robyn rede, 123.
Romaynes, 245.
Rome, 198.

Saba, 152. Realm of Baltazar (Ps. lxxii. 10).

Sabyn Sprynge, 123.
Salamon, 59. (Matt. i.)
Sale, 37 (note). Salah (Gen. xi).
Salmana, 198. Country seen from Mount (cf. Ps. lxxxiii. 11).
Salome, 62 (note). 3rd husband of Anne.
Salome (Salomee), 140. The doubting midwife.
Samary, 350. Samaria.
Samaryan, 269. (?)
Sampson, 68.
Samuel, 63 (note), 68.
Sara, 68.
Sarazyn, 230.
Sarug, 37 (note). Serug (Gen. xi).
Sathan, 9.
Sathanas, 25.
Sawdyr Sadelere, 123.
Scottlonde, 199.
Scriba, 201.
Sella, 29 (note). Zillah (Gen. iv).
Sem, 36 (note).
Sephore, 93. One of the Virgin's handmaidens (cf. Pseudo-Matthew).
Seraphyn, 2.
Seruasius, 62 (note). Geruasius by mistake, grandson of Elyud brother of Elizabeth mother of John. (*Legenda aurea.*)
Sesar, 231.
Seth, 29 (note). (Gen. iv.)
Shem (Seem), 36.
Sophonye (Sophosas), 61. Zephaniah.
Spayn, 198.
Stevyn Sturdy, 123.
Susanne, 93. One of the Virgin's handmaidens (cf. Pseudo-Matthew).
Sybile Sclutte, 377.
Sybyly Slynge, 123.
Sym Somnore, 127.
Symeon Justus, 162.
Symeon Zelotes, 270.
Symme Smalfeyth, 123.
Symon, 62 (note), 250.
Symon leprows, 243.

Tarys, 152. Realm of Melchizar
Tharsis (Ps. lxxii. 10).
Thadeus, 251.
Thare, 37 (note). Terah (Gen. xi).
Thomas of Ynde, 15.
Thom Tynkere, 123.
Tobie, 63 (note). Tobias for Tobit (Tobit i).
Tyffany Twynkelere, 123.

Veronica, 296.

Walys, 199. Wales.
Whatt at þe Welle, 123.
Wyttsunday, 15.

Ynde, 15.
Ypotan, 153. Realm of Jasper (source of name not found).
Ysaac (Isaac), 3.
Ysaias (Ysaie), 57.
Ysakar, 65. (Isakar, 73.) High priest.
Ysakar, 63 (note). Father of Anne ('Achar' name given in Pseudo-Matthew. Isakar and Nazaphat parents of Anne in pedigree.—Harl. 6148).

Zabulon, 198. Country seen from the Mount.
Zabulon 'þe devylle of helle', 58. 'Terra Zabulon et terra Nephthalem ... populus qui sedebat in tenebris vidit lucem magnam.' In the 'Gospel of Nicodemus', cap. xiii. 9, the text in this form is applied to the Harrowing of Hell. Zabulon is thus associated with the 'regio umbrae mortis.' Zabulus occurs with the meaning 'devil' in *De Nugis Curialium.*
Zakarye, 116.
Zebedeus, 63 (note).
Zebee, 198. Country seen from the Mount (cf. Ps. lxxxiii. 11).
Zelomy, 140. The believing midwife.

GLOSSARY

This glossary cites, in most cases, the first occurrence only of the word in the text, and it includes only those variations in spelling which seem significant with regard to the date or the locality of the composition. Initial *ff* is given as *f*; *ȝ* follows *g*; *i* and *j*, and *u* and *v* are taken together.

a, 50, *have*; abene, 175, *have been*; abrowth, 281, *have brought*; affaylid, 40, *have failed*; ago, 186, *have gone*; asavyd, 176, *have saved*.
abey, 260, *obey*.
abhomynabyl, 100.
abought, 117, *about*; aboute, 39; abowth, 20.
abyl (v.), 72, *enable*.
accende (v.), 201, *kindle*.
acorde (v.), 58, *agree*.
ad, 14, l. 471, ? for *as*.
adon, 371, *adown*.
adrad, 68, *afraid*.
advowtrye, 203; adultrye, 9.
affendyn, 37, *offend*.
afferde, 196, *afraid*.
afflyght, 208, *afflicted*.
affye (v.), 357, *trust*.
aglottys, 227, *ornamental tags*.
agryse (v.), 36, *horrify*; agreseth, 307, *terrifieth*.
aȝen, 11, *again*; agayn; ageyn, 6.
aȝens, 2, *against*; ageyns, 21.
alkende, 184, ageyn nature and alkende, *kind*.
allether, 12, oure allether, *of all of us*; alderers, 322; althere, 217; altheris, 191.
almyght (adj.), 328, *almighty*. See myht, 359, l. 123, *mighty*.
alow, 253, *approve, accept*; alowyht, 33; alowe, 56, *permit*.
alwyse, 292, in alwyse, *in every way, inevitably*.
alyaunce, 211, syb of alyaunce, *related in blood*.
alye, 136, *kindred*.
alythe (v.), 297, *alight*. See lyght.
amat, 274, make ȝou amat, *dismayed*.
an, 76, *have*.
anameryd, 125, *enamoured*.
angelle (pl.), 1, *angels*; aungelle (pl.), 197.
anodyr, 64, *another*.

anow, 21, *enough*: inow, 253.
antecer, 80, *ancestor*.
anvemynyd, 67, l. 124, an vemynyd, *have poisoned*.
apayed, 59, *pleased*.
apparens, 255, to ȝour apparens, *in appearance to you*.
apperyd, 182, l. 122, ? *apportioned*.
applyande, 229, *applying*.
applyed, 239, *apply it*.
appose, 2, *confront, examine*; apposyn, 195.
aqwhyte, 49, *pay, acquit*; aqwhyte (pp.), 239.
arde, 196, *hard*.
are, 40, *hare*.
arere, 124, *raise*.
aresyn, 14, *arisen*.
arn, 30, *are*; ar, 95. See be.
arneys, 264, *harness, armour*.
arryn, 294, *torment, irritate*.
arwe, 40, *arrow*.
aseyth, 100, *satisfaction*.
ashyrlyng (adj.), 170, ? *a shrilling*. See shyrle, 113, *shrill*.
askuse, 2, *excuse*.
askyght, 144, *asketh*.
aslake, 224, oure deth ȝe may aslake, *cause to cease*. See slake.
asmatryk, 178, *arithmetic*.
asoyle, 34, *solve*.
assedually, 360, *assiduously*.
assumpte (? pt.), 354, ? *rose to heaven*; (pp.), 355.
ast, 50, *hast*.
astat, 11, *state*.
asyse, 53, þe last asyse, *judgement*.
atast (v.), 27, *taste*.
atent (sb.), 3, *entent*.
athreste, 179, *athirst*.
atreyd, 324, *vexed, afflicted*.
atwynne, 197, *in twain*.
auantoryslv, 310, *by chance*.
aventure, 75, per aventure, *chance*.
auerte (sb.), 80, ? *averter*.

avexit, 348, *vexeth*.
augrym, 178, *arithmetic*.
aunge, 80, 115, *angel*.
avoyd, 123, *stand aside, depart*.
autere, 65, *altar*.
avyse (sb.), 47, *advice, counsel*.
avysement, 247, *advice*.
awe, 9, 203, vndyr þer awe, *power*.
awyse (v.), 47, *advise*.

baftys, 170, *behind*.
bake, 48, *prepare, make ready*.
balke, 317, be brook or balke, *mound*.
bane, 314, *destroyer*.
baptyme, 188, *baptism*; baptyze, 318, l. 1371.
bargany, 253, *bargain*; cf. gracy, 75, towaly, 260.
barne, 147, *child*; baron, 154, 157, 172.
barrany, 65, *barren*.
barynes, 68, *barrenness*.
bat, 11, *debate*.
batte, 277, *stroke, blow*.
bawmys, 58, *of balm*.
baye, 170, be bestys baye, *stall*.
bayle, 273, *jurisdiction*.
bayn, 163, *ready*.
be, 3, *by*.
be, 2, *to be*; bene, 2; beyn, 228; byn, 218.
be, 14, *are*; byn, 24; beth, 43. *See* arn.
be, 165, *been*.
bede (pt.), 33, *offered*.
bede (pp.), 167, *awaited*.
bedellys, 226, *heralds*.
bedene, 2, *forthwith*.
beetys, 19, *beet plants*.
begchis, 355, *bitches*.
beglete, 155, l. 125, ? *for* begete, *begotten*; cf. vnbegete, 258. *See* glete.
begownne, 110, *begun*.
behete (v.), 350, *promise*.
behovyth, 135, *belongs*.
belde, 17, *build*.
beleve, 170, *remain*; belave, 370; belevyth, 328.
belthe, 371, *evil*. *See* note in Dr. Greg's *Assumption of the Virgin*.
belyff, 171, *quickly*; blyff, 2; belyve, 192.
bemys, 107, *beams*.
bende (sb.), 153, *bond*; 169, *band*.
berde, 247, 361, *woman, lady*.
bere, 22, *bear*; berun, 351; bar (pt.), 70; bore (pp.), 69; beryght, 137, *beareth*.
berst, 181, *breast*; bryst, 12; brest, 14.
beryed, 10, *buried*; burry, 215.
beryels, 16, *graves*.

berynt, 294, 296, *for* beryn it. *See* levynt, reysynt.
beschop, 87; bushop, 4.
beseche, 65; beseke, 121, 358.
beseyn, 230, *beseen, arrayed*.
bestad, 69, 150, 305, *appointed, placed*.
betake, 65, *commit*; beteche, 63; betake (pp.), 363.
bete (sb.), 23, *bite*.
bete, 170, on bete, *unbeaten*.
bethwen, 92, *between*.
bett, 40, *better*.
bety, 268, *beat*.
bey, 228, *buy*; bye, 58.
bey, 227, *boy*.
blaberyn, 154. *chatter*.
blad (pp.), 172, *bled*.
blake, 2, 19, blewe and blake (colour of flowers), *yellow*; cf. blayke, *Pearl*, l. 27.
ble, 169, *complexion, colour*; bryth as ble, 17.
blewe, 2, *blue*; bloo, 311, *livid*.
blomyght, 136, *bloometh*.
blyff, 12. *See* belyff.
blyn, 313. *cease*.
blysse (v.), 44, *bless*; blyssyd, 4.
bobbyd, 307, *buffeted*.
bone, 37; boun, 137, *boon, prayer, request*; 25, *command*.
bone (adj.), 27, þi byddyng bone, ? *good, gracious*.
bonet, 228, *cap*.
boorde, 328, his burryenge boorde, ? *his burial feast*.
boot (pt.), 26, *bit*.
boot (sb.), 26, *remedy*; bote, 4.
borwe (v.), 37, *save*.
borwe (sb.), 68, *pledge*.
borys, 297, *bored holes*.
bothers, 25; ʒour bothers, 25; here botherys, 71; *of both*
bountevous, 225, *bounteous*.
bourgh, 135, bourgh and cety, *town*.
bow, 21, *bough*.
bowne, 157, *ready, willing*; 248, *destined*.
bowth, 11, *bought*.
boyst, 329, *box*.
brace (sb.), 248, *embrace*.
brake, 19, *fern*.
brast, 15, *burst*; breste, 68.
braunce, 359, *branch*.
brayde, 218, *short space of time*.
brede, 40, *breadth*.
breganderys, 264, *body armour for foot-soldiers*.
breng, 258, *bring*.

Glossary

brent, 2, *burnt*; brenne, 248, *burn*; brynnyng, 188.
brest, 321, 339, brest ful bolde, *dauntless breast*.
brethellys, 287, *abandoned wretches*.
brethere, 364, *brothers*; bretheryn, 77.
brewe, 38, *bring about*.
bronde, 47, *brand, torch*; 173, *sword*.
bronston, 287, *brimstone*.
brook, 36, 197, brennyng brook, *hell*.
brothel, 205, *abandoned woman*.
brouth, 10, *brought*; brout, 10; brougth, 15.
brybory, 228, *bribery*.
brybour, 173, *scoundrel*.
brymmys, 152, *brinks*, by brymmys browne.
bryth, 1, *bright*; bryght, 33.
buske, 148, *set out, go*.
buxom, 46, *obedient*; buxhum, 94.
byche, 205, byche clowte. *See* bychyd.
bychyd, 368, *cursed, execrable*; bygyd, 368. *See* bicched, N. E. D.
bylle, 36, *list*.
bynne, 150, *manger*.
bysmare, 132, *shameless creature*.
byth (v.), 22, *bite*; boot (pt.), 26.
byware, 54, *beware*; bewhare, 54.

cadace, 227, *cotton wool*.
calabere, 228, *a kind of fur*.
call, 178, l. 24, be call, ?*by repute*. Cf. callyng, *proclamation*, *Cleanness*, l. 1362.
calsydon, 244, *chalcedony* (Rev. xxi. 19). *See* Note at end of Glossary.
cammaka, 154, *a fine fabric*.
care (v.), 363, *lament*.
careyn, 43, *carcass, corpse*.
carpynge, 147, *talking*.
cast (sb.), 121, *intention, purpose*.
cast (v.), 116, *intend, purpose*.
catel, 176, *property*.
celestly, 96, 367, *celestially*.
cenacle, 15 *See* List of Proper Names.
cevyle, 179, canon and cevyle lawe, *civil*; sevyle, 228.
chalys, 259, þese chalys, *chalice*.
chare, 303, 324, *drive away*.
chargyng (sb.), 89, *charge, command*.
charle, 131, *churl*.
chase, 20 (r. w. place), ?*for* chaste (v.), *discipline, correct*.
chastement, 228, *chastisement*.
chaumbyr, 39, *chamber*; chawmere, 107.
chauncel, 64, *chancel*.
chavyl bone, 33, *jawbone*.
chef, 228, l. 114, ?*chief*.
chene, 269, *chain*.

cheryse, 213, *cherish*.
chese, 48, *choose*; ches, 100, *chose*; chose (pp.), 133.
cheselys, 50, *pebbles*.
chete, 214, *sheet* (*corrected to* schete).
cheve, 151, *fare, succeed*; cheuith, 300, *befalls*.
cheverelle, 227, *kid leather*.
chevesauns, 228, *resource, shift*.
childely, 76, *childlike*.
chownys, 170. *See* schelchownys.
chyldere, 26, *children*; childyr, 74; childeryn, 7.
chyse (adj.), 170, *choice*.
clappyd, 316, clappyd in cloth; *for* clyppyd, *clasped, wrapped*. *See* N. E. D.
claryfieth. 96, *illumines*
clenche, 356, *affirm, insist*.
clepe, 63, *call*.
clergyse, 178, *learning*.
cleue (v.), 48, 143, *stick*.
cleue (v.), 156, *split*; clovyn (pp.), 40.
clos, 119, l. 98, ? in clos.
clowdys, 374, *clods*.
clymbyn, 170, ? A barn . . . wolde clymbyn kynge and knytys.
clyne, 106, *incline*.
clynge, 48, my hert doth clynge and cleue as clay, *shrink*
com, 29, *come*; cum, 63; comyn, 15; cam (pt.), 8; com (pp.), 65.
comawnde (v.), 278, *commend*.
comberaunce, 211, *trouble*.
combros, 325, *difficult*.
comeryd, 98, *cumbered, overwhelmed*.
comfortation, 108, *comfort*.
comfownys, 236, 298, 357, *commons, people*.
compellyd (pp.), 42, coupellyd (*coupled*). Emended by J. M. Manly on authority of Kittredge.
con, 163 (r. w. son), *can*; kan . . . thanke, 249; cone . . . thanke, 274, *thank*.
conclave, 13, *locked chamber*; his dede conclaue, 13, *tomb*; my conclave, 97.
concludyd, 275, *overcome in argument*.
conseyve, 246, l. 442, ?*for* constreyn. *See* p. 262, l. 913.
consorcyte, 358, *fellowship* (*consortium*); N. E. D. consorce, earliest quotation 1512.
contenawns, 254, 265, be contenawns, by *gesture, signs*.
contewnyng, 244, *continuing*; contenue, 232.

convycte, 9, *prove in error*; convycte (pp.), 359, *overcome*.
coors, 214, *corse*.
corde, 241, of corde, *accord*.
cordewan, 227, *Cordovan leather*.
cornys, 32, *corn*.
coryous, 173, *elaborately wrought*.
cost, 173, helle cost, *region, country*.
cost, 32, croft and cost, ? *field and hill-side*.
costyous, 227, *costly*.
cote, 88, *cottage*.
coverte, 132, *covering*.
covetyse, 198, *covetousness*.
counawnt, 279, *covenant*.
countyrfe, 226, *contrive*.
cownde, 288, *could*.
cowthe, 96, *known*.
craggyd, 355, *rough*.
credyl, 129, *cradle*; cradly, 179.
crenseyn, 227, *crimson*.
creppe, 36, 353, *creep*; crepe, 37.
cressetys, 254, *a form of lantern*.
crowch, 328, *cross*.
croyse, 147, *cross*.
cunnyng, 2, *knowledge*.
curat, 64, *pastor*.
curryd, 154, cruel and curryd, *for cursyd, curst, evil-tempered*.
curteys, 152, *courteous*.
curyng, 228, 364, *covering*.
cus, 81, *kiss*.
custom, 108, a custom ocupacion, *customary*.

dalyawnce, 127, *dalliance*; dalyauns, 203, *trifling*; 342, *talk*.
dampne, 9, *condemn*; dempt (pp., r. w. ment), 293.
datys, 21, *dates* (fruit).
daungere, 26, *danger*; 233, *power*.
dawe, 272, 275, bryng o (a) dawe, *slay*.
dawe, 279, the day dawe, *dawn*.
day, 236 (r. w. say), 313 (r. w. away), *die*; deye, 25; deyd (pt.), 222, *altered to* dyyd.
declinande, 229, *declining*.
dede (v.), 1, *did*; dyd, 7; dude, 129.
dede, 320, *death*.
defame (sb.), 127, *ill report*.
defendyd, 300, *for* offendyd. See defense *for* offence, N. E. D.
defens (v.), 58, *defend*; dyffende, 74.
defens (sb.), 121, *without defens, opposition*.
delacion, 234, *delay*.
delayde, 144, *mitigated*.
dele, 213, every dele, *bit*.
delve, 29, *dig*; dolve (pp.), 321, *buried*.

delyre, 193, *act madly*.
dene, 289, *din*.
dent, 153, *blow*.
depart (v.), 65, 204, *part, go asunder*; deth me xulde departe, 211, *dismiss*.
deppe, 36 (r. w. scleppe, creppe), *deep*.
deprave, 189, *decry*;· dyspravyn, 324.
dere (v.), 54, *injure*.
derth, 139, *dearness, costliness*; dyrthe, 175.
derwurthy, 193, *precious*.
desideracyon, 357, *desire*.
dete, 96, *ditty, song*.
detent (pp.), 362, *kept away*.
deve, 156, *stun*; devid, 322.
devele (pl.), 139 (*see* aungelle), dewelys, 371, *devils*.
devcyde, 229, *withdraw*; devoydyng, 240, *casting out*.
devyr, 112, *duty*.
devys, 8, *mechanical contrivance*; devyse, 132, bettyr it is to telle þe trewth devyse, *for þe trewe devyse. device, trick*.
diffynicion, 100, *limitation*.
diplois, 204, in deploydo, *doublet*. In a statute of 1433 of the Cistercian Order ecclesiastics are bidden to abstain 'a gestatione diploidorum'—among other secular garments.
displesaunte, 45, to þi displesaunte, ? *for to þe d., displeasing*.
dissponsacion, 81, *betrothal*.
do, 169, *cause*; 226, I gan þer do play, *I did cause there to be played*; 275, we han þe do sowth, *we have caused thee to be sought*.
dobbelet, 227, *doublet*.
dodemvsyd, 368, ? *amazed*.
dompnesse, 117, *dumbness*.
domys, 64, *judgements*; 178, *prescriptions*.
dowcet, 21, *sweet*.
dowe, 3, 166, *dove*; dove, 43.
dowm, 118, *dumb*; dum, 193.
dowse, 83, *sweet*.
dowth (sb.), 77, *doubt*; doute, 9; dowhte, 85; dowt, 173, *fear*; no dowth of, 77, 88, *without doubt*.
doyl, 42, *dole, sorrow*.
doyst, 261, *dost*.
dragonys, 272, 287, *dragons*.
drawe, 40, *draw*; drowe (pt.), 225; drawe (pp.), 246.
dredynge, 26, deth dredynge, *fear of death*.
drepe (v.), 161 (r. w. slepe), *drop*.
dresse, 61, *prepare*.

Glossary 387

dreve (pp.), 297, *driven*; drevyn, 60.
drewe, 32, 377, *morsel.*
drey (adj.). 159, *dry*; dreye (v.), 216.
dreynt (pp.), 39, *drowned.*
dronkesheppe, 377, *drunkenness.*
drynchyng (sb.), 38, *drowning.*
dur (inf.), 261, *dare.*
dure, 340, *hard.*
dwelle, 101, *abide.*
dwere (sb.), 15, *fear.*
dyet, 178, *diet, regimen.*
dyght, 40, *bring about, make ready*; dyth, 16; dyht (pp.), 31; dyth (pp.), 238; dyght (pp.), 162, *put, removed.*
dygne (sb.), 173, for dygne of my dygnyte þei haue of me dowt, ? *worth.*
dylexcion, 78, *love.*
dyng, 27, *strike, knock.*
dyngne, 154, *worthy.*
dyrknes, 164, *darkness.*
dyscres, 10, 211, *fall off, decrease.*
dysprave, 267 (r. w. rave), *disprove*; dysprevyd, 293; dyspravyn, 324. *See* deprave.
dyssend, 365, *descended.*
dystrye, 235, *destroy*; dystroyt, 231, *destroyeth.*
dyswary, 354, *doubt.*
dyvide, 241, *divided.*

echon, 19, *each one.*
edyfy (v. intr.), 237, *grow, prosper,* Mannys sowle in blys now xal edyfy; edyfied (pp.), 76, *built.*
eerly, 240, *earthly.*
effectuously, 352, *effectually.*
efne (sb.), 261, *heaven.*
egal, 260, *equal*; 255, eqwall.
elde (sb.), 65; (adj.), 255, *old.*
empere, 190 (r. w. manere), *emperor.*
empere, 307, *empire.*
empres, 108, *empress.*
enbrace, 87, 208, *embrace.*
encheson, 257; incheson, 108, *cause, reason.*
enclosyd, 227, ? hosyn enclosyd.
ende, 262, an ende to make, *agreement, compact.*
ende (adj.), 165, *for* hende, my savyour ende.
enform, 36, *inform*; 85, *explain.*
engynes, 226, *contrivances.*
enjonyd, 108, l. 340, *enjoined*; *or see* injouyid.
enjoyd, 117, *was joyful, rejoiced.*
ensure, 82, 112, *assure.*
entent, 5, *intention.*
enterly, 254, *entirely.*

er, 277, *before, ere.*
erbe, 19, erbys, 11, *herb.*
erde, 281, *heard.*
erde, 265, *earth.* See herd.
erdon, 263, *errand.*
erraunt, 192, *wandering, departing.*
erste, 98, *sooner.*
erte, 191 (r. w. smerte), *for* arte.
erthelech, 19; erthly, 27, *earthly.*
erthqwave, 307, *earthquake.*
ete, 20, *eat*; ete (pp.), 254 (r. w. fete), *eaten*; etyht, 251, *eateth.*
eternalyte, 269 (r. w. paraclyte), *eternity.*
ethe, 136, *easy.*
eve song, 120, *evensong.*
everich, 3; every, 13, *every*; everychon, 10, *every one.*
eueryth, 228, *or* eneryth, ? *inherit.*
evy, 46, *heavy.*
evyn-crysten, 77, *fellow Christian.*
evysum, 338, *heavy.*
exorte, 65, 241, *request, entreat.*
exort, 99, *issue, spring forth.*
expowne, 14; expownd, 85, *expound.*
exys, 254, *axes.*
ey (pl. ?), 75; eyn, 13, *eyes.*
eyd (sb.), 303, *heed.*
eye, 114, sen at eye; cf. *at hand.*
eyʒil, 302, *vinegar.*
eylight, 211, *aileth*; heylith, 131.
eylsum, 82, *wholesome.*
eynes, 98 (r. w. peynes), *eyes.*
eyre, 230, *heir.*
eyte, 122, *eight*; eyted, 76, *eighth.*

facyon, 370, *face.*
fad (pp.), 21, *fed*; fade, 159 (r. w. glade).
fade, 24, 51, *fade*; ffrom feyth nevyr þou ffade, 348, *grow weak, fall away.*
falfage, 34 (*for* falsage), *falsehood.*
falle (pp.), 98, *fallen.*
falle, 195, 327, ? *cause to fall*; felle (v.), 58, *fell, cast down.*
falsed, 9, *falsehood.*
fame (v.), 131, *defame.*
famyt, 97, *famished.*
fast (pt.), 226, *fasted.*
fawe, 274 (r. w. lawe), *fain, glad.*
fawte, 64, *fault*; for fawth of, 259, *default.*
fay, 14, in fay; faye, 299, *faith.*
fe, 173, ffryth and ffe, *property.*
fekyll, 329, *treacherous.*
felas, 67, 275; felawus, 228, *fellows, equals.*
fele, 21, 50, 345, *many*; in fele degre, 9.
felle, 177, 320, *skin.*

Glossary

fenaunce, 210, *end.*
fend, 58; fynde, 9, *fiend.*
fende (v.), 248, *protect.*
fenne, 157, 248, *fen, marsh.*
ferder, 231, *further.*
fere, 84, *companion*; in fere, 46, *together.*
fere (sb.), 19; fer, 23, *fear*; fere (v.), 64, *to fear*; ferying, 68 (*gerund*).
fere, 7, *far.*
fere, 15, 68; feyr, 2; fyer, 18; fyre, 19, *fire.*
ferforthe, 118, so forforthe ffor þat (*ecce enim ex hoc, for behold from henceforth*), *for* lo *forsooth*; cf. Wycliffite version, Luke i. 49.
ferly (adj.), 15, *marvellous.*
ferne (r. w. sterne), 158, be feld and ferne, ? *fern.*
fers, 76 *fierce.*
ferth, 161, *forth, forwards.*
ferthe, 19, *fourth.*
feryth, 96, *causeth to fear, aweth.*
fest (v.), 372, *fasten.*
fest (sb.), 15, 63, *feast*; ffeste and food, 326; *feast or fast*, 303, l. 894.
fett (v. inf.), 215, *fetch*; fettyn (3 pl. pres.). 371.
fewte, 329, *fealty.*
feyn, 231, *feign, make a pretence.*
feyn, 328, The naylis gun his lemys feyn, ? *make faint. See* feynnesse.
feyn, 227, *fine.*
feynnesse, 39, *faintness, weakness.*
feyt, 304, *feet.*
fise, 357, ? *term of abuse.*
flamke, 160, *flank.*
flem (v.), 262, *drive away.*
flescly, 24; flesly, 24; fleschly, 76, *fleshly.*
flokkys, 227, *stuffing for doublet.*
flomjordon, 8, *river Jordan.*
florens, 157, *florins.*
flowe (sb.), 38, *flow, current*; flowe (sb.), 319, *flood.*
flyth (sb.), 2, *flight.*
flytt (v. inf.), 176, *save, cause to flee*; flytt (pp.), 248; flytt (inf.), 315; flitte, 340, *flee, depart.*
fode, 305, now xal wepynge me fode and fede, *supply food to.*
fol, 310; fole, 295, *full.*
fole, 284 (r. w. to þe), *folly.*
folwe, 11, 36, *to follow*; folwygh, 202, *followeth.*
fon, 10, *foes.*
fond (v.), 22, *try, attempt*; fonde, 28, l. 386, leve spowse now þou fonde, *findest, provest.*

fonge (v.), 36, *begin, undertake.*
fonnyng, 284, *foolishness, madness.*
fonnys, 340, *fools.*
fonnyst, 32, *art mad.*
food, 140, *person, creature*; 185, *child, offspring.*
footmayd, 64, *waiting-maid.*
fop, 276, *fool.*
fordere, 226, *further.*
forfare (v.), 42, *be lost, come to destruction.*
form faderys, 256, *forefathers.*
forme, 202, *form of a hare.*
formere, 49, *former, creator.*
fors, 316, with fyne fors, *by main force*; 32, I ʒeve no fors, *I attach no importance to.*
forsake, 2; forsakyn (inf.), 36; forsake (pp.), 264, *avoid, forsake, deny.*
fortyfye, 226, his deth to fortyfye, *secure, ensure*; 292 (v. intr.), *resist, stand firm.*
foulyng, 286, *wretch.*
fourtene, 77, *fourteenth.*
fowth, 76, *fought.*
foyson, 58, *abundance.*
fraught, 129, *freight, burden.*
fre (adj.), 3, *noble.*
freke, 27, *man, fellow*; 303, þe fend þat freke.
frelnes, 101, *frailty.*
frelte, 195, *frailty.*
frensche, 110, *French.*
fryght, 149 (r. w. kyth), *for* fryth.
fryth, 138, *wood, forest.*
fullich (adv.), 58; fulleche, 263, *fully.*
furryd, 228, furryd with ermyn, calabere or satan, *lined.*
fyftene, 77, *fifteenth.*
fynyaly, 102, *finally.*
fynyte, 82, for *infinite.*
fyth, 228 (r. w. delyte); fytyn (inf.), 298, *fight.*
fytt, 176, *tune, song.*

gadere (3 pl.), 9; gaderyn (inf.), 21; gaderyd (pp.), 15, *gather.*
galaunt (sb.), 125, *gallant.*
gan, 2; gonne, 147; gun, 10; gunne, 319 (aux. v.), *did. See* gynne.
gast (2 sg.), 323, *gavest. See* gyff.
gate, 26; ʒatys, 28, *gate(s).*
gate, 46; gatt, 327, *way*; al gatys, 291, *at all costs.*
geawnt, 13, *giant.*
gef (inf.), 231, *give. See* gyff.
geneologye, 61; genealogye, 62, *lineage, pedigree.*
gent (adj.), 127, *pretty.*

Glossary

gentyl, 4, *noble*; jentylle, 59.
gentyllys, 1, *gentles.*
gerlys, 171 (knave gerlys), *children, young people.*
gerth (r. w. byrth), 175, on ground nere on gerth, ? *garth.*
geste, 155, *event, tidings.*
geste, 161, *guest.*
gesyne, 141, *childbed.*
gett, 302, *fashion, manner.*
geyne, 368, *meet, oppose.*
glabereris, 368, ? globaris, *gluttons; or for* glavereris, *deceitful talkers.* See Dr. Greg's *Assumption of the Virgin.*
glathe, 161, 222, *welcome, gladden.*
glent (pt.), 361, *shone.*
glete, 155, *mud, slimy matter.* See Note at end of Glossary.
gleterynge, 152, *glittering.*
gleyvis, 254, *glaives, weapons.*
glose (sb.), 8, *gloss, commentary*; 373, *explanation.*
glyde, 26, *crawl*; 155, *go, pass.*
gnaggyd, 355, *knotted, tied up.*
godhyd, 81, *godhead, deity.*
godnysse, 94, *goodness.*
gomys, 169, 355, *men, fellows.*
gonge, 319, in helle gonge, *pit, latrina.*
gouernayl, 361, *government, direction.*
grace (sb.), 9; gracy, 75; gracys, 79; gras, 82, 94, *grace.*
graffe, 314 (r. w. saffe), *grave.*
grame (v. inf.), 24, 202; gramyd (pp.), 133, *make angry, enrage.*
grame (sb.), 2, *wrath, anger.*
gramercy, 30; gromercy, 242, *thanks.*
gramly, 368, *fiercely.*
gratulacyon, 357, *thanks.*
graunt (pp. r. w.), 209, *granted.*
graue (pp.), 214, 223, *buried.*
grecys, 77, *steps, stairs* (Psalms 119-133, Wyclif. Version, The Song of Grees *or* Greces)
grede, 171, 334, *cry out.*
grees, 74, *steps.* See grecys.
greff (r. w. sheff), 32; greve (r. w. preve), 143, 151, *grief*; at greff, 253, *amiss.*
gres, 26; gresse, 19, *grass.*
gret (pt.), 115, *greeted, saluted.*
grote (sb.), 88, *groat.*
ground, 14; grounde, 34; grownd, 32; grownnde *or* growunde, 316, 347, *ground.*
grownde (pp.), 139, *grounded.*
grugge (v.), 215, *grudge.*
grw, 161, *Greek.*
grylle (adj.), 217, *fierce.*

gryse, 149, grysly as a gryse, ? *pig, boar.*
grysly, 149, *horribly.*
gryth, 7, *peace.*
gyff, 98; gyf, 226; gef, 221, *give*; geve, 252; gevyth, 77; gast (2 sg.), 323.
gyldyn, 69, *golden.*
gynne (1 pl.), 16; gynnyth, 1; gun (pt.), 10, *begin.*
gynne (sb.), 39, *gin, trap, plan of destruction.*
gynny, 156, ? *scribal error; altered by reviser to* wonyt.
gyse (sb.), 11, *way, manner of life.*
gysse (v.), 360 (r. w. wysse), *guess.*

ʒa, 38, 67, *yea, yes.*
ʒardys, 86; ʒerde, 5, *rod(s).*
ʒede (v. pt.), 182, *went.*
ʒelde (v.), 91, *yield, give.*
ʒemanry, 1, *yeomanry.*
ʒenge, 139, *young*; ʒonge, 182.
ʒer (pl.), 8; ʒere (pl.), 4; ʒerys, 72, *years*; yer, 374.
ʒerd, 278, *garden.*
ʒevene, 165; ʒevyn, 167, *even.*
ʒewys *for* jewys, 307.
ʒiff, 102; ʒyff, 64; ʒyf, 252; ʒef, 272, *if.*
ʒit, 13; ʒitt, 30; ʒyt, 10, *yet.*
ʒonge, 44, *youth*; ʒough, 76; ʒoughʒ, 183.
ʒyf (v. 1 sg.), 29, *give*; ʒevyn (inf.), 32; ʒeve (inf.), 54; ʒevyn (pp.), 21; ʒovyn (pp.), 77; ʒove (pp.), 73; ʒaff (pt.), 103; ʒove (pt.), 144; ʒovyn (pt. pl.) 338.
ʒyft, 31, *gift.*
ʒyng, 71; ʒynge, 59, *young*; cf. ʒenge.
ʒys, 276, *yes.*

[Words having an adventitious initial aspirate, not otherwise needing explanation, are not included.]

ha, 34, *have*; han, 33; haht, 318, *hath.*
hald, 293, *hold*; halt, 293, *holdeth.*
halpe (v.), 301, *help.*
halse, 300, *embrace.*
halwe, 54, *keepholy, hallow*; halwyd, 96.
hand-maydeʒe, 118, *handmaiden.*
haras, 138, hous of haras, *stable for a stud of horses.*
hardely, 314, *certainly.*
harrow, 199, *the devil's cry.*
harwere, 150, *harrower.*
hat (sb.), 11, *hate.*
hattyht, 147, *is called.* See hyght.
hay, 154, *net for catching wild animals, or hedge.*

hede, 10, *hide*; heyde, 41; hede (pp.), 159, *hidden*.
hedyr, 67, *hither*; hider, 361.
hefly, 240, *heavenly*.
hele, 211, *health*.
helpyngys, 116; cf. Wyclif. Version, 1 Cor. xii. 28, aftirward graces of heelingis, helpyngis, . . . kyndis of langagis.
hem, 2, *them*; þem, 2.
hem, 159, 228, *him*.
hende, 5, *gracious, gentle*; hynde, 57, 113; hende, 358, *near at hand*; fer and hynde, 113, *far and near*; hendyr, 81, *nearer*.
hens partyng, 31, *death*.
hent, 87, *take, hold*; hent (pp.), 11.
herborwe (sb.), 137, *shelter*; (v.), 227; herboryd (pp.), 138.
herd, 264, *earth*.
herde, 147, *herdsman*.
here, 2, *their*.
here, 4. *her*; hire, 71.
here, 227, *hair*.
herne, 371, sere sathan in the herne, *corner*.
hert, 40, *hart*.
hertely, 36, 122, *hearty*; hertyly, 31, *heartily*.
heryght, 327, *heareth*.
hese, 1, *his*; *more usual form* his.
hest (v.), 40, *promise*; hestyd (pp.), 85.
hete (sb.) (r. w. mete). 175, *hit, blow*.
hevy (v.), 93, *make heavy*; hevyin, 363, *grow heavy*.
hey, 44, *high*; hey3, 131; hie3, 135; hy, 11; high, 58.
heyd (sb.) (r. w. dede), 49, *heed*.
heyn, 169, *villain, wretch*; cf. Chaucer, C. Y. Tale, 1319.
heyued, 76, *lifted*.
ho, 68, *who*; hoo, 52.
ho, 100, byddyth him ho, ?*bid him cease*.
hoberd, 169, 302, *a term of reproach*, ? *clown*
holond, 227, *holland cloth*.
hond, 22 (r. w. husbond), *corrected from* hand.
honge, 303, *hang*; hyng, 12; henge (pt.), 338; hynge (pt.), 320; hangyn (pp), 236.
hont (v.), 154, *hunt*.
hool, 16, *whole*.
hors, 246, 272, *horses*.
howeth, 89, *oweth, ought*.
howlott, 169, *owl* (*here used for woman*).
howte, 172, *hoot*; howtyn, 169.

hurle, 368, *strive, contend*.
hy (v.), 6, *hast*; hy3, 27.
hyde, 303, *skin*.
hyght, 6; hyth, 15; hygth, 69; *be called*; hyth, 116, *was called*. *See* hattyht.
hy3ht, 7, 87, 165, in hy3ht, *in haste*; on hyght, 107, ? *in excelsis*.
hylle (adj.), 248, *ill*.
hylle (v.), 33, 238, *cover*; hylte, 374, *hidden*.
hyn, 298, *him*.
hynde. *See* hende.
hyth, 25, *promised*.
iche, 86, *each*; ilke, 170.
ichon, 135, *each one*.
jematrye, 178, *geometry*.
jentylle, 59, *noble*; 109, *gentle*.
ierarchie, 98, *hierarchy (of angels)*.
jewus, 301 (r. w. þus); jewys, 301 (r. w. þis), *Jews*.
i-fownde (pp.), 149, *found*.
i-knowe (pp.), 132, *known*.
ilke, 170, *each*.
inbassett, 70, *embassy*.
incheson, 108, *occasion, cause*; encheson, 251.
indeploydo, 204 (Latin st. dir.). *See* diplois.
indute, 193, *clothed*.
inhabith (pp.), 354, *settled, established*.
injouyid, 118, *rejoiced (exultavit)*; enjoyd, 117; injoyeth, 358; injoye, 373. *See* enjonyd.
inow, 253, *enough*; 21, anow.
inportable, 272, *unbearable*.
i-nvm (pp.), 149, *taken*.
irke (v.), 168, *grow weary*.
juge, 116, hese juge, ? *See* Intro., p. xliii.
i-wys, 7, *certainly*; i-wus (r. w. þus), 122; wys *for* i-wys, 73, l. 32.
kachyd, 274, *caught*.
kage, 153, 156, *scaffold, pageant, seat of honour* (*Prompt. Parv.* cage, catasta); cf. myn hy3 cage, *Hymns to the Virgin and Christ*, ed. Furnivall, E. E. T. S; castell and cage, *Castle of Perseverance*.
kan, 140, þat kan 3our good, *know what to do*.
kayser, 173, *emperor*; caysere, 190.
ken (v.), 117, *acknowledge*.
kend (sb.), 148, *kind, nature*; kendely, 30, *natural*; kende, 80, *race*. *See* kynde.
kepe (sb.), 171, *care*.
kepyng knyght, 34, *keeper, protector*.

Glossary

kerchere, 49; kerchy, 296, *kerchief.*
kest (pt.), 278, *cast.*
knad, 355, craggyd knad (Halliwell, *knife*), ? *for* knag, *stump*; hence *bludgeon.*
knawe (r. w. awe), 79, 355, *know*; knawe (r. w)awe), 284, *known*; knove, 267 (r. w. lowe), *know.*
knawe, 177, 272, *gnaw.*
knelande, 355, *kneeling*; knelende, 67.
knett (pp.), 326, *knotted, fastened*; knyt, 17, *joined.*
knowlych (v.), 60, *acknowledge*; knowlage, 115.
kure (v.), 49, *cover.* See curyng.
kynde, 119, *generation (progenies)*; 39, *species.*
kynde (adj.), 149, *natural, native.*
kyngdham, 229, 350, *kingdom*; kyngham, 282.
kynnys, 110, in no kynnys wyse, *by no means.*
kyrke, 168, *church*; cherch, 55; chirch, 121.
kyth (sb.), 149, kynde in our kyth, *kindred.*
kythe (v.), 170, *make known, show*; kydde (pp.), 45; kyd (pp.), 125, *well known.*
kytt (v.), 345, *cut.*

lacche (v.), 26, *gain, receive.*
lak, 129, withowtyn lak, *without fail.*
lake, 288, in lake, *lake of Hell*; 324, 333, 334, 370, *pit, grave*; 359, lake of lyonys, *pit.*
lappe, 343, so nere our lappe, *so near our persons.*
lappyd, 117, *closed, bound.*
las (r. w. was, pas), 14, *less.*
las (sb.), 25, *net, snare.*
lasse, 19, *lest.*
lat (imp. s.), 47; late (imp. p.), 65, *let.*
latyng, 243; lettyng, 4, *hindrance.*
laue (r. w. haue, raue, saue), 91, whyte as laue. *See* Note at end of Glossary.
lave (r. w. knawe), 355, *law*; lawe, 4; lay, 55, 377 (lede no lecherous lay), *rule of life, conduct.*
lawhȝ, 125; lawgh, 141, *laugh*; lowh (pt.), 21.
lay, 154, the lesse lay, *laity, ignorant people.*
lech (v.), 179, 192, *heal. cure.*
lechory, 42; lychery, 228, *lechery.*
ledys, 173, *peoples.*
lees, 328; les, 331, *falsehood, deceit.*

lef (v.), 32; leff, 164, *leave.*
lefful, 82, *lawful.*
leke (adj.), 289, *like*; lycke, 274.
lely, 90, lely whyte, *lily.*
lemys, 163, *limbs*; lyme, 160.
lenage, 183, *lineage, pedigree*; lynage, 58.
lende (v.), 160, *land, arrive.*
lenger, 81, *longer.*
lenyall, 60, *lineal.*
lere (v.), 24, *learn.*
lese (inf.), 48. *lose*; lore (pp.), 56; lorn (pp.), 42.
lesse than (conj.), *unless.*
lest (v.), 20 (*corrected from* last, r. w. best); leste, 68; lestyght (3 sg.), 92, *last, endure.*
lest, 126, 179, *list, listen*; leste (imp.), 84; lyst (imp.), 45.
lesyng, 4, *deception.*
lete (v.), 5, *cause*; lete, 28, late, lett, 72, *let.*
lete (v. pt.), 16, *leave.*
leve (v.), 233, *believe.*
leve (inf.), 26, *live*; levyn (inf.), 26; lyff (inf.), 30; levyth (3 sg.), 47; levyd (pp.), 71.
leve (adj.), 28, *dear*; levyr, 30, *rather.*
levers, 106, *livers.*
levyn, 146, *lightning.*
levyng, 29, 31, *living, life.*
levynt, 258, i. e. levyn it, *leave it*; cf. reysynt.
levyth (pp.), 256, *left.*
lewd, 62, *ignorant*; lewdness, 144, *ignorance.*
leyke (v.), 148, *go quickly.*
leyn (inf.), 276, *lay, pledge*; leyn (inf.), 324, *lay, cast down.*
leysere, 184, *leisure*; 298, leysere seyng, *slowly, mournfully* (N. E. D. *leisurely*).
lofsummere (adj.), 151, *more lovesome.*
lofte, 76, on lofte, *aloft, on high.*
logge, helle logge, 26, 176, 305, 319, *prison of hell.*
lokygh, 154, *locketh.*
lokyn (pp.), 17; loky *for* lokyn, 26, *locked.*
longe (v.), 36, *belong*; longyht (3 sg.), 57; longygh (3 sg.), 199..
loof, 195, *loaf*; 345, loff; ? 91, laue.
losel, 33, *scoundrel.*
loth (sb.), 329, *evil*; loth (adj.), 338, *wicked.*
loveday, 103, *day of agreement.*
lovnesse, 357, 359, *lowness, humility.*
lowlyte, 329, ? *for* lewte, *loyalty* (r. w. fewte).

lowte (v. 3 pl.), 53; lowth (inf.), 156, *bow.*
loyn (pp.), 13, 97, *lain*; loyn (1 pl. pt.), 274, *lay.*
lullay, 129, *lullaby.*
lullyd (v. int. pt.), *slept peacefully.*
lurdeyn, 41, *lazy person.*
lyberary, 80, we xal lerne you þe lyberary of oure lordys lawe, *the whole of, the canon.*
lyce (v. 3 sg.), 85 (r. w. servise), *lies.* See lyggyst.
lyche, 154 (r.w. dich); lycke, 274, *like.* See leke.
lyfflad, 1, *life.*
lyfte (pp.), 75, *lifted.*
lyfte (adj.), 294, *left (hand).*
lyggyst (2 sg.), 150, *liest*; lyght (3 sg.), 140; lyce (3 sg.), 85; lyth (3 sg.), 7, *lies*; lyne (inf.), 214 (3 pl.), 172.
lyght (v. pt.), 320, I lyght out of, *lit, alit*; lyth (pp.), 5; (inf.), 241; lyhtyn (inf.), 157.
lympe, 160, *lump.*
lymyd, 56, handys lymyd; cf. *later expression* lime-fingered, *given to pilfering.*
lynacion, 178, *measuring.*
lyste (sb.), 161, 309 (r. w. resquest), *desire, liking*; lyst (v.), 67.
lyst, 235, *lest.*
lytenyth (3 sg.), 96, *lighteneth, illumines.*
lyth (sb.), 17; (adj.), 68, *light.*
lyty *for* lytyl, 147.

maculacion, 133, *spot, stain,* fig. *defilement* (1st example N. E. D.).
make (sb.), 19, *mate.*
make (inf.), 1; makyn (inf.), 3, *make*; made (pt.), 1; mad (pt.), 58; makyght (3 sg.), 172.
males, 9, *malice.*
mankende, 53, 60, *mankind, men*; 163, 302, *human nature.*
mankynne (r. w. begynne), 148, *mankind.*
mansclawth, 290, *manslaughter, murder.*
many, 87, *for* man.
marryn (inf.), 153, *mar, destroy.*
massage, 43, *messenger.*
massage, 108, *message.*
mast (v.), 219, *mayest.*
mavmentryes, 371, *idolatries.*
mayn (sb.), 22, *power.*
maystrye, 179, *mastery*; maystryes, 199, 226, *cunning tricks.*
me, 220, *men.*

meche, 68, *great (much).*
mede (sb.), 16, *reward*; mede (v. inf.), 50.
medele (v.), 34, *have to do with*; 148, *mingle.*
medylerth, 26, *earth, world.*
meke (v.), 97, *make meek, temper.*
mekell, 159, *great*; mekyl, 18, 125; mykyl, 172.
melle (v.), 184, *declare, utter.*
melle (v.), 18, *mingle.*
melyon, 266, *million.*
men (v. inf.), 87, *mean.*
mende (sb.), 7, *mind* (haue mende = *know*); mendys, 226, *thoughts, memories.*
mene, 108, l. 327, *mine, my.*
mene (sb.), 166, 364, *mediator.*
mene (sb.), 369; meny, 254, *company, gang.*
mercyabyl, 100, *merciful.*
mere (sb.), 162, 329, *boundary,* hence *way, place.*
merke (v.), 195, *darken.*
merveyl (adj.), 90; mervaylle, 224, *marvellous.*
mete (sb.), 8, *food, meal.*
meth, 147 (r. w. breth, deth), *moderation, gentleness.*
methowut, 274, *methought.*
meve (v.), 192 (r. w. greve); meef (r. w. breff), 229, *move.*
mo (adv.), 26; moo (adj.), 109, *move.*
mokador, 179, *bib, napkin.*
molde (sb.), 1, *earth, world.*
mon (sb.), 110, *moan.*
mon (aux.), 277, *may, shall.*
monyth, 106, *month.*
moote (sb.), 4, *dispute.*
morny (adj.), 97, *mournful.*
mote (aux.), 45, *may*; moty *for* mot I, 29.
mothalle, 278, *moot hall, judgement hall.*
mow (aux.), 99, *may*; mowne (pl.), 374, *may, can.*
mowe, 302, *grimace.*
mownth, 294, *mount.*
mullinge, 151, *term of endearment* (1st example N. E. D.).
mure, 358, 370, *demure, modest.*
muse (v.), 131, *think, consider.*
muste (sb.), 353, *new wine.*
mut *for* mot, 65, 118, *may.*
myche, 153, on a myche? *See* Note.
myche, 356, *much.*
myght (sb.), 35; myghtys, 73; myth, 1, 226, *might.*

Glossary

myht (v.), 31; myth, 2, *might*.
myht, 359, *mighty*. See almyght.
mynstrelle (pl.), 152; mynstralle (pl.), 174; menstrelle (pl.), 176, *minstrels*.
myrable, 360, *wonderful*.
myrke (adj.), 161, *dark*; myrkenes (sb.), 217, *darkness*.
mysse (sb.), 38, *wrong, injury*.
mys (v.), 7; mysse, 44, *miss*.
myscheve (v.), 127, *to come to grief, go astray*; myschevyd (pp.), 100, *in misery*.
myschyf (sb.), 18, *evil plight*; myschevys, 98, *miseries*.
mystyr, 132, *kind*.

nale, 54, at þe nale, *at the ale (house)*.
nat, 25; natt, 56, *not*. See noth.
ne, 20, *nor*. See nere.
negramauncye, 178; nigramansye, 266, *witchcraft, necromancy*.
nemene (v.), 164 (r. w. evene), *name*; nempne (r. w. hevynne), 47.
nere, 40, 228, *nor*.
nesch, 28, *soft, tender, weak*.
nest, 107, *next*.
neyhand (adj.), 162, *near at hand*.
neyth (v. inf.), 226, *draw near*.
neyther (. . . nor); 32; neythyr (. . . nore), 53, *neither*. See nowther.
nome (pp.), 88, *taken*. See i-nvm.
norche (inf.), 196, *nourish*; norchyth, (3 sg.), 77.
norsshere, 225, *nourisher*.
norture, 29, 358, *nurture*.
noth, 11; nott, 67; notht, 109; notwh, 299; notȝ, 305; nought, 24; nowght, 41; nowth, 11, 97 (r. w. wrought); notwth, 233, 267, *not*.
novelle, 346, *news*.
nought, 75; nowht, 38; nowth, 20, l. 126; not, 68, *nought*.
nowther (. . . ne), 56; nother (. . . ne), 87; noyther, 253, *neither*. See neyther.
nowthty, 34, *worthless*.
noyis, 76, *noise, sound*.
noyous, 76, *grievous*.
nyghe (v.), 156; nyhyn, 359, *draw near*. See neyth.
nyn . . . nyn, 62; nyn, 117, *nor*.

o, 2, *one*; oo, 17; on, 21.
obecyon, 281, non obecyon of errour, *obstacle* (Fr. *obicion*. See Godefroy, *Dict. de l'anc. langue française*). Not in N. E. D.
obeschyauns, 357, *obedience*; obeschaunce, 367; obedyens, 57, 75.
oble, 255, *sacramental wafer*.
oblocucyon, 62, *bad delivery* (N. E. D. in this case only); or *interruption* (Halliwell, Latin obloqui, *to interrupt a speaker*).
ocapye, 352, *occupy*; occapyed, 81.
odyr, 264, *other*.
ogyl, 368, *shudder*.
olyff, 186, *in life, alive*; on lyve, 323.
onethys, 137, *with difficulty*.
onpossible, 180, *impossible*.
ony, 96, *honey*.
onyth, 227, *at night*.
or (conj.), 48, *before*.
ordenaryes, 79, *ecclesiastical officials*.
ore, 71, *grace*.
ore, 299, *for* oure, *our*.
ostage, 137, *hostelry*.
outh, 359, 367, *aught*; owght, 81; owgth, 86; owught, 115.
outrage (v. int.), 127, *commit excess, transgress*.
ouyr (v.), 8, *hover*.
ovyrdon (adv.), 21, *exceedingly*.
ouyrlede, 245, *domineer over* or *lead astray*.
ovyth, 46, *behoveth*.
owe (adj.), 25, *own*.
owth, 79, *out*; owughte, 131 (r. w. aboute).
owtrage (adj.), 35, *violent*.
owyght, 92, *oweth, should*. See howeth.
oyle, 262, oyle of mercy.

pace (v.), 12, *pass*.
pad, 154, *toad*; padde, 175.
page, 13, *slip for* pagent.
page, 174, *fellow, knave*.
paphawk, 154, 169, *suckling* (only instances given in N. E. D.); Halliwell *parrot* (? *for* papjay, popinjay).
paramoure (sb.), 44, *darling*, (Abraham, of Isaac)
parayl, 231, *equipment, ways and means*; 253, *apparel*.
parlement (of hefne), 82, 264, *Divine Council*.
parochonerys, 64, *parishioners*.
partabyl, 259, *capable of sharing*.
pas (sb.), 74, *pace, step*.
pasche, 244, *Easter*.
passage, 106, 108, *period of pregnancy*.
passent, 345, we went with hym evyn passent, *journeying*.
pawsacion, 81, *pause*.

Glossary

pay, 43, *pleasure, satisfaction*.
payd, 72, *pleased*.
pelle, 158, 232, *fur, furred cloak*.
pende, 169, prevyn pychyn and topende; 302, with stronge peyn yt gynnyth to peynde, *press, pinch*, (N. E. D. dial.).
pepyr, 20, *pepper*.
perchyn, 225, *pierce*; perysch, 181.
perdure, 239, *continue*.
pere (v.), 123, *appear*.
pere (sb.), 20, *pear*.
persevere, 106, *for* perservere, *preserver*.
pertly, 1, *briskly, readily*.
perverte, 232, *perverteth*.
perysch, 181, *pierce*.
pet, 371, *pit*.
pete (sb.), 9, *pity*.
pete (v.), 25 (r. w. ete), *put*; 155 (r. w. wete), *put* or *pity*.
peusawns, 245, peusawns of pepyl, *crowd, number of*.
peys, 223, *weight*.
pillid, 355, *bald*.
plasmacion, 180, *fashioning, creation*.
playn, 207, telle us þe playn, *the plain case*; 237, *openly*.
pleand, 62, *acting*; pleyand, 355.
plentevously, 35, *plenteously*; plentevous, 227.
plesande, 229, *pleasing*.
plesawns, 17, *pleasure*.
pleson, 101 (r. w. reson), *pleasure* (not in N. E. D.).
plete (v.), 175, *plead*.
pleyn (v.), 197, *act*.
pleyne (adj.), 157, poundys pleyne, *full*; 356, 370.
pleynge fere, 107, *playfellow*.
pleynyn, 112, *complain*.
ply, 232, *work at, carry out*.
plyght (v.), 144, *fold*.
plyth (v.), 2, *promise*; plyghtys (1 sg.), 170; plyght (pp.), 56.
plyth (sb.), 12, *plight, state*; plyght, 37.
poer (sb.), 282, *power*; pore, 119.
poer (adj.), 245, 249, *poor*; pore, 24; power, 144, 249; porys, 99, *of the poor*.
popetys, 169, *dolls, puppets*.
portature, 30, *figure, form*. (See N. E. D. portraiture.)
portys, 50, *gates*.
postelis, 372, *apostles*.
povert, 64, *poverty*.
pouste, 17, *power*; pooste, 339.
powndys, 157, *ponds for fish*.

praty, 96, *pretty*.
pray, 300, to delyuere man fro þe develys pray, *action of preying, capture* (earliest instance in N. E. D. 1523); pray, 204, *prey*.
prayr, 89 (r. w. her), *prayer*.
prayt, 233, *prayeth*; preyand, 355, *praying*.
precepte (pp.), 255, *prescribed* (earliest instance in N. E. D. 1531).
pref (v.), 289 (r. w. belef), *prove*.
prendyd, 175, ? *taken*; Halliwell, *pricked* (? *for* prened).
prerogatyff, 107, *prerogative, peculiar right*.
present, 235, *now, at this place*.
prest, 11, *ready*.
presyn, 346, *prison (altered from* preson *to rhyme with* aresyn).
pretende, 74, *aim, direct*; 358, *hold out, offer*; 359, ? *make an attempt*; 362, pretendist, *intendest*.
prevyde (v.), 102, *look* (meaning not given in N. E. D.).
prevydens, 81, ? *for* provydens, *provision*.
prevyn, 154, 169, *put to the proof, test*.
prise, 37, 149, of prise, *excellent, precious*.
promyssyon, 58, *promise*; promiscion, 350.
pronunciation, 62, *delivery, declamation*.
properyd, 230, I am properyd, *mine is the property, the right*; N. E. D. proper (v.), *to make master of*.
prose, 231, *story, narrative*.
provaylys (v. 3 pl.), 226, *for* prevaylys, *avail, benefit*.
prow, 111, 308, *advantage*.
prune, 154, *trim, preen*.
prynsesse, 252, *princes*; prysis, 355.
psalmus, 96; psalmys, 74, *Psalms*.
pshalmodyeth, 260, *sing psalms*.
punchement, 100. *punishment*; punchyth, 68, *punish*.
punche, 328, *pierce*.
purvyauns, 228, *provision*.
pyan, 20, *peony*; pepyr, pyan; cf. peper and piones ... hote spices, *P. Pl.* B. v. 312.
pychen, 169, *stab. See* pyth.
pygth, 96 (r. w. with), *pith*.
pyke (v.), 91, *pick*.
pylis, 199, *towers*.
pylle, 277, whele and pylle, *a game*.
pyn (sb.), 142, *suffering*; 211, pyne.
pynde (pp.), 28, to pynde, *wasted*.

pynne (sb.), 129, *centre of the target*; cf. prycke, 40; pynne, 197, *peak*.
pynne (v.), 24, 251, *shut up, confine*.
pyth, 12, *thrust*; pyth (pp.), 1, *placed, fixed*; pyght (pp.), 6, 144, 196; pyht (pp.), 18.

qwalle, 60, *whale*; whallys (gen.), 340
qwan, 241, *when*; quan, 109.
qwart (sb.), 211, *health*; qwerte, 191, 344 (adj.), *whole, sound*.
qweche, 259, *which*; qwhich, 216; qwyche, 331. *See* weche.
qwed (adj.), 13, *evil, wicked*.
qwedyr (v.), 114, *quiver*.
qweke (v.), 27 (r. w. freke), *quake*; qwake, 15.
qwelle (v.), 12, *kill*.
qwelp (sb.), 45, helle qwelp, *whelp*.
qweme (v.), 101, *unite, reconcile*; queme, 361, *please, be agreeable*.
qwenys, 171, *women, queans*.
qwens, 75, *whence*.
qwere, 269; quer, 365, *choir*.
qwere, 102, *where*; qwher, 278.
qwethynge, 335, quyk and qwethynge, *speaking* (earliest quotation of this phrase N. E. D. 1529).
qwhethe, 204, *bequeath, consign*; qwethe, 213.
qwhyl (conj.), *while*.
qwy, 112, *why*.
qwyght (adj.), 336 (r. w. syght), *free, clear*; (v. inf.), 16, 39, *requite, reward*.
qwyke, 96; qweke, 227; quyk, 335, *quick, alive*.
qwyle (sb.), 65, *time*.
qwyppys, 294, *whips*.
qwyte (adj.), 19, *white*.

race (v.), 185, *break asunder*.
raftys, 170, *beam, pole, shaft (of a spear)*. *See* rakynge.
rake (v.), 173, rake þem on rought, *rake, sweep*.
rakyl (adj.), 21, *hasty, unstable*.
rakynge, 170, with rakynge raftys, *dashing, violently moving*.
ransake, 143, *search carefully*.
rape (sb.), 169, *violence*.
rape (sb.), 218, *haste*; (v.), 369, I renne I rappe; (imp.), 373.
rapely, 374, *swiftly, quickly*.
rappe (v.), 173, *smite, hit*.
ravaschyd, 348, *transported*.
ray (sb.), 170, a reed ray, *aphetic for* array *or* ray, *dance* (dance of blood).

ray (adj.), 233, ray tabardys ... ray hodys, *made of ray, i. e. striped cloth*.
rebate (v.), 69, *abate*.
reclyne (v.), 132 (r. w.), *to return to earlier condition* (quoted in this sense N. E. D.).
record (sb.), 40, record of my boy (absol. phrase), *my boy bearing witness*.
recure (v.), 82, *win, obtain*; recuryd (pp.), 226, *recovered, restored*; recuryn (inf.), 281, *restore to life*.
reddure, 355, *rigour, severity*; redrure, 239, ? *for* reddure.
rede (sb.), 5, *counsel*; reed, 23.
rede (v.), *counsel, advise*.
redolent, 80, *sweet-smelling*; redolens (sb.), 241, *perfume*.
reducyd, 263, ? *brought back again*.
reed, 48 (r. w. god), *for rod*.
reed (adj.), 170, *red, blood-stained*.
regyon, 80, Regina of regyon, *rule, government*; 102, *region*.
rejoyse (v.), 56, *enjoy*; (sb.), 377, *joy, pleasure*.
relacion, 82, *regard, consideration*; 38, *report*.
relefe (v.), 356, *remain*. *See* relevys.
relese (v.), 195, *relieve, alleviate*.
releve (v.), 220, *relieve, assuage*.
relevys (sb.), 81, *remains*.
rem (r. w. bedleem), 146, *realm*.
remeffe, 229, *remove*.
renne, 10; ronne, 10, *run*.
replye (v.), 256; replyeth (imp. pl.), 360, replyeth and assedually wachith, *apply*.
repreff, 32; repreve, 127, *reproof*.
reprevable, 232, *conducing to the reproof of*.
rere (v.), 56, *raise up*.
resch (sb.), 161, resch and root, *rush*; rosch, 28.
rescu (sb.), 106, *rescue*.
rese (v.), 372, *rose*; reson (pp.), 312, *risen*.
responcyon, 355, *answer, response* (earliest example N. E. D. 1502).
respyt, 113, *delay*.
rest (v.), 228, *remain*.
resydens, 182, *seat, position*.
reve (v.), 156, *spoil, plunder*; 302, *take (away)*; 322 (pp.), wyttys ben revid.
revere, 353, *river*.
revyfe, 231, *revives*.
rewe (sb.), 10, *row*; rowe, 35.
rewelerys, 236, *rulers*.

Glossary

rewly (adj.), 10, *rueful, piteous*; ruly, 12.
rewthe, 99, *matter for regret*.
reynenge, 16, *sovereignty*.
reynes, 227. *cloth of Raines, linen made at Rennes*.
reysynt, 246, 299, *for* reysyn it, *raise it. See* levynt.
roberych, 260, *rubric, stage direction*.
rochand, 287, I reyne as a rochand, ? *ruler* (Halliwell), *or cognate with* roch (= *rough*).
rought (sb.), 173 (r. w. dowt, abought), *rout*.
rowel, 169, *spur* (I ryde on my rowel = *I ride spurring*).
rowncys, 170, *horses, hackneys*.
rowne (v.), 374, *utter*.
rowse (v.), 83, *proclaim, make public*; 97, *utter*.
rowte (sb.), 35, *company, assembly*; rowthte, 176.
rowth (v.), 167, 281 (impers.), 369, *reck, care* (*past tense form with present meaning*).
rubbe (v.), 375, ?.
rustynes, 42, rustynes of synne, *corruption*.
ruyne, 104, *fall, overthrow*.
ryff, 3, his chylde þer offered xuld be upon an hylle full ryff; þe busshop toke here iii maydonys ryff; 92, I wedde here ryff; (adv.), *readily*.
ryght, 34; ryte, 2; ryth, 1, *right*.
ryghtwysnes, 100, *righteousness*.
ryme, 227, *match*.
rys, 20, gentyl rys, *rice* (*choice rice*).

sa, 119, ? *so*.
sacryd (pp.), 255, *consecrated*.
sad, 1, *serious*; 154, *firmly fixed* 223, *heavy*; sadly, 41, *resolutely*.
sage (adj.), 8, 105, *wise*; 117 (sb.), *wise woman*.
sage, 93 (r. w. age), in old sage, *saying*, cf. *Pearl*, l. 226, saghe; sawys, 80, *sayings, speech*.
salver, 166, *healer*.
saluse (v.), 361, *salute*.
same, 19, in same, *together*.
satan, 228, *satin*.
sawe, 267 (r. w. have), *save*.
sawtere book, 95, *Psalter*.
sayll (v.), 40, *assail*.
scappys, 197, *slips, mistakes*.
schadu (v.), 105, *shadow*.
schal. *See* shalle.
schamly, 362, *shamefully*.

schape (v.), 132, *escape*; shape, 313.
schapman, 252, *chapman, trader*.
schapyn, 317, *cause, bring about*.
scharlys, 171 (r. w. gerlys), *churls*.
schedyn (v.), 156, *shed*.
schelchownys, 170, MS. schel chownys *altered to* schel chowthys, ? *for* selcouthys, *marvels*; possibly in the first case a mistake for scheltrownys, *troops*.
schende, 46, *injure, destroy*; schent (pp.), 110, *disgraced*.
schep, 139, *cheapness, abundance*.
schet (pp.), 28, *shut*; shytt (pp.), 248; shytte (pr. t.), 215.
schet (pt.), 40, *shot*. *See* shete.
schon, 52, *shoes*; sho, 25, *shoe*.
schonde (sb.), 317, *shame, disgrace*.
schrowde, 25, *garment*.
scle, 46, *slay*; scloo, 46, 171; slo, 33; slen, 147; slewe (pt.), 34; scle (pt.), 41; sclow (pt.), 28; slawe (pp.), 9; slayn (pp.), 3.
scleppe (v.), 36, *slip*.
sclepyr, 93, *slippery*.
sclyde, 216, our sorwe doth slake and sclyde, *slip away, be forgotten*.
scorn, 220, *shorn, rent*.
scowte, 128, *a term of abuse*. 205/145
se (v.), 2, *see*; sene, 4; seyn, 52; saughe (pt.), 162; sey (pt.), 277, 311; say (pt.), 330; sowe (pt.), 253; seyn (pt.), 351: sene (pp.), 2; sayn (pp.), 34, 323; seyn (pp.), 81; seyne (pp.), ? 152.
se (sb.), 18, *seat*.
se, 359, l. 123, *for* so *or for* be = *by*.
se, 103, l. 193, ?.
seche, 246, *such*. *See* soch, suych.
seche (r. w. leche), 152; seke, 153, *seek*; south (pt.), 7; sowth (pt.), 8; sought (pp.), 30; sowte (pp.), 102; sought, 303, *attacked, visited*; han de sowth, 275 (*see do*).
seke, 65, *sick*.
sekyr, 38, *safe, sure*.
sel, 258, 276, *time*; seyl, 274.
selkowth, 146, *strange*.
selph, 93, *self*
semlant, 153, *appearance*.
sen, 298, *saint*.
sen (conj.), *or* son, 66, *since*. *See* syn *and* sythyn.
sende, 103 (r. w. kende), *sent*.
sengler (adj), 360, *singular*.
senstere, 357, ? *seamstress* (disparaging term for women).
senues, 297, *sinews*.

sequens, 65, *sequence, chant.*
ser, 369, *withered, sere.*
serge (sb.), 273, *search.*
serteyn (sb.), 241, a serteyn of chylderyn, *a number.*
seryattly, 357, *one by one, seriatim.*
sese, 220, *cause to cease.*
sesyd, 112, *seized, in possession, established.*
sethe (adv.), 118, *since.*
settynt, 275, settyn it, *set it.* See levynt, reysynt.
seuer (v.), 358, 370, *promise, assure, declare.*
severe (v.), 232, *sever.*
severe (adj.), 366 (r. w. cure), ? *sure*; sewre, 242.
sevyle, 228, *civil law.* See cevyle.
sew (v.), 13, *proceed, go with speed*; 108, 247, *petition, sue; follow*, 289 ; suenge, 29 ; sewyng, ʼ256, *following.*
sewe (sb.), 377, *sow* (?).
sewte, 182, *suit, cause, action*; hold the sewte, *defend the action.*
sey, 25, *say*; seyn, 40, 52, l. 22 (or *see*); seyt, 299; sythe, 235, *saith*; seyand, 355, *saying*; ȝe haue me herd seyd, 30.
seyd, 125, *seed.*
seyn, 228, *sign.*
seyne (r. w. tweyne, regne), 152, ? *for* sheyne, *bright.* See shene.
seynge, 76, l. 123 ; 311, st. dir., ? *seeing.*
seyse, 228, seyse nere sessyon, *assize* (form not given in N. E. D.). *See* syse.
shadyr, 148, *shudder.*
shaftys, 154, *spears.*
shalle, 58, 202, 207, 348 ; shal, 204 ; schal, 229, 262 ; sal, 41 ; shalt, 50 ; schulle, 4 ; shulle, 8, 30 ; shul, 195 ; schulde, 48 ; schuld, 256 ; and throughout the *Assumption* Play *for usual* xal, xul, xulde.
shamfastnes, 105, *shyness.*
shede, 266, *sheath.*
sheff, 32, *sheaf.*
shendynge (sb.), 162, *harm.*
shene (adj.), 146, *bright.*
shenshipp, 45, *shame.*
shete (v.), 40, *shoot*; schet (pt.), 40; shote (pt.), 40; shet (pp.), 212.
shetyng (sb.), 41, *shooting.*
shewyght, 147, *sheweth*; shewyth (3 pl.), 80, l. 235, ? *cause to appear.*

shove (v.), 229, *show.*
showe (v.), 33 (r. w. loue), *shove, thrust* ; 310 (r. w. anow).
shray (s.), 170 (r.w. clay) ? *clamour.* See N. E. D. scry. Cf. day *for* die.
shrevyn (pp.), 193, *shriven.*
shrewe (sb.), 125, *wretch*; brede a shrewe, 194, *something troublesome, vexatious, mischief.*
shrewe (v.), 124, *curse.*
shrynkyd (as a snayle), 198, *shrank, drew back.*
shulderyn, 172, *shoulders.*
shynand, 167, *shining.*
shyne (sb.), 146, *radiance* (earliest instance in N. E. D. 1529).
shyrle (r. w. fylle), 113, *shrill.*
skore (v.), 120, *scour.*
skorgyd, 165, *scourged.*
skye, 147, out of Jacob xulde shyne a skye, ?; in a bryght skye, 351, *cloud.*
skylle, 32, *reason* : skele, 361; it is skyl, 92, *reasonable, proper.*
slake, 216, *grow less, come to an end,* cf. *Pearl*, l. 942.
slauth, 376, *sloth.*
slawdyr, 124 ; *for* slawndyr (64), *slander*; sclaundyr, 125.
slawe (r. w. lawe), 236, *slow.*
sle, 7, *slay.* See scle.
sleytys, 146, sleytys slyȝ, *skilful devices*; slithtis, 367 ; 200, *artifices, wiles.*
slought, 340, *slothful.*
slye, 9, *cunning*; slyȝ, 146, 156, *clever.*
smyght, 48, *smite*; smyth, 253; smyth, 266, *smiteth*; smet (pt.), 277 ; smete (pp.), 266.
snarle, 368, *catch in a snare, strangle.*
snelle, 113, *quick, ready* (i. e. to help mankind).
sneveleris, 368, *snivellers, whiners.*
soch, 307, *such.* See seche, suych.
socurraunce, 207, *succour* (not in N. E. D.).
soferauns, 76, *endurance* ; 271, *sanction.*
sofreynes, 81, *masters,* term applied to audience by Contemplacio and in the *Assumption* Play. See sovereynes.
sofron, 66, *suffer, allow*; soferyd (pp.), 42, *endured.*
sokelyng, 254, *clover.*
solempne (adj.), 15, *solemn.*
solennyte, 293, *solemnity.*
somowne, 123, *summon.*

son *for* sen, 66, *since.*
sond, 22 *send.*
sonde, 28, 46, 109, 132, *message, dispensation, ordinance;* 87, 169, *messenger.*
sondyr, 40, on sondyr, *asunder, in pieces.*
sool, 227, *sole of the foot.*
sor (sb.), 144, *injury, ailment.*
sore (adv.), 297, *violently* ; 5, *greatly.*
sorwatorie, 306, ? (Halliwell, *place of sorrow*).
sorwe (sb.), 8, *sorrow, trouble*; sorwyn, 154, (? pl.).
sorwyth, 68, *sorroweth.*
soserye, 284, *sorcery.*
sote, 4, *sweet.*
sotely, sotylly, 253, *subtly, stealthily.*
sottys, 153, *fools.*
sotyl (adj.), 195, *cunning.*
sotylte, 9, *trick, crafty device.*
sovereynes, 71, *masters, sirs*; souereynes, 354. *See* sofreynes.
sought, 303. *See* seche.
sowe (*or* sewe), 229, *follow.*
sowe, 253 (r. w. knowe), *saw.*
sowkyn (v.), 7, *suck* ; sokyn (pp.), 24.
sownd (adj.), 32, *reasonable* ; 34, *sound*; 134, on sownd, *unsound, ill-founded.*
sownde, 35, Thy seruauntys saue lord fro synful sownde, *swoon, torpor.*
sowre (adj.), 154, *sour.*
sowyght, 67, *seweth.*
spede (sb.), 16, 71, *help, helper.*
spede (v.), 30, *assist, cause to succeed*; sped (pr. t.), 20, *hasten* ; spedly sped (pp.), 148, *accomplished.*
spedful, 82, *helpful.*
spedly, 148, *successfully, prosperously.*
spekyn, 30, *speak* ; all maner langage hem spak, 15 (refl.); cf. bot than hym spake Gallerone to Gawayne, *Anturs of Arth.*, N. E. D.
spelle (sb.), 115, *language.*
spelle (v.), 147, *relate, preach.*
spere (v.), 27, *shut off, bar* ; sperd (pp.), 59, 288, *fastened.*
spetously, 268, *despitefully.*
splayed, 228, *spread out.*
spoused, 354, *wedlock* ; 365, *espousal.*
spowsage, 83, *espousal, marriage.*
spowsyng, 85, *marriage.*
sprede, 35, *spread* ; sprad (pp.), 21.
sprynge (v.), 4, *spring, arise* ; spreng (pt.), 364 ; spronge (pp.), 60.
spryte, 191, *spirit* ; sprytt, 60.
spylle, 12, *destroy* ; spilly (1 sg.), 355 ; spylt (pp.), 41, *shed.*

stage, 35, strete and stage, *raised platform for spectators* ; 93, xul dwelle with ʒow in stage, ? *now, at this time, straightway*; cf. N. E. D. stage, *a period of time.* See stownde.
starkly, 116, *stoutly.*
statis, 355, *dignitaries.*
sted (sb.), 23, with-inne a sted ; þis steed, 134 ; as if used for point in time instead of point in space, cf. *on the spot*; stede and place, 43, *situation.*
steke (v.), 154, *stab, kill.*
stent (v.), 368, *stop, bring to a stand.*
steracle, 197, *spectacle, show.*
sterre, 7, *star.*
steryd (pp.), 25, *excited, instigated* ; styrth, 138, *stirreth.*
stevene, 146, *sound* ; 350, *speech* ; stewyn, 90, *voice, speech.*
stey, 335, *arise, ascend* ; stye, 351.
stody (sb.), 75, *study, acquisition of learning*; in a cold stodye, 207, *state of abstraction.*
stomachere, 227, *waistcoat.*
stomble, 28, *stumble* ; stumbyll, 161 ; stomele, 197.
stondyn (v.), 10, *stand*; stonde, 52 ; stound, 87 ; stant, 138, *standeth* ; stonde (pp.), 40.
stondynge (that), 179, 205, 214, *it being the case that.* (This absol. use not given in N. E. D.)
stotte, 205, *term of contempt for a woman.*
stow (v.), 204, *stop.*
stownde, 12, *time*; þis stownde, *now* ; in stownde, 156, *at this moment, now.*
strekyn, 157, *passed, come forth.* See stryke.
strengere, 229, *stronger.*
strenght (sb.), 75, *strength* ; strenghthis, 78 ; strenthis, 163, *powers.*
strengthe (v.), 217, *strengthen.*
streyte, 26, *straight, directly* ; streyth, 140.
stryke, 153, *go, make one's way.*
sty, 26, helle sty, *pigsty.*
stye, 160, ouer þis stye, *path* (cf. AS. stīg).
stynkygh, 222, *stinketh.*
styrte (v.) (r. w. herte), 231, *go.*
styward, 7, *steward.*
submyt (pp.), 260, *submissive.*
sudary, 332, *napkin.*
suerd, 3, *sword* ; swerd, 2.
sumdele, 341, *somewhat, to some extent.*
sumtyme, 26, *formerly.*

Glossary 399

supportacion, 122, *countenance, support.*
sustyr, 107, *sister.*
sute, 193, folwyth þe fowle sute of þe devyl, *train.*
sutere, 190, *follower.*
suych, 8, *such*; suech, 76; swyche, 68; sweche, 239; seche, 246; soch, 307.
swage (v.), 35, *abate*; 170, *cease.*
swap (sb.), 7, *blow.*
swappynge (adj.), 172, *slashing.*
swelle, 75, *swallow, devour.*
swem (sb.), 65, *grief*; 101, *a pity.*
sweme (v.), 361, *faint, be overcome.*
swemful, 64, *full of grief.*
swemynge (sb.), 74, *grief, emotion.*
swete (v.), 32, *sweat.*
swetyng, 151, *darling.*
swoot (sb.) 26, *sweat.*
swowne (v.), 12 (r.w. stownde), *swoon*; swownde, 139; swonge, 299 (st. dir.); swuonyng, 298.
swynk (sb.), 26, *toil.*
swynke (v.), 32, *toil.*
swythe, 38, *quickly.*
sybb (adj.), 48, sybb blood, *kindred*; sybbest blood, 213; sybbe, syb, 126, *akin.*
syeng (sb.), 35, *sighing.*
sygnes, 80, *signs*; syne, 77, 146; synge, 370.
sygnyfure, 188, *signification, sign*; signifure, 340.
syhyn (v.), 863, *sigh. See sythe.*
sympyl, 193, *of little value.*
syn (prep.), 118; (conj.), 165, *since. See* sen, sythe, *and* sythyn.
syndony, 311, *fine linen* (sindon, word in Vulgate for linen cloth used by Joseph of Arimathea).
syne (sb.), 77, 146, *sign. See* sygnes.
syre, 61, a grym syre, *lord, master.*
syse, 27, *sighs.*
syse, 11, *assize*; in syse, 161, *in state, in proper fashion. See* seyse.
syt. 101, *is fitting, seemly*; syt, 26, *sitteth. See* syttyn.
syth, 2, *sight*; syght, 33; syht, 39; syte, 52; sygth, 223; sythte, 273; syghtys, 76, *for* fyghtys.
sythe (v.), 334, *sigh.*
sythe, 235, *for* seythe, *saith.*
sythe (adv.), 62, *afterwards*; (prep.), 165, *since*; syth (conj.), 240, *since, as*; sythe þat, 110, *since.*
sythym (adv.), 260 *for* sythyn, *afterwards, then.*

sythyn (conj.), 258, *since, as*; sythyn þat, 130, *since. See* syn *and* sen.
sythys, 349, *times.*
syttyn (v.), 18, *sit. See* syt.
syttynge, 101, *suitable, fitting.*

take (imp.), 39, *take*; take (pp.), 41; tan (inf.), 38; tan (pp.), 13; takyght (imp.), 171; take, 92, l. 308, *give*; toke (pt.), 2; tokyn (pt. pl.), 345, *gave.*
takke (v.), 288, *tack, fasten*; takkyd (pp.), 297, *nailed.*
takylle, 40, *weapon, bow.*
taske, 318 (r. w. aske), to taske a wynk, *tax, levy,* i. e. *seize, take.*
tast (v.), 142, *touch, examine.*
taught, 162; tauht, 354, *taught.*
tee (v. inf.), 29, *go.*
tekele (v.), 227, *tickle*; tekyl (adj.), 126, *ticklish.*
teme (sb.), 361, *theme,* sayd us this teme, *spoke to us to this effect.*
tempte (v. pt.), 226, *tempted.*
tende (v.), 43, *intend*; 369, *attend*; tent (imp.), 364, *take care of.*
tendyrly, 83, *carefully,? for* tentily.
tene (sb.), 6, *malice, anger. See* trey.
tene (v.), 85, *rex*; 368, *suffer vexation.*
tent (sb.), 86, take tent, *take heed.*
terme, 9, *space of time*; 92, terme, tyme and tyde.
termynable, 232, To seuere ryth and wrong in me is termynable; 272, In ȝow alle jewgement is termynabyle. Hence 'to be terminable in' = to be derived from, to have its ultimate source in. (Not in this sense in N. E. D.)
terrewth, 102 (*for* trewth), *truth.*
teryeng, 72, *tarrying, delaying*; tary (imp.), 87; tery (subj.), 278.
testyficacion, 61, *testimony, witness.*
teyl (v.), 26, *till, obtain by tillage.*
teynt, 368, *attainted, corrupt, guilty.*
the (v.), 29, *in phrase so* moty (= mot I) the, *prosper, thrive*; then, 87.
thedom, 131, evyl thedom, *bad luck. See* the.
thedyr, 65, *thither.*
therkeness, 96; thyrknes, 270, *darkness.*
therlys, 171, *thralls, serfs.*
thes (r.w. pres, *press*), 61, *these*; theys, 57.
thes, 231, l. 24, *this*; þese, 259. *See* chalys.
tholyn (v. inf.), 153, *suffer*; tholyd (pp.), 172.

thore (r. w. more), 74, *there.*
thorw, 17; thorwe, 9; thour, 108; thourghe, 57; throwh, 31, *through.*
thouht (sb.), 35; thouth, 11; thought, 28; thowth, 57, *thought.*
thralle (sb.) (r.w. alle), 325 = thraw, *period, space of time.*
thrawe (pp.), 233, *thrown*; throwe (pp.), 24.
thredde, 364, *third*; thryd, 2; thrydde, 75.
threttene, 76, *thirteenth.*
threwth, 100, *for* trewth, *truth.*
throwyht (3 sg.), 67, *throweth.* See thrawe.
throwys, 139, *throes.*
thrylle (v.), 8, *pierce*; thyrlyng, 15; thyrlyd (pp.), 268.
thryste (sb.), 98; thrust, 302, *thirst.*
thy, 112, what for thy, *that.*
thyk, 25, þi fals fablis þei be ful thyk, *plentiful, 'thick'.*
thynke, me thynke (r. w.), thynkyth, thynkeht (v. impers.), 32; thynkyht, 40; thynkygh, 173, *methinks.*
thynkyth (3 pl.), 106, *think*; thynkys (imp.), 120.
thyrknes, 270. See therkeness.
to, 111, l. 72, *two.*
to-breke, 48, *break open.*
tokenynge, 146, *sign, betokening.*
too (sb.), 126, *toe*; ton (pl.), 131.
toost (sb.), 173, (not) wurth a toost, *a piece of toast.*
to-pende (v. inf.), 169. See pende.
to-pynde, 28. See pynde.
tormentry (sb.), 184, *tormenting.*
tose (v. inf.), 373, *' tease out ', search out.*
tother, 53, þe tother, *other.*
to-torn (pp), 27, *torn in pieces.*
towaly, 260, *towel.*
trace (sb.), 50, *track*; returnyth his trace, *comes back.*
tras (v.), 13, *trace, follow a course*; 249, on of ȝow my treson xal tras, *contrive, work out.*
trast (v.), 84, *trust.*
tray (v.), 319, *betray.*
trayn, 324, *deceit.*
trekyl, 64, *trickle.*
trenyte, 17, *Trinity.*
trepett, 174, I ȝeve hym ... a trepett, *I trip him up.*
tretable, 201, *tractable, yielding.*
trewth, 2, *truth*; trowth, 56, l. 160.
trey, 6, 97, *anger, spite.* See tene.
tribus, 63, *tribes.*
tron, 8, *throne.*

trow (v.), 33, *think, consider.*
trowth, 56, l. 145, *troth.*
trus (v.), 171, *truss, pack.*
tryne, 81, *threefold, triune.*
tryste (v. 3 sg., r. w. myste), 161, *trusteth.*
tundyr, 40, *tinder.*
turtelys, 64, *doves.*
twey, 99; tway, 345, *two.*
tweyners, 117, here tweyners (g. pl.), *of them two.*
twynne, 303, *part in two, cleave asunder.*
tyde (sb.), 1, *time.*
tydandys, 289, 323, 330, *tidings*; tydyng, 3; tydyngys, 43.
tyght, 222, *readily, promptly*; tyth, 16; as tyght, 34, *as quickly (as possible)*, i. e. *at once.*
tymbre wryth, 136, *carpenter*
tythyng, 3, *tithing, paying tithes.*
tyxt, 226, *text.*

þar (v.), 257, *need.*
þare, 66 (r. w. bare), *there.*
þenge, 2, *thing*; thyng, 19.
þenke (v.), 1, *think, purpose*; thynk, 7.
þese, 259, þese chalys, *this.*
þo, 1, *those.*
þurowe, 1, *through.*

valure, 340, for your valure, *for your worth*, i.e. *to redeem you*; 341, *value.*
varyable, 232, *contrary to.*
varyauns, 6, withoute varyauns, *without deviation, straightway*; varyauns, 232, *divergence.*
vathe, 299, *interjection, ? what*; cf. whath, 231.
vengeabyl, 100, *revengeful, prone to vengeance.*
venym (adj.), 192, *venomous.*
verament, 14, *truly, verily.*
verdyth (r. w. lyth, *light*), 339, *verdict.*
verray (adj.), 2, *true*; verry (adj.), 60; veray (adv.), 183, *truly.*
vertuis (adj.), 256, *virtuous.*
veruent, 216, *fervent*; veruently, 15, *flamingly.*
veryfye, 114, *confirm as true.*
vesage, 34, *visage, face.*
vesyte (v.), 98; vycyte, 240, *visit.*
vesytacion, 49, *visitation.*
vetaylys, 244, *victuals.*
veyn (adj.), 34, *worthless.*
vmbyl, 260, *humble, lowly.*
vnbegete (pp.), 258, *unbegotten.*
vndowteful, 75, *free from doubt.*

Glossary 401

vndryd, 97, *hundred.*
vnhede (r. w. dede), 24, *uncover.*
vnhende, 176, *discourteous, unknightly.*
vnkende (r. w. wynde), 7, *unnatural* ; 99 (r. w. ende).
vnknowlage, 113, *lack of knowledge.*
vnlosne (v.), 237, *unloose, untie.*
vnpynne, 306, *unfasten.*
vnstable (*altered from* vnstabyl, r. w. able), 203, *unreliable, inconsistent.*
vnterest, 167, *uttermost.*
vntey (r. w. wey), 224, *untie.*
vnthende (adj.), 32, *poor, meagre.* See the.
voydnes, 116, *emptiness.*
vyl (sb.), 263, *for* wyl, *will.*

wace, 265 (r. w. place), *was.*
wake (v. int.), 17, *to exist fully, be active* ; 252, *watch* (int.) ; wakyn the way, 314, *watch* (tr.), *guard.*
wakyn, 161 ; *awake* ; cf. ore he be wakyn, *Torrent of Portugal*, N. E. D.
walkyn, 19, *welken, sky.*
walterid, 317, *weltering, steeped in.*
wan (v.), 226, *won* ; wonnyn (pp.), 146.
wantruste, 212, *mistrust, lack of hope.*
war (adj.), 4, be war, *take note* ; ware, 187, *informed, aware* ; be whare, 54, *beware.*
wardeyn, 91, *guardian.*
ware (sb.). 187, *goods.*
wark (sb.), 307, 1. 1046, *work. See* werk.
warly, 310, *warily.*
wasch (pp.), 348, *washed.*
wast (pp.), 27, *wasted.*
watt, 274, *fellow* ; cf. *Richard the Redeless*, wattis, *people.*
wawys, 42, *waves.*
way, 111, do way, *leave off.*
wayted, 169, *kept watch* ; wayten, 314, *watch, spy.*
weche (v.), 262, 312, *watch.*
weche, þe weche, 12, *which* ; þe which, 3 ; wich (þat), 44 ; wheche book, 95 ; whiche, 164.
wede, 25, *garment.*
wedyr (sb.), 307, grett wedyr, *storm.*
wedyr, 113, *whither.*
weldygh, 90, *wieldeth, disposeth* ; weldyth, 17.
wele (v.), 188, *will. See* wole.
welsom, 27, *wild, unpleasant* ; whylsum, 191.
welthis, 95, ryches and welthis.
wem, 4, *stain, pollution.*
wene, 2, *ween, believe* ; wend (pt.), 107.

wene, 4, withoutyn wene, *doubt.*
wepone, 265 (st. dir.), *weapons.*
wepyng, 28, wepyng dale, *hell.*
werch (v.), 55 (r. w. cherch), *work* ; werke, 19 ; werkyn, 86 ; werkyht, 55, *worketh* ; werke, 236, l. 158, *worketh.*
werd, 1, *world* ; werlde, 26 ; werdl, 172.
werdly, 10, *worldly* ; wurdly, 54.
were, 137 (r. w. to be), *weary.*
werke (v.). *See* werch.
werk (sb.). 16, *work* ; werk wylde, 136, *difficult, tiresome business. See* wark.
werm, 19, *worm, creeping thing* ; 25, *serpent* ; worm, 26.
weryd (pt.), 328, awey he weryd þo ffyndys, ? *exorcized.*
weryn, 42 ; ware, 65 ; wore, 82 ; worn, 122 ; where, 240, *were* ; wore, 318, *wert.*
wese, 159, *wise.*
wete (v.), 30, *know* ; wetyn, 267 ; *See* wyst, woot.
weyll (adv.), 37, *well.*
weyth, 223, *weight.*
weyys, 159, this weyys (r. w. reyse), *this way.*
whanhope, 11, *despair.*
whan, 307, *wan, gloomy.*
whar, 277, whar, whar, *a hunting cry* (cf. modern ' Ware fox ').
whath, 231, *what. See* vathe.
whe, 235, *we.*
whight, 55. *creature, man* ; wyght, 55 ; wythe, 115 ; wyhgte, 135 ; whyt, 138 ; whyht, 176 ; whith, 213 ; wytys, 307 ; whithtys, 327.
whith, 72, *with* ; whit, 240 ; *for usual* w^t or *with.*
who, 206, *woe* ; whoo, 34.
whonde (v.), 115 (r. w. honde), *hesitate.*
whow, 65, *how* ; whov, 164.
whyght, 89, *white.*
whyle (sb.), *time* ; wheyle, 230. *See* qwyle.
whylsum, 191. *See* welsom.
whysshe, 251, *wish.*
withsytt, 200, *oppose, withstand* ; withsett (2 sg.), 200.
wo (adj.), 14, *miserable.*
woke (sb.), 4, *week.*
wole (v. sg.), 17, *will* ; wele (sg.), 88 ; wole (pl.), 292 ; wyl (pl.), 1 ; wul (pl.), 67, 332 ; wolyn (pl.), 291 ; welyn (pl.), 292 ; wolne (pl.), 258 ; wole, 88, *for* wolde.
won (v.), 355 (r. w. on), *wane, cease,*

slacken, *or for* wonde, *hesitate See.* whonde. *See* Dr. Greg's Note in *Assumption of the Virgin.*
wond, 87, *wand, rod*; whande, 87.
wone (sb.), 17, *habitation, place*; won, 137, *habitation, house.*
wonyn (v.), 149, *dwell*; 146, wonyght, *dwelleth.*
wood (adj.), 41, *mad.*
woolde, 41, *wold, ground*; wolde, 154, 327; 154, on wolde *altered to* ouer mowlde.
woot (v.), 26, *know*; wot, 35. *See* wete.
worde, 183, *world. See* werd.
wore, 88, þerfore now wole I · so 'my staff holde I · þis jurny to wore. *were to,* i. e. *arrived. See* to (adv.), 'expressing motion resulting in arrival', N. E. D.
woundyr (adv.), 21, *wondrously*; 138 (adj.), *wonderful.*
wrake (sb.), 129, *injury, harm*; 189, 288, 375, *persecution, vengeance, injury*; 86, thi wurde xal I werkyn withoutyn wrake, ?.
wrech (sb.), 194, werke us mech wrake, both wrech and woo, *punishment, destruction*; wreche, 327.
wrecche (sb.), 26, *wretch, outcast.*
wreke (v.), 148, from woe us wreke; 306, from wo I wole hem wreke, *protect or rescue*; wrokyn (pp.). out of whoo ȝet art not wrokyn, *rescued* (cf. *Chronicles of Robert of Brunne*, ed. Furnivall, l. 16076, 'Christendam fro Paen wreche', *rescue, defend, vindicate*); wreke (pp.), 171, *avenged*; wrokyn (pp.), 133, *avenged.*
wrete (pp.), 3, *written*; wretyn, 53.
wrethe (sb.), 26, 209, *wrath.*
wrothe (adj.), 2, *wroth.*
wrougth (pp.), 2, *wrought, created*; wrouth, 9; wrowth, 17.
wry (v.), 229, *turn aside.*
wryth (sb.), 136, *see* tymbre wryth.
wrythe (v.), 28, *twist.*

wul (adv.), 330, *well, very (usual* wele).
wul (v.), 67, 332. *See* wole.
wurth (adj.), 158, *valuable.*
wyght (adv.), 40, *strongly, actively*; 310, wyth, *swiftly, readily*; 328, l. 10, *swiftly* or (adj.) *strong*; whyt as thought, 279 (adj.), *swift.*
wyght (sb.), 55, *creature. See* whight.
wyghtly, 152, *nimbly.*
wyk (adj.), 25, *wicked*
wyl (adv.), 16, *well. See* wul.
wyndand, 369, wyndand wod (Halliwell prints wynd and wod), ? *raving mad*
wynne, 164, to jherusalem fast now wynne, *make for Jerusalem*; 146, *gain, win*; wonnyn (pp.), 146, *won.*
wys, 73, *for* i-wys.
wysse (v.), 94, *direct, guide.*
wyst (v.), 112, 310, *knew*; 113, I myght wel a wyst, *known. See* wete, woot.
wyte (v.), 115, *blame.*
wyth (adv.), 310. *See* wyght.
wytte, 65. *for* wylte, *wilt.*
wyttys, 155, *wits.*
wytys, 307, *wights. See* whight.

xad (v. inf.), 259, *shed.*
xal (sg.), 1, *shall* (x forms are the standard forms for this text, except in the *Assumption* Play. *See* shalle); xul (pl.), 6; xuln (pl.), 298; xulde (pt.), 3, *should.*
xamefullest, 273, *shamefullest.*

ye, 87 (*for usual* ȝe), ye.
yif, 356, *if.*
yis, 87, *yes.*
yit, 356, *yet.*
yon (adj.), 159 (*for usual* ȝon), *yon.*
yong, 158 (*for usual* ȝonge), *young.*
yow, 160, 188 (*for usual* ȝow), *you.*
yowr, 161 (*for usual* ȝoure), *your.*
ys, 87, *is.*
yt, 87, *it.*

NOTES

Calsydon. p. 244, l. 374. The reference is possibly to the hardness of chalcedony (cf. 'chalcedonius ... sculpturis resistere dicitur', Bede, *On the Apocalypse*, and 'The calcydone whiche harde is in to grave', *Court of Sapience*); or possibly to the fact that chalcedony was held to typify those who show forth the light that is within them when called upon to give public testimony of their faith (cf. 'Chalcedonius ... habet fulgorem sub dio non in domo. Quo demonstrantur hi qui ... cum vel doctrinae vel aliis sanctorum usibus in servitute ad publicum procedere iubentur, mox quid fulgoris intus gesserint ostendunt', Bede). The path was to lead to the Trial and Death; it is the way taken 'for mannys love'.

glete. p. 155, l. 125. Miss H. M. R. Murray suggests that 'glete' may be connected with the verb 'gleit' or 'glete', N. E. D. = to glitter. In this case the meaning of the passage may be: 'He may put (in) us wits to recognize in the flesh, by the gleam (of the star), God's noble son.'

laue. p. 91, l. 275. Halliwell gives 'lave' = *washed*. Miss Murray suggests that 'laue' may be a Northern form of 'loaf'. Cf. 'whyt as payndemayn', *Sir Thopas*.

on a myche. p. 153, l. 73. The word 'myche' has been altered by the reviser to 'mysse'. 'on a myc' is still quite clear, and the full form can be deduced from the rhyme words 'rycħ' and 'dycħ'. Miss G. D. Willcock suggests 'miche', *a loaf of bread*, N. E. D. (cf. *Rel. Ant.* ii. 192: 'with-oute wyn and miche', where the word rhymes, as in the text, with 'riche' and 'diche'). The word would here be used with reference to the sacramental wafer, and the missing verb must have been some word equivalent to the 'beleuen' that has replaced it. The meaning is borne out by 'þer-inne sette here sacrementys' in the following line.

PROVERBS, SAYINGS, ALLUSIONS

p. 17. I am fadyr of myth
My sone kepyth ryth
My gost hath lyth
And grace with-alle.

p. 25. I walke as werme with-outyn wede
Awey is schrowde and sho.

p. 28. Schort lykyng xal be longe bought.

p. 29. ȝe must delve and I xal spynne.

p. 31. þat alle ȝour here levyng
May be to his plesyng
And at ȝour hens partyng
To come to good place.

p. 54. By-ware of custome ffor he wyl dere.

p. 58. Of Regalle lyff xal come suche foyson
þat a clene mayde modyr xal be.

p. 64. So xulde euery curat in þis werde wyde
Ȝeve a part to his chauncel i-wys
A part to his parochonerys þat to povert slyde
The thryd part to kepe for hym and his

p. 65. And þo þat departe in sorwe god make þer metyng glad.

p. 67. Aftere grett sorwe · evyr gret grace growyht.

p. 70. Aftere grett sorwe · evyr grett gladnes is had.

p. 110. Olde cokwold þi bow is bent
Newly now after þe frensche gyse.

p. 111. Many a man doth bete þe bow
Another man hath þe brydde.

p. 128. But ȝit sum mede and ȝe me take
I wyl with-drawe my gret rough toth
Gold or sylvyr I wol not for-sake
But evyn as alle somnorys doth.

p. 132. In ffeyth I suppose þat þis woman slepte
Withowtyn alle coverte whylle þat it dede snowe.

p. 140. Whan women travayl grace doth growe.

p. 202. þe hare fro þe fforme we xal a-rere.

p. 227. In trost is treson.

p. 252. Mony makyth schapman.

p. 288. Ffor alle his barfot goyng · fro me xal he not skyp.

p. 314. þow þer come both jakke and gylle.

p. 326. Ffor mede doth most in every qwest
And mede is mayster bothe est and west
With mede men may bynde berys.

p. 356. Whoso clyme ouer hie · he hath a foule falle.

p. 357. It wolde cause the comownys to ryse
And rathere the devyl sle hym · than we schulde that abide.

jolyere than þe jay, p. 154; lyth as ro, p. 327; shrynkyd as a snayle, p. 198; whyte as laue, p. 91; whyte as swan, p. 56; wroth as wynde, pp. 7, 315.